LF

Microwave Devices

The Wiley Series in Solid State Devices and Circuits

Edited by

M. J. Howes and D. V. Morgan

Department of Electrical and Electronic Engineering, University of Leeds

---

Microwave Devices

Edited by

M. J. Howes and D. V. Morgan

University of Leeds

Microwave Devices

Device Circuit Interactions

Edited by

M. J. Howes
D. V. Morgan

Department of Electrical and Electronic Engineering, University of Leeds

A Wiley—Interscience Publication

JOHN WILEY & SONS

Chichester · New York · Brisbane · Toronto

Reprinted April, 1978.

Library of Congress Cataloging in Publication Data:
Main entry under title:

Microwave devices.

'A Wiley—Interscience publication.'
1. Microwave devices. 2. Semiconductors.
3. Microwave circuits. I. Morgan, D. V.
II. Howes, M. J.
TK7876. M5267 621.381'3 75-15887

ISBN 0 471 41729 7

Typeset in IBM Century by
Preface Limited, Salisbury, Wiltshire, England and
printed in Great Britain by the Pitman Press, Bath

Contributors

L. F. EASTMAN	*Cornell University, Ithaca, New York, 14850, USA*
W. HAYDL	*Institute fuer Festkoereperphysik, 7800 Freiburg, Eckerstr., L, West Germany*
M. J. HOWES	*Electrical and Electronic Engineering, The University of Leeds, Leeds LS2 9JT, UK*
K. KUROKAWA	*Bell Laboratories, Murray Hill, New Jersey, USA*
J. S. LAMMING	*GEC/AEI Hirst Research Centre, Wembley, Middlesex, UK*
I. W. MACKINTOSH	*Royal Radar Establishment, Malvern, Worcestershire, UK*
D. V. MORGAN	*Electrical and Electronic Engineering, The University of Leeds, Leeds LS2 9JT, UK*
H. W. THIM	*Institute fuer Festkoereperphysik, 7800 Freiburg, Eckerstr., L, West Germany*
P. WEISSGLAS	*Microwave Institute Foundation, Fack S-10044, Stockholm, Sweden*

Series Preface

The Oxford Dictionary defines the word revolution as 'a fundamental reconstruction'; these words fittingly describe the state of affairs in the electronic industry following the advent of solid state devices. This 'revolution', which has taken place during the past 25 years, was initiated by the discovery of the bipolar junction transistor in 1948. Since this first discovery there has been a worldwide effort in the search for new solid state devices and, although there have been many notable successes in this search, none have had the commercial impact which the transistor has had. Possibly no other device will have such an impact; but the commercial side of the electronics industry stands poised, awaiting the discovery of new devices as significant perhaps as the transistor.

Research and development in the field of solid state devices has concerned itself with two important problems. On the one hand we have device physics, where the aim is to understand in terms of basic *physical concepts* the mode of operation of the various devices. In this way one seeks to optimize the technology in order to achieve the best performance from each device. The second aspect of this work is to consider the important contribution of the circuit to the operation of a device. This problem has been called *device circuit interaction*. It is a great pity that in the past these two major aspects of the one problem have been tackled by separate groups of scientists with little exchange of ideas. In recent years, however, this situation has been somewhat remedied, the improvement being due directly to the very rigorous system specifications demanded by industry. Such demands constantly require greater performance from devices, which can only be brought about by coordinated team work.

The objective of this new series of books is to bring together the two aspects of this problem: device physics and device circuit interactions.

The editors wish to thank most warmly the authors contributing to this series. They also wish to thank Professor J. O. Scanlan (University of Eire, Dublin), Dr R. D. Pollard (University of Leeds), Professor A. E. Ash (University of London) and Professors L. F. Eastman and J. Frey (Cornell University) for their advice and encouragement in preparing this first volume. Finally we would like to thank our wives Dianne and Jean for their infinite patience and valuable assistance in checking the manuscript.

M. J. HOWES
D. V. MORGAN

University of Leeds
January 1975

Preface

This, the first volume in this series of books, is concerned with microwave devices. It is perhaps rather fitting that this topic should be chosen to initiate this theme of integrating device physics and device circuit interaction into a coherent text, since the microwave area, perhaps more than any other, illustrates the important role the circuit has to play on device performance. Indeed many of the so-called advances in microwave device performance during the past 5—10 years have concerned themselves with the optimization of the circuit to suit the device in question.

This book covers the major microwave devices which are making significant contributions in present day research and systems applications. Chapter 1 provides a brief résumé of the various historical stages in the development of devices. Chapter 2 is concerned with devices which utilize the transferred electron effect. Chapter 3 deals with the avalanche and baritt devices. The fundamental concept involved in both these devices embody the 'transit time' phenomenon outlined by Shockley as early as 1948. Chapter 4 deals with the veteran of solid state devices — the transistor. However, the practical devices capable of operation in the microwave region, although similar in principle to their ancestors, have superficially changed beyond recognition. Both the bipolar and the field effect transistors are dealt with in this chapter. Chapter 5 typifies more than any other the way in which this book departs from those with similar titles. It deals in depth with microwave circuits and considers the many and varied circuit problems. Many of these solutions originate from the author concerned. Chapter 6, 'Microwave Amplifier Circuit Considerations', relies heavily on the physical concepts introduced in Chapters 2 and 3. Here, however, the stress is on the realization of amplifiers rather than oscillators. In Chapter 7 we have a comprehensive and precise account of the 'Applications of Microwave Solid State Devices'.

We hope to achieve, by coordinated co-authorship of leading experts in the respective fields, a varied and balanced review of past and current work. The books in the series will cover many aspects of device research and will deal with both the commercially successful and the more speculative devices. Each volume will be an in-depth account of one or more devices centred on some common theme. The level of the text is designed to be suitable for the graduate student or research worker wishing to enter the field of research concerned. Basic physical concepts in semiconductors and elementary ideas in passive and active circuit theory will be assumed as a starting point.

M. J. HOWES
D. V. MORGAN

University of Leeds
January 1975

Contents

CHAPTER 1

The Development of Solid State Microwave Devices

M. J. HOWES and D. V. MORGAN

1.1 INTRODUCTION

During the past two decades we have witnessed a very rapid growth in the field of telecommunications. This growth has thrown together many research topics which hitherto were moving along their separate paths: electromagnetic wave propagation, antennae, digital techniques and microwave power generators, to name but a few. Each of these topics was being studied intensely and yielding new basic information which in turn stimulated more research. A common theme in all this work has been the production and manipulation of electromagnetic energy. In this book we are going to be concerned with microwave devices and, in order to put this into its context, we show in Figure 1.1 the full spectrum of electromagnetic radiation starting with ULF and ending with gamma rays at the high frequency end. Many parts of this spectrum are now being used in communication systems, the oldest of these being the optical region which has provided us with our day to day communications. It is rather interesting that the recent discovery of the LASER has opened again the interest in optical frequencies for long-distance communication systems.

1.2 MICROWAVE DEVICES

The first practical use of electromagnetic wave in man-made communication systems originated at much lower frequencies since Marconi — the innovator -- had to start in a region of the spectrum where the technology had progressed to a point where energy sources were available.

As more and more demands were made on the frequency spectrum in terms of carrier frequency and bandwidth, fundamental power sources and amplifiers operating at higher and higher frequencies were required. Consequently a number of microwave tubes were developed to take advantage of the large bandwidth available in this part of the spectrum. Many of these devices still retain a tight grip on the market and this is

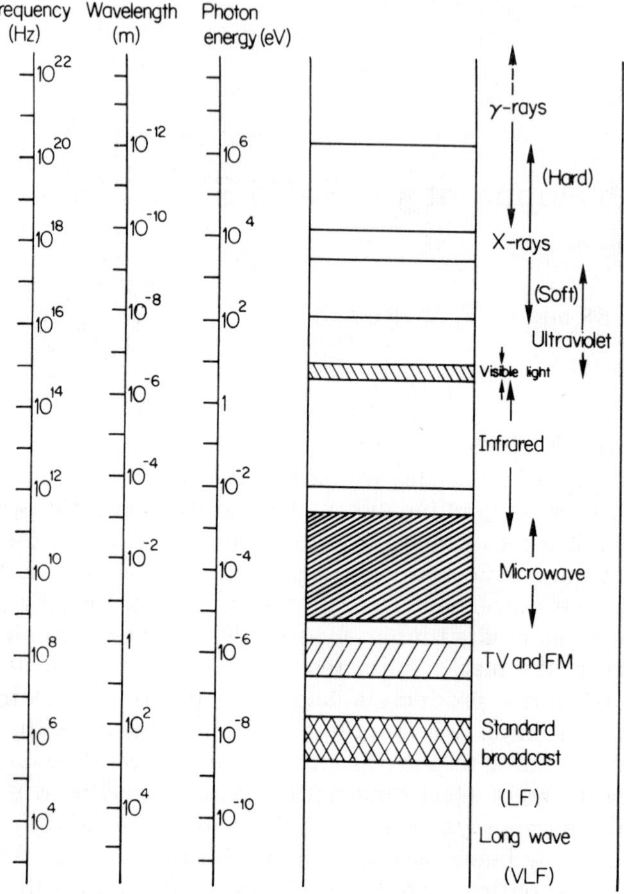

Figure 1.1

likely to be the true in the foreseeable future, particularly in the cases of weather radars (high power pulse magnetrons), distance measuring equipment (microwave triodes) and air traffic control transponders for example. It seems that, despite the relatively larger effort put into solid state devices, tube technology is at this time more advanced in terms of frequency coverage, power handling, reliability and a number of other performance areas. However, at lower power levels solid state devices are either in the process of replacing microwave tubes or are opening up new applications mainly because of the relatively larger power units required by conventional microwave tubes. Among the commercially viable new applications are short-range portable radar systems, microwave field sensors and flowmeters.

1.3 SOLID STATE MICROWAVE DEVICES

The discovery of the bipolar junction transistor and the subsequent replacement of thermionic valves in the field of electronic circuits has reshaped our thinking in almost all of the regions of applications. It is therefore not surprising that the electronics industry at the present time awaits the advent of a suitable high-powered solid state microwave device to replace the large and cumbersome thermionic counterparts. The development of solid state sources has in general been of two kinds. In the first category we have the steady progress resulting from the pushing up of the upper limiting operating frequency of devices such as the bipolar and field effect transistors (Chapter 4). In the second category we have the more erratic progress resulting from the discovery of new devices working on fundamentally different principles; one example was the discovery of the transferred-electron effect (Chapter 2).

The search for microwave solid state devices can be traced back to the paper published by William Shockley in 1954.[1] In this paper Shockley speculates on the idea of a two-terminal device with a negative resistance arising from transit time effects. The first example he considered was a 'minority carrier delay diode' (Figure 1.2) here a p^+-n-p(or an n^+-p-n) structure was suggested in which minority carriers injected at the p^+-n junction then exhibit a transit time delay in drifting to the other p-n junction. An alternative structure discussed by Shockley again consists of a p-n-p device but in order to achieve unipolar action he suggests operating the device in a punch-through

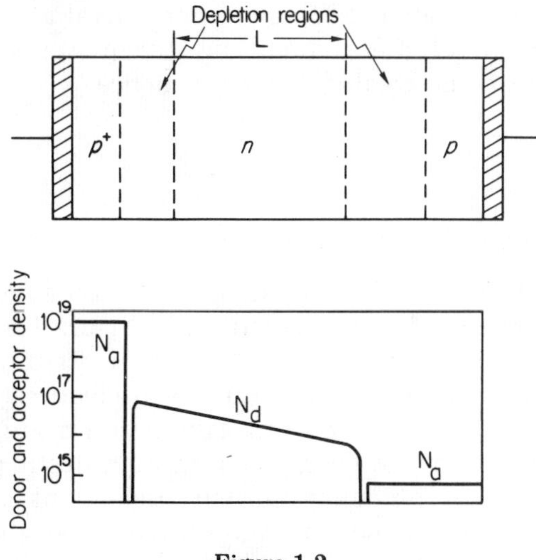

Figure 1.2

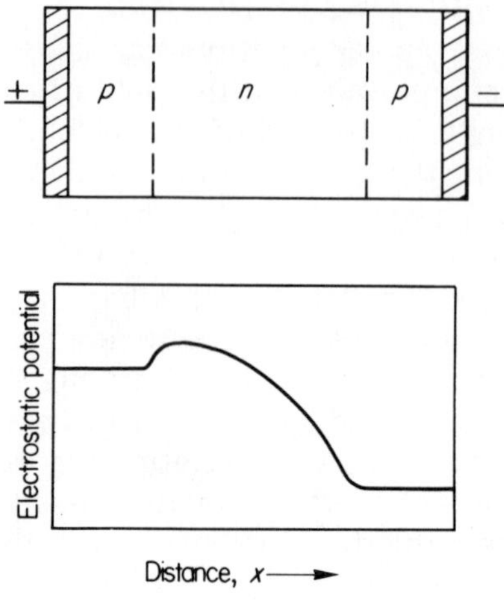

Figure 1.3

mode (Figure 1.3). These two proposed structures bear an uncanny resemblance to the more recently developed baritt diodes to be discussed in Chapter 3.

In the same paper Shockley speculates on the possibility of producing a two-terminal device with negative differential resistance by simply subjecting a piece of uniform semiconductor to a high field, which he argued might show deviations from Ohm's law, such that the carrier velocity would decrease with increasing field (i.e. produce a region of negative differential mobility). This theory[1,2] is founded on the idea that if holes can lose energy to phonons at a certain maximum rate P_{max} then under these conditions the power supplied by the electric field must be no greater than P_{max} hence

$$eEv < P_{max}$$

or

$$v < |P_{max}/eE|$$

so that the drift velocity will decrease with increasing field giving rise to a velocity field curve of the type shown in Figure 1.4. This idea was developed by Kroemer[2] and in turn led unsuccessfully to an experimental search for practical high frequency oscillators. The reason for this lack of success was the incorrect assumption made in the analysis that the only mechanism for energy loss was to acoustic phonons.

A possible way of obtaining current oscillations arises from the negative mass concept. To illustrate this idea, consider the energy versus wave vector curve arising from the simple Kronig Penney model for a one-dimensional solid (Figure 1.5). The effective mass ($m^* =$

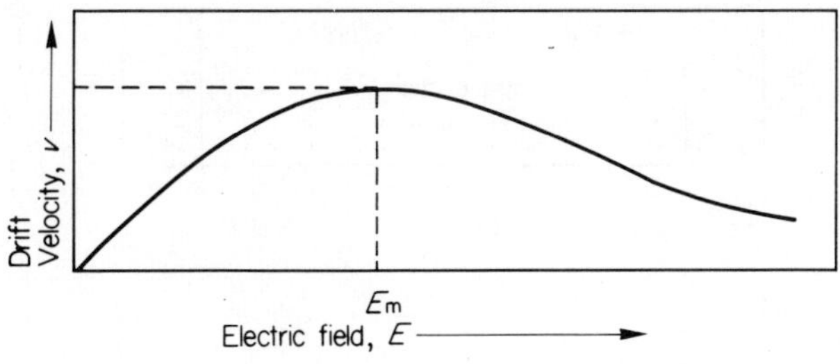

Figure 1.4

$h(\partial^2 E/\partial k^2)^{-1}$) is positive for the lower end of the band ($k < k_1$) and negative when $k > k_1$ where k_1 is the inflexion point of the E-k curve. In this simple picture an electron accelerated by an electric field will take energy from the field when the mass is positive ($0 < k < k_1$) and conversely will give up energy to the field when it passes through into the negative effective mass region ($k > k_1$). The electron would therefore move in the opposite direction until it returned to the positive mass region. Thus the electrons should oscillate to and fro in the sample, giving rise to current oscillations at the terminals. In practice this simple picture does not hold because avalanche multiplication and electron scattering limit the energy the electron may gain from the field. This concept is interestingly still alive today and has initiated some work on the so-called 'superlattice oscillator',[3] In terms of the Kronig Penney model the idea is to build into the solid a superlattice of spacing $D \gg a$. Thus the band edge (π/D) could be brought down to low k values which could be reached before the onset of velocity saturation or avalanche multiplication.

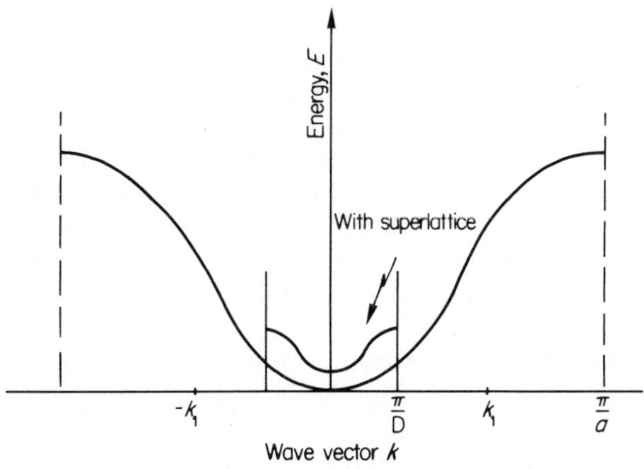

Figure 1.5

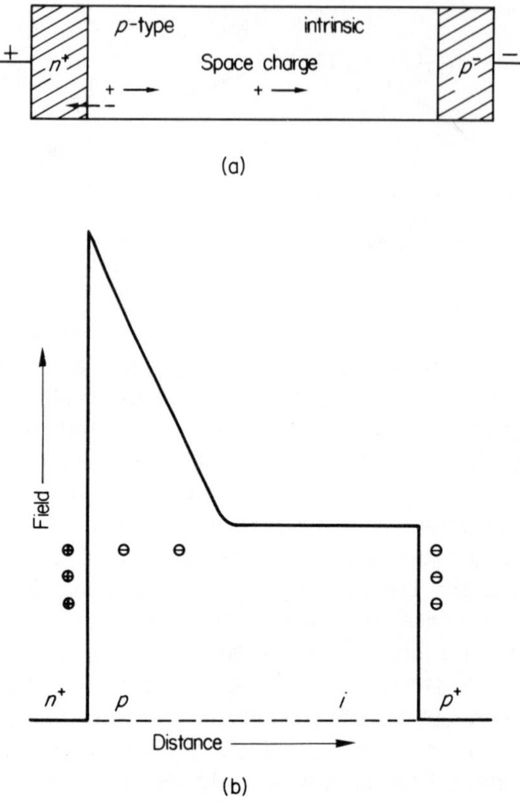

Figure 1.6

An interesting, though somewhat novel, two-terminal negative resistance device — the tunnel diode — was discovered by Esaki in 1957.[4] The measured negative resistance was observed in the forward bias characteristics of very narrow germanium *p-n* junctions (i.e. high doping) and is the result of field emission (tunnelling) across the narrow depletion region. Sommers[5] in 1959 suggested a microwave application of the device which initiated a period of intense research activity to realize some of these potentialities. In retrospect, however, the device has been disappointing, primarily as a result of the low output power resulting from the small current and voltage excursion across the negative resistance region.[6]

W. T. Read in 1958[7] proposed a multilayer diode (Figure 1.6) for generating microwave power. The device operates on a combination of impact avalanche breakdown and electron transit time effects and, for this reason, has been called an IMPATT device (IMPact Avalanche and Transit Time). Hole electron pairs are generated at the n^+-p interface when the electric field in this locality exceeds that required for avalanche multiplication. The free electrons generated then drift rapidly towards the n^+-region under the influence of the field, whilst the associated holes move to the right and drift across the intrinsic region at a velocity appropriate to the saturation velocity of the semiconductor material. It follows that, if this drift current can be made to occur whilst the superimposed a.c. voltage across the device is negative, then the device will behave as an *a.c. negative resistance* and deliver power at a frequency approximately equal to the carrier transit time across the intrinsic layer; the negative resistance exists by virtue of a phase lag between the applied voltage and the resulting current. This structure was in essence the same 'transit time' concept suggested by Shockley which owed its origin to much earlier work on thermionic valves. Experimental evidence for the first practical avalanche transit time-device was reported some seven years later in 1965 by Johnston, De Loach and Cohen[8] who achieved pulsed power outputs from a silicon *p-n* junction driven into avalanche. Since this early work, advances have been so rapid that today these avalanche devices are established as one of the most important microwave solid state power sources and the topic will be dealt with in detail in Chapter 3.

One of the most outstanding technical accomplishments of the past decade has been the practical realization of a bulk semiconductor microwave device — the transferred electron oscillator. As with the Read diode, the transferred electron effect originated as a theoretical concept. In 1961 Ridley and Watkins[9] suggested the possibility of obtaining bulk negative resistance in certain semiconductors by the transferred-electron effect. This concept involves the exploitation of a certain type of band structure in order to obtain differential negative conductivity. The effect is shown schematically in Figure 2.1. Here the conduction band of an *n*-type semiconductor contains two minima, a central valley centred at $k = 0$, together with a satellite valley. If the electrons in the central valley have a high mobility and those in the satellite a low mobility then it is possible that, as the central valley electrons are 'heated' by the application of an electric field across the device, hot electrons will at some critical field be transferred into the low mobility valley giving rise to the negative differential mobility. The current voltage characteristic of such a device (depicted in Figure 2.2) will contain a differential negative resistance region and in principle could, with a suitable circuit, be used for supplying microwave power. It was left to Hilsum[10] in 1962 to show by detailed calculations that GaAs and GaAs—GaP alloys had band structures which were suitable to exhibiting the transferred-electron effect. Although bulk negative resistance has also been observed in compounds such as InP, ZnSe, CdTe and InAs, so far only GaAs, GaAs—GaP and InP have produced useful amounts of power.

The picture we have described so far has an important omission and this was the subject of the now classical paper, 'Specific negative resistance in solids', written by B. K. Ridley[11] in 1963. Ridley showed that a bulk semiconductor exhibiting an n-type current voltage characteristic would be unstable and would break up into domains (i.e. regions of high and low electric field) and the characteristic shown in Figure 2.2 could not be observed directly. The domain formation process may be compared with the all-too-frequent motoring situation in which cars are travelling all at roughly the same speed along a motorway which contains a region of zero visibility. Cars passing through such a region will have their mobility seriously reduced and there will be, as a result, a region depleted of cars. In the case of electrons in semiconductors, the bunching or domain formation will occur at high field irregularities in the solid (i.e. points where the field exceeds the critical field corresponding to that in Figure 2.2). For simplicity, assume that the dipole layer (domain) forms near the negative terminal of the voltage source. The phenomena described above contained the key to understanding the experimental observation reported some two years later by J. B. Gunn.[12] Gunn, working on electrical noise in semiconductors observed that microwave noise powers of the order of watts were emitted by GaAs and InP samples subjected to pulsed electrical fields of several thousand volts per centimetre. When short samples (<0.2 mm) were used this noise changed to coherent microwave oscillations with frequencies corresponding to the transit time frequency. Although many early explanations were considered for the 'Gunn effect', it was Kroemer[13] who, in 1964, pointed out that the observations were in full agreement with the earlier predictions of Ridley and Watkins, and Hilsum. To this day the term 'Gunn diode' is used to describe transferred-electron devices in general, regardless of whether they utilize the Gunn effect proper.

The final device, which we will mention briefly, is the TRAPATT diode, a device which perhaps typifies the philosophy of this volume and indeed the whole series. The trapatt mode of operation of avalanche devices discovered at RCA in 1966, involves an extremely complex interaction between the device and the microwave circuit, details of which are given in Chapter 3. Some idea of the design difficulties can be realized, however, when one considers the trapatt amplifier which requires harmonic and subharmonic tuning as well as initiation through the impatt mode. Despite the sophisiticated nature of this device and its associated circuitry, however, the trapatt is a leading contender for phased-array radar system amplifiers because of its potential of meeting the requirements of high peak powers (>100 W), high duty cycles (1%—20%), high efficiency (>25%) and a bandwidth of at least 15%.

REFERENCES

1. W. Shockley, 'Negative resistance arising from transit time in semiconducting diodes', *Bell System Tech. J.*, 33, 799—826 (1954).
2. H. Kroemer, 'Zur theorie des germaniumgleichrichters und des transistors', *Zeits. f. Physik*, 134, 435—450 (1953).
3. L. Esaki and R. Tsu, 'Superlattice and negative differential conductivity in semiconductors', *IBM J. Res. Dev.*, 14, 61 (1970).
4. L. Esaki, 'New phenomenon in narrow germanium *p-n* junctions', *Phys. Rev.*, 109, 603 (1958).
5. J. S. Sommers, 'Tunnel diodes as high frequency devices', *Proc. Inst. Radio Eng.*, 47, 1201 (1959).
6. J. O. Scanlan, 'Analysis and synthesis of tunnel diode circuits', Wiley—Interscience, London, 1966.
7. W. T. Read, 'A proposed high frequency negative resistance diode', *BSTJ*, 37, 401 (1958).
8. R. L. Johnston, B. C. De Loach, and B. G. Cohen, 'A silicon diode microwave oscillator', *Bell System Tech. J.*, 44, 369—372 (1965).
9. B. K. Ridley and T. B. Watkins, 'The possibility of negative resistance in semiconductors', *Proc. Phys. Soc. (London)*, 78, 293—304 (1961).
10. C. Hilsum, 'Transferred electron amplifiers and oscillators', *Proc. I.R.E.*, 50, 185—189 (1962).
11. B. K. Ridley, 'Specific negative resistance in solids', *Proc. Phys. Soc. (London)*, 82, 954—966 (1963).
12. J. B. Gunn, 'Microwave oscillations of current in III—V semiconductors', *Solid-state Commun.*, 1, 88—91 (1963); also 'Instabilities of current in III—V semiconductors', *IBM Journal Res. Dev.*, 8, 141—159 (1964).
13. H. Kroemer, 'Theory of the Gunn effect', *Proc. IEEE* (correspondence), 52, 1736 (1964).

CHAPTER 2

Transferred-Electron Devices

LESTER F. EASTMAN

2.1 INTRODUCTION

There has been a need for a solid state negative conductance at microwave frequencies for effective oscillation and amplification. Such a negative conductance was discovered as a bulk effect of gallium arsenide, indium phosphide and a few other compound semiconductors. This bulk effect, a property of the material, is the reduction of electron drift velocity with increasing electric field above a threshold value.

Early theoretical papers by Ridley and Watkins[1] and by Hilsum[2] predicted such a reduction of electron drift velocity with increased electric field, due to the gradual transfer of a fraction of the electrons from a high mobility state to a low mobility state. The first, independent experimental observation of transit time oscillations due to this property of GaAs and InP was then reported by Gunn.[3] Kroemer[4] showed that the theoretical transferred-electron effect and experimental Gunn effect were one and the same. Finally, Copeland[5] reported computer and experimental studies yielding oscillations that were not transit time limited and thus effectively used the bulk property of this phenomenon.

During the past decade these GaAs devices have been operated over a frequency range from below 1 GHz to above 100 GHz. Continuous powers from a few milliwatts up to over 2 watts, with efficiencies from a few per cent to 15% have been generated. Pulsed powers from one watt to six kilowatts with efficiencies from a few per cent to 30% have been generated. These devices have wide tuneability, low noise, very fast turn-on in pulsed operation, life expectancies of decades, simplicity and low cost; so they are in wide use in microwave electronics systems.

This chapter presents the physical basis, the electronic modes of operation, the epitaxial growth and processing of materials, the ultimate power and efficiency performance capabilities, as well as brief mention of the noise performance, and tuneable operating of the transferred-electron effect.

2.2 PHYSICAL BASIS OF TRANSFERRED-ELECTRON EFFECT

The electron energy versus electron wave number for GaAs and InP are shown in Figure 2.1 with crystal directions, effective masses and energy values. The electrons are normally in low energy, low effective mass states near the lowest minimum in the conduction band. When a low value of electric field is applied, the electrons drift, having a high mobility near 8000 $cm^2/V\ s$. As the electric field value is increased, the electrons become more energetic and begin to interact strongly with phonons in the crystal.[6,7] The main electron energy loss mechanism is due to the generation of optical phonons. The rate of electron energy loss to optical phonons is dependent on the electron energy. Below about 0.03 electronvolts energy this polar optical phonon energy loss rate is zero, but rises rapidly above 0.03 electronvolts energy to a peak value at a few times this electron energy and then gradually reduces. As the electric field is raised to 3—3.5 kV/cm in GaAs, and 10—10.5 kV/cm in InP, a significant fraction of the electrons begins to have energy exceeding the value having the peak loss rate. As these electrons become more energetic, they lose less energy to the lattice. The strong electric field keeps supplying these electrons with more and more energy which is not lost and they 'run away' to higher energy levels. When this field-dependent significant fraction of the electrons in GaAs reaches the energy value separating the two conduction band minima shown in Figure 2.1 of 0.35 eV, electrons transfer to the many empty states near the next higher minimum. In this upper region, the electron masses are high and thus the electron mobility and drift velocity are low. As a result, the average drift velocity gradually reduces

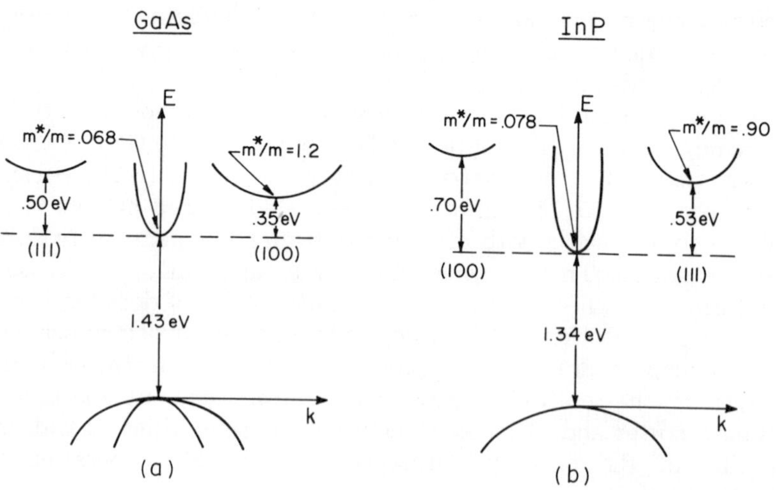

Figure 2.1

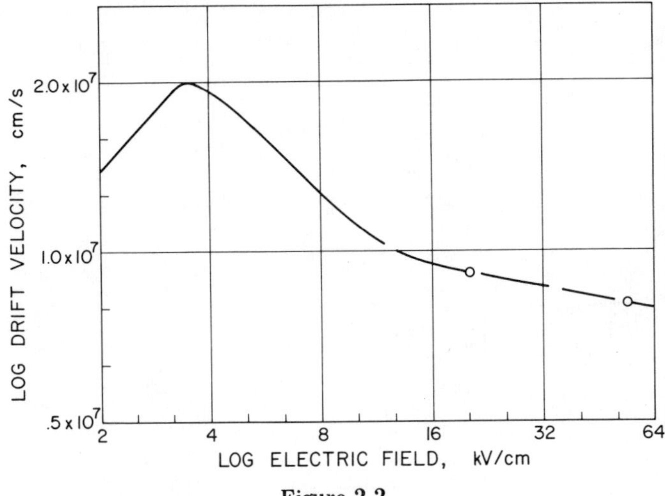

Figure 2.2

as the electric field rises above the threshold value of about 3,500 V/cm..

Figure 2.2 shows a log—log plot of electron drift velocity in GaAs versus electric field for room temperature and for about $2 \times 10^{15}/\text{cm}^3$ ionized impurity density. The results of the Monte Carlo computer calculations of Ruch and Fawcett[8] are used up to four times threshold electric field, the time of flight measurements of Houston and Evans[9] are used from about six to fifteen times threshold electric field, indicated by the section between circles, and the high field extrapolation agrees with approximate transit time results of avalanche device tests. Ruch and Fawcett calculated the effects of changes in ionized impurity density and temperature on the $v(E)$ curve. The calculated peak velocity rises to about 2.15×10^7 cm/s for pure material at room temperature and drops to 1.75×10^7 cm/s at $2 \times 10^{17}/\text{cm}^3$ ionized impurity density. The velocity from about two times threshold field on up in electric field is nearly independent of ionized impurity density. Hilsum[10] has recently shown that the room temperature low-field mobility in GaAs depends on donor density as

$$\mu = \frac{\mu_0}{(1 + \sqrt{N/10^{17}})} \tag{2.1}$$

Variations of electron velocity values in GaAs with temperature can be directly inferred from device current variations with temperature. In Figure 2.3, the experimental variation of peak current and valley current (the latter at bias voltage twelve times the threshold value) are shown as the temperature is raised from room temperature. The

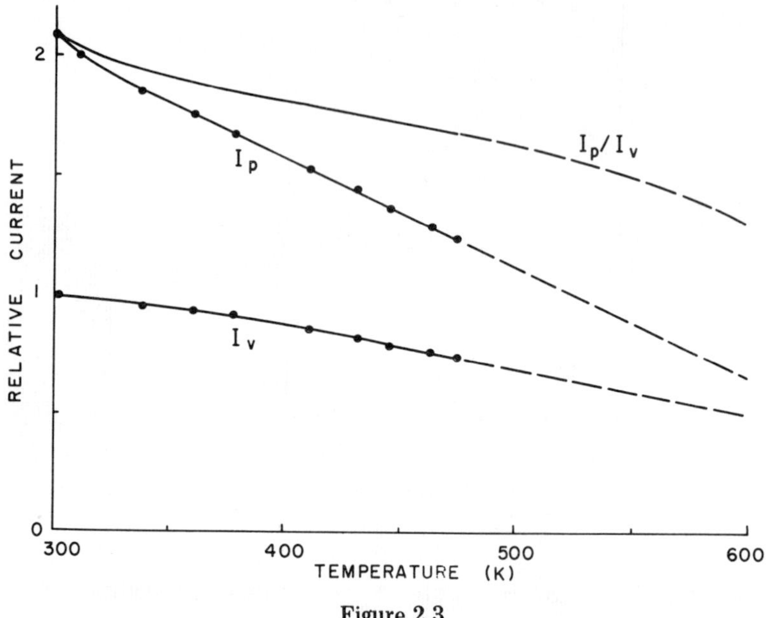

Figure 2.3

resulting current peak to valley ratio variations are also shown. The values are normalized to the room temperature value of the valley current. The extrapolation of the straight line variations at high temperature gives about 380 °C as the temperature at which the current peak to valley ratio goes to unity. The peak current varies more rapidly up to 80°C above room temperature than it does above this value, while the valley current varies less rapidly below this value than it does above. The particular device used has $1.7 \times 10^{15}/\mathrm{cm}^3$ electron concentration and 40 μm thickness, but other uniformly doped devices with higher and lower doping times length products behave similarly except for room temperature current peak to valley ratio values as shown later in Figure 2.7. The peak current value varies from its room temperature value I_{p0} as

$$I_p \cong I_{p0}(1-2.7 \times 10^{-3}\ \Delta T) \quad \text{for} \quad \Delta T < 80\ \text{K} \tag{2.2a}$$

or

$$I_p \cong I_{p0}(0.98-2.25 \times 10^{-3}\ \Delta T) \quad \text{for} \quad \Delta T > 80\ \text{K} \tag{2.2b}$$

The valley current varies from its room temperature value I_{v0} as

$$I_v \cong I_{v0}(1-1.15 \times 10^{-3}\ \Delta T) \quad \text{for} \quad \Delta T < 80\ \text{K} \tag{2.3a}$$

or

$$I_v \cong I_{v0}(1.03-1.83 \times 10^{-3}\ \Delta T) \quad \text{for} \quad \Delta T > 80\ \text{K} \tag{2.3b}$$

where ΔT is the rise in temperature above room temperature.

The variation of low field mobility, from its room temperature value μ_R, in lightly doped GaAs, with temperature is

$$\mu = \mu_R \frac{300}{T} \cong \mu_R (1-3.3 \times 10^{-3} \Delta T) \tag{2.4}$$

InP has a room temperature peak velocity near 2.5×10^7 cm/s, a high-field valley velocity near 0.6×10^7 cm/s and a threshold electric field of 10.5 kV/cm. The experimental variation of InP peak velocity with temperature is

$$v_p \cong v_{p0} (1-1.2 \times 10^{-3} \Delta T) \tag{2.5}$$

A transferred-electron device is simply a piece of n type GaAs, or other appropriate compound semiconductor, with plane, parallel electrodes on two opposing faces. An electric field is imposed on the material by applying a voltage between the electrodes. The area of a cross-section of the device, taken to be in a plane perpendicular to the electric field, is normally the same at any location between the electrodes. The electric conduction current density caused by the electric field is simply the product of the electron charge times the electron density times the electron average drift velocity.

By considering continuity of electric conduction current and the relation between electric field and applied voltage, a profile of electric field can be deduced for each profile of electron concentration (and profile of cross-section area) along the path between electrodes. If all values of electric field in the device are well below the threshold value of 3,500 V/cm, the nearly linear dependence of electron velocity on electric field, related by mobility, gives simple results. The region with the lowest electron concentration ,or lowest product of electron concentration and area, if area varies, has the highest electric field. When the electric field in any portion of the device is raised above the threshold value of 3,500 V/cm, the mobile carriers can develop space charge layers that can accentuate any electric field variations. Two different profiles of donor concentration, N, variation with distance are shown, along with electric field profiles, in Figure 2.4. The electron density profile is the same as the donor density profile when the electric field strength is below the threshold value and for extremely short times after the electric field exceeds the threshold value. When the electric field exceeds threshold, the region above threshold rapidly develops a space charge profile that in turn accentuates any electric field variations. Thus careful study of the region with field above threshold is needed. In Figure 2.4(a) in the region where the electric field is rising above threshold, in the direction of electron drift, an accumulation of excess electrons begins to develop rapidly. In the

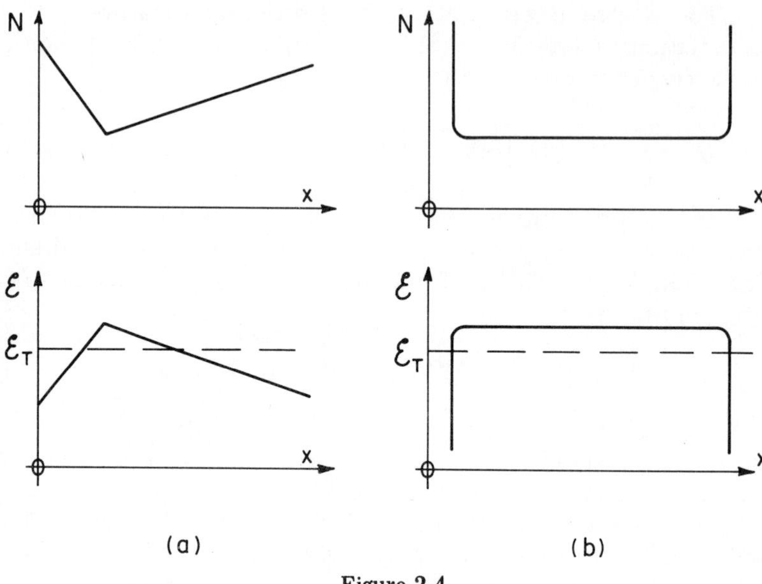

Figure 2.4

region just ahead of it where the above-threshold electric field is dropping, in the direction of electron drift, a depletion of electrons begins to develop rapidly. This space charge dipole disturbance moves along with the average electron drift velocity as it develops.

It is possible to show the reasons for excess electron accumulation and depletion in these adjacent regions by considering the velocity—field characteristic applied in these regions. In the region where the above-threshold electric field is rising in the direction of electron drift, the electrons move more slowly as one moves to the right because higher fields cause lower velocity. Thus the electrons at threshold electric field are overtaking electrons at higher electric fields causing electron accumulation. As the electron accumulation develops, the excess negative charge causes the electric field to vary more steeply, causing an even faster rate of electron overtaking and accumulation. Conversely, in the region where the above-threshold electric field is falling in the direction of electron drift, electrons move more quickly as one moves to the right because lower fields cause higher velocity. Thus the electrons ahead draw away from those behind causing an electron depletion that steepens the (negative) slope of the electric field variation with distance here also. A rather large maximum of electric field, called a dipole domain can thus quickly develop at such a localized above-threshold field maximum. This domain is ultimately limited in field strength and size by the limit of the applied voltage. When fully developed, this domain with its space charge layers travels at about 1×10^7 cm/s.

If the initial slope of the electric field versus distance on one side of the maximum is much smaller than that on the other side, a significant difference in performance results. The much smaller slope of electric field can cause a much smaller difference in relative drift velocity and the accumulation or depletion process can be initially slowed down appreciably. Indeed, it can be made negligible during the period of time that a microwave oscillation has device electric fields above threshold.

The formation time of the accumulation and depletion space charge layers depend on the density of donors as well as on the slope of the electric field variation with distance. The higher the donor density, the higher the flow of charge, to form the space charge layers more quickly. At higher operating frequencies, yielding shorter periods of oscillation and thus shorter periods above threshold, higher donor density is possible with any given electric field profile.

In Figure 2.4(b) the electric field profile is shown as rising nearly abruptly on the left and falling nearly abruptly on the right with a flat region between. This situation occurs with uniform doping and uniform area in transferred-electron devices. In this case an accumulation layer easily forms in a rapid manner on the left-hand side, but no appreciable depletion layer forms in the adjacent region. The accumulation layer moves from the left boundary into the uniformly doped region at about 1×10^7 cm/s. There must be a depletion layer, on the right-hand side of the uniformly doped region, but it is formed at the interface of a much more heavily doped region. The electric field inside this heavily doped region is below threshold, so the depletion layer does not move into it, but is nearly stationary.

2.33 MODES OF OPERATION

One of the most generally useful modes of oscillation is that mode where only an accumulation layer, and not a depletion layer, forms at the negative electrode. If the time that the device voltage exceeds the threshold value is properly limited by the circuit each cycle of oscillation, the doping profile is relatively flat, and the microwave voltage swing is not severely limited by overloading the oscillation, this mode is assured. This mode was termed 'limited space-charge accumulation' (LSA) by Copeland when he dealt with layers with thickness exceeding the accumulation layer transit distance during the above-threshold portion of each cycle of voltage oscillation. Only the depletion layer is 'limited' in such operation, since the accumulation layer quickly forms, moves into the active layer with a magnitude fixed by the device instantaneous voltage, and is eliminated each cycle when the device voltage goes below the threshold value. Some other workers in the field have used the term 'accumulation-layer mode' to describe this operation near transit time frequencies, although it was Copeland who developed the critical criterion that the electron concentration in the active layer required a limited range of circuit-controlled, above-threshold operating time during each cycle, to prevent the depletion layer, regardless of device length.

If the equivalent circuit of the microwave cavity used with the device in this mode is an inductor with a nearly fixed value of inductance over a wide frequency range, the circuit-controlled, above-threshold operating time is

$$\tau_A = 2\pi\sqrt{LC} \tag{2.6}$$

where L is the circuit inductance and C is the capacitance the device would have if it had no free electrons in its active layer. It is also possible to limit the time spent above threshold each cycle by making the device quite thin, as was shown by Camp and coworkers.[11] In this situation the device voltage is quenched to a value below the threshold value when the accumulation layer reaches the positive electrode.

At room temperature, the depletion layer is limited, in GaAs with a flat doping profile, to a negligible size if the product of electron concentration in the active layer times the above-threshold time each cycle is less than a value just over 2×10^5 s/cm^3. In order to allow the fast onset of oscillations in this mode, especially at the start of a bias voltage pulse applied to a very thick device, this product must also be more than a value just less than 1×10^5 s/cm^3. At higher operating temperatures, where the negative slope of $v(E)$ reduces, both limits of this product are raised. Conversely, at lower operating temperatures, both limits of this product are lowered.

The time spent below threshold can be calculated separately, as a portion of an R_0, L transient, as approximately

$$\tau_B \cong \frac{L(1 - I_v/I_p)}{(R_0 V_T / R_0 I_p)(V_b/V_T - I_v/I_p)}$$

$$\cong \frac{L}{2.5R_0 \left(\dfrac{V_b}{V_T} - 1/2\right)} \tag{2.7}$$

where L is the total circuit inductance, R_0 is the device low field resistance, V_b/V_T is the ratio of bias voltage to threshold voltage and I_p/I_v is the current peak to valley ratio.

A resonator yielding a fixed inductance value for L is the multi-axis radial cavity (MARC),[12] shown in Figure 2.5. This resonator has a radius of the order of $\lambda/8$ or less and has strong output coupling by virtue of the close spacing of the diode and the coaxial output line. The equivalent circuit of this cavity is given in Figure 2.5(b). The wave trap has two very low characteristic impedance radial sections, separated by a higher impedance section, and the length of these sections is set to

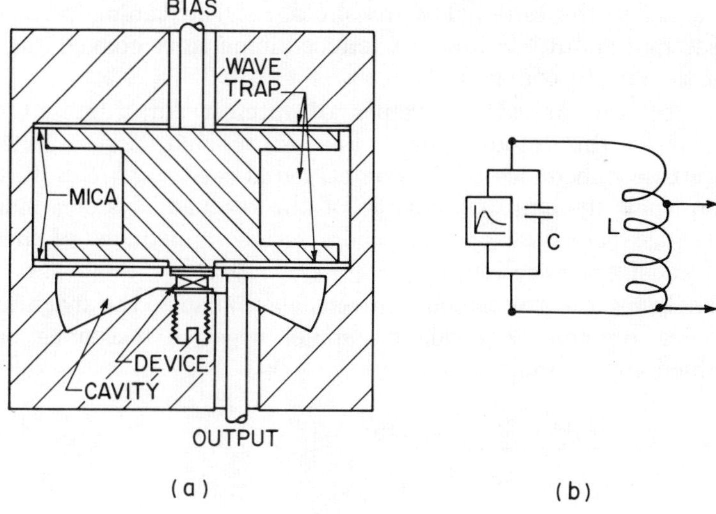

Figure 2.5

avoid half-wave resonances at the fundamental and the second harmonic frequencies.

The relaxation wave shape[13,14] universal in the heavily coupled, low Q LSA operation is shown in Figure 2.6. Figure 2.6(a) gives the current and voltage wave shapes with no special tuning of wave shape to optimize efficiency, while Figure 2.6(b) is with 'tuning' to cause increased efficiency. In both cases the time above threshold, τ_A, is

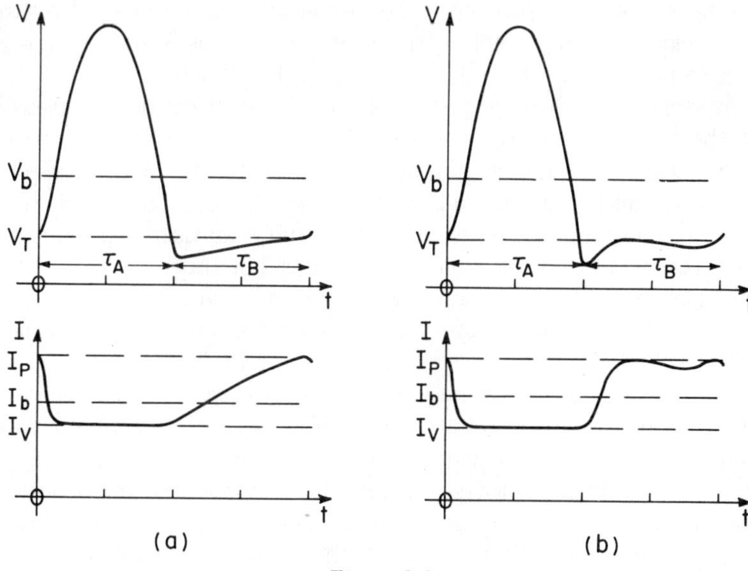

Figure 2.6

nearly equal to the time below threshold, τ_B. The 'tuning' involved is a short section, about $\lambda/8$ long, of reduced impedance coaxial line at the start of the output coaxial line.

Some of the impedance values of interest for a pulsed device operating in this mode might be: $R_0 = 1.5\ \Omega$; $Z_{0c} = 20\ \Omega$, the characteristic impedance of the conical radial cavity; $\sqrt{L/C} \cong 40\ \Omega$, the low-frequency dynamic impedance of the resonant system, including the cartridge parasitics; and the characteristic impedance of the short coaxial transformer section, $Z_{0T} \geqslant 20\ \Omega$.

By analysing the voltage and current wave shapes, the magnitude of the device dynamic equivalent parallel negative resistance can be determined approximately as

$$|R_N| \geq \frac{8R_0(V_b/V_T - I_v/I_p)}{(I_p/I_v - 1)} \tag{2.8}$$

for special tuning of wave shape for high efficiency. If this harmonic tuning is absent, the constant 8 rises to 12. Thus if $V_b/V_T = 5$ and $I_p/I_v = 2$, $|R_N| \geqslant 56\ \Omega$, for $R_0 = 1.5\ \Omega$ and harmonic tuning.

This same analysis yields the highest possible efficiency of

$$\eta \leq (0.9)\frac{(I_p/I_v - 1)}{(I_p/I_v + 1)}\left(1 - \frac{V_T}{2V_b}\left(1 + \frac{I_v}{I_p}\right)\right) \tag{2.9}$$

for special tuning of wave shape. The constant 0.9 falls to 0.6 if the tuning is absent. The best wave shape for high efficiency occurs when the time spent below threshold nearly equals that above threshold. Thus the best low duty cycle efficiency for $V_b/V = 5$ and $I_p/I_v = 2$ is expected to be 25.5% with special tuning, and 17% without.

If temperature or doping gradients, or non-uniform device areas are present the I_p/I_v ratio will be reduced, lowering efficiency. This current ratio is a critical 'efficiency' parameter, just as the product of electron concentration and the above-threshold time, τ_A is a critical space charge control parameter. In an unloaded cavity, an oscillating device in this LSA (or accumulation layer) mode has bias current that closely approximates the value I_v at higher pulse bias values. Uniformly doped devices have I_p/I_v values, at several times threshold voltage, ranging from the velocity peak to valley ratio of 2.0×10^7 cm/s:0.81×10^7 cm/s $\cong 2.45:1$ for extremely thick devices, to much lower values of I_p/I_v for accumulation layer transit time devices. Figure 2.7 shows the reciprocal of the best values of I_p/I_v found experimentally on uniformly doped devices plotted against the reciprocal of the product of electron density and device thickness l. The square intercept point is the ratio of valley to peak velocity. A theoretical result, calculated on the basis of an increased valley current due to an in-phase, or positive

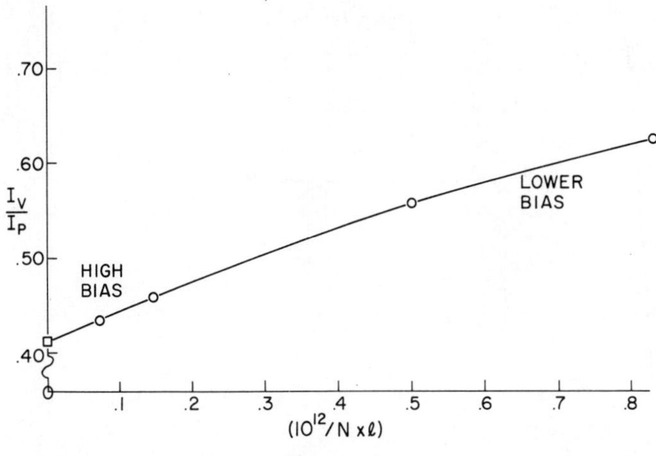

Figure 2.7

conductance, component of current, above threshold, induced by the accumulation layer transit, agrees approximately with this experimental result. Figure 2.8 shows the bias current versus bias voltage on a uniformly doped, unloaded device, with $N \times l$ product of $6.8 \times 10^{12}/\text{cm}^2$, in both polarities. Figure 2.9 shows this curve for an unloaded device with $N \times l$ product of only $1.5 \times 10^{12}/\text{cm}^2$ and with a small variation of electron concentration across the active layer.

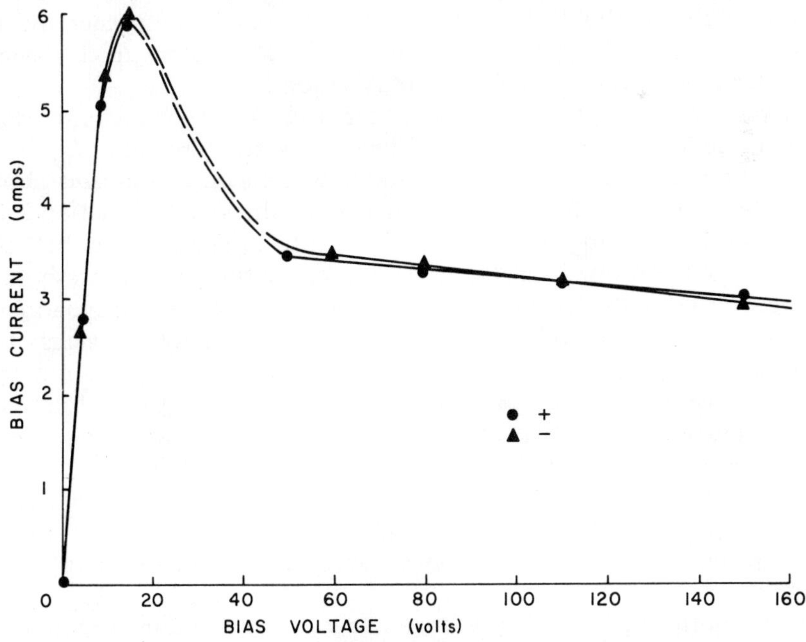

Figure 2.8

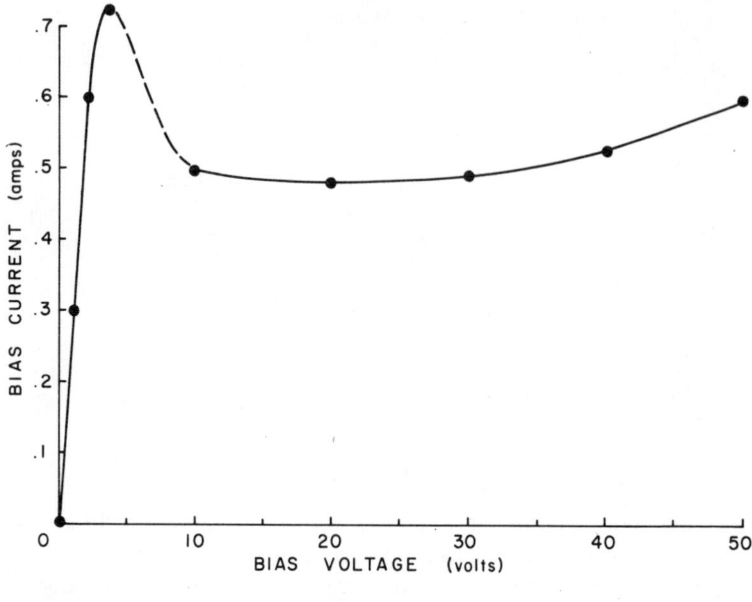

Figure 2.9

If the electron concentration on one side of a device is 1.25 times the electron concentration on the other side of the device, and the device area is uniform, the I_p/I_v ratio is reduced from its best value by the ratio: $1/1.25 = 0.8$, so that the usual $I_p/I_v = 2$ ratio on thick devices becomes $I_p/I_v = 1.6$ with this doping variation.

A variation of area or temperature across the device would also cause a similar reduction of this critical efficiency-determining ratio, I_p/I_v.

The period of oscillation in this mode is the sum of the time above threshold (Equation 2.6) and the time below threshold (Equation 2.7). The time below threshold is usually equal to or smaller than half the period of oscillation. At higher bias voltage, the time below threshold is reduced, causing the frequency to rise. Figure 2.10 shows the experimental values of the period of oscillation in pulsed operation versus the reciprocal of $(V_b/V_T) - 1/2$, yielding a straight line to confirm this approximate theory. In c.w. operation, the tuning with bias follows a similar curve, but there is additional upward frequency tuning due to device temperature rise. At a fixed bias voltage, there is a drop in the time spent below threshold when the temperature of the device is raised. The main effect is the relatively fast drop in I_p, due to the drop in v_p (Equation 2.2) with increased temperature. This drop in I_p in Equation 2.7, as well as the smaller effect of the reduction of I_p/I_v, are large in comparison to the very small rise in V_T. When temperature rises there is usually a variation of time spent below threshold, with

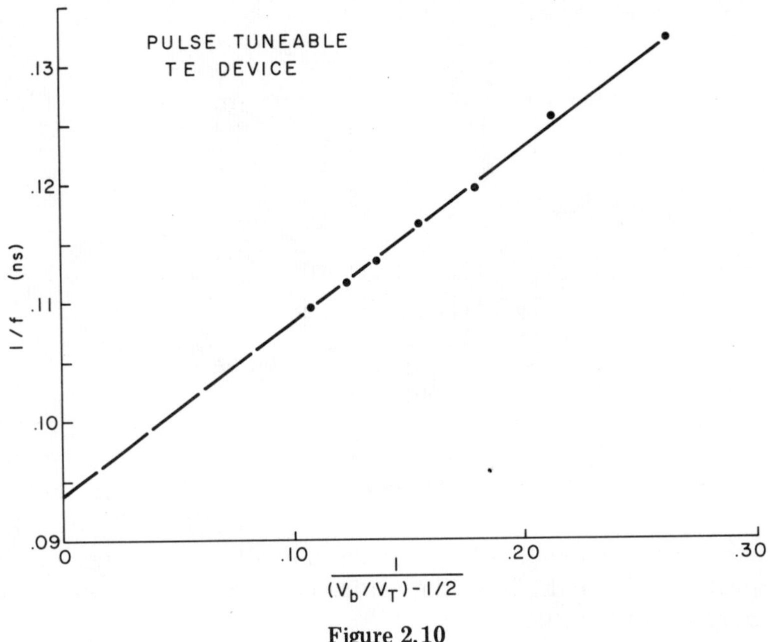

Figure 2.10

temperature of

$$\tau_B \cong \tau_{B0}(1-2.8 \times 10^{-3}\ \Delta T) \tag{2.10}$$

where τ_{B0} is the value of the time spent below threshold at room temperature and ΔT is the change in temperature from this value.

It is possible to use voltage tuning with a frequency sensor and a feedback loop to stabilize the frequency of oscillation. It is also possible to store 10—50 times as much energy in an auxiliary cavity to stabilize the oscillation against frequency drift. Figure 2.11 shows a simple cavity that is useful for stabilizing c.w. devices, operating near the transit time frequency so that starting transients are less critical. The inner $\lambda/4$ coaxial line, between the device and the output coax, is useful as an impedance transformer to optimally load the oscillation. The characteristic impedance of this line is less than 50 Ω, so that raising its characteristic impedance from a very low value to a higher value raises the series loading resistance presented to the device, thus raising the output power. The outer $\lambda/4$ coaxial line, between the output coax and the short, stores energy in proportion to the reciprocal of its characteristic impedance, for a given output power. Since much of the stored energy is in this portion of the circuit, changing its length, or using a perturbing screw in this line, can be used to tune the operating frequency. Such a cavity has little or no second harmonic loading at the output, unless special provisions are made. It should be

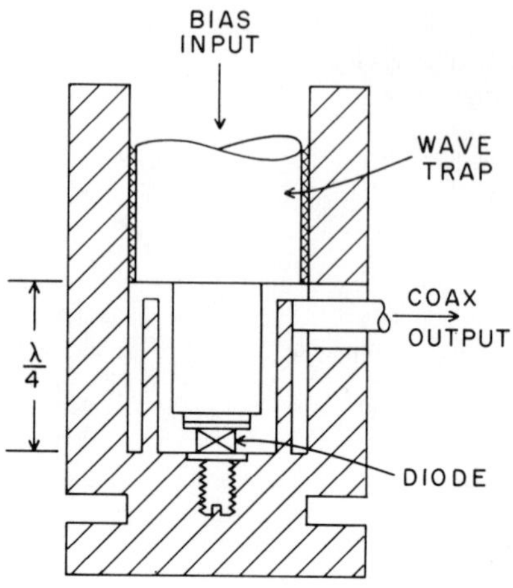

Figure 2.11

noted that the compact, folded coaxial line allows the device to have good heat sinking, and allows the centre conductor of the output coax to be at ground potential.

Historically, the high field domain mode of oscillation was discovered first, in experiments by Gunn. He had rather high electron concentration, for the operating frequency, had alloyed contacts that often yield higher than usual fields at the boundaries, and had no sure way of assuring uniform resistivity in the primitive GaAs and InP material available to him. As a result of this historical fact, a majority of the research and development people in the transferred-electron device field have assumed that domains were always present. Indeed, as well as some researchers analyzing domains that are not present, some force domains to be present by forcing high doping gradients. 'The early discovery of high-field domains has set the development of transferred-electron oscillators back several years' is an often-heard quote. At the present time several laboratories are fully aware of the means and advantages of eliminating domains in order to achieve more efficient and more reliable device performance.

When the electron concentration or area of a transferred-electron device is much smaller near the negative electrode than it is near the positive electrode, domains can form. In addition, moderate to heavy average electron concentration and long circuit-controlled times above threshold also tend to ensure domain formation. If such a domain forms near the negative electrode, it moves toward the anode at about 1×10^7 cm/s and is rather difficult to eliminate in transit. The electric field in front and in back of the domain is just about half the threshold value and these regions are positive resistance in series with the domain in an equivalent circuit. As device voltage rises, the domain both gets wider and has higher field and the device current reduces. Thus the domain is both a voltage variable capacitor and a negative resistance in parallel. Since the domain has a significant amount of parasitic positive resistance in front of it, as well as behind it, the power lost in this resistance lowers the efficiency of the domain mode oscillation compared with that of the accumulation layer mode. The above-threshold portion of the oscillation can be either transit time limited or it can be circuit limited in the domain mode, as it was with the accumulation layer. If the microwave voltage swing in the domain mode is allowed to go below threshold, the R_0, L transient behaviour is again present. If the device voltage rises to very high values with high N,x l product in the domain mode, avalanche breakdown occurs. To prevent breakdown, the following inequality must be approximately met:

$$(N \times l)(V_b/V_T - 1/2) \leq 2.5 \times 10^{13}/\mathrm{cm}^2 \tag{2.11}$$

One significant role that the high-field domain is beginning to play is in microwave logic elements. Figure 2.12 shows the physical and electric field geometry — as well as the characteristic of a three-terminal Gunn domain logic element. The active layer is grown on a semi-insulating substrate and the main electric field is along the interface. The gate is back-biased to partly deplete the channel. The drain-source current can then be raised to the point where a peak in electric field under the gate triggers domain formation and thus substantially lowers the device current. Such a device used as a logic element has nearly 2:1 drop in terminal current when a domain is in transit. The active layer thickness times the electron concentration should be $1 \times 10^{12}/\mathrm{cm}^2$ or more to cause such a substantial current drop at domain initiation. When the active layer thickness and the gate width is about 1 μm and the source drain separation is 10 μm, rates as high as 5 Gbit/s have been obtained.[15] Several logic circuits have already been developed,[16] and

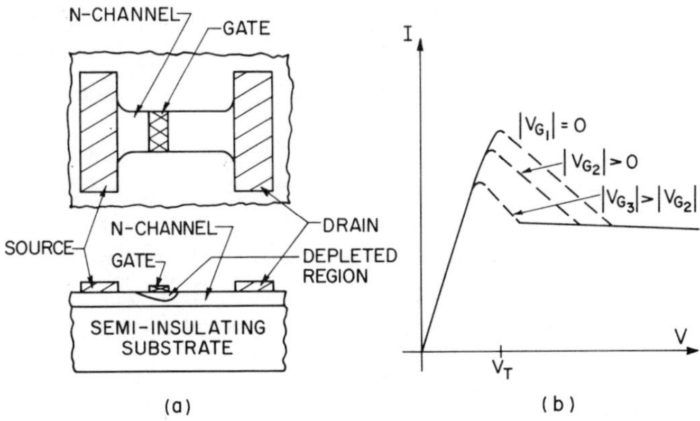

Figure 2.12

there may be a considerable development activity in this area in the near future. A book on various Gunn logic[17] has also recently been published.

It is also possible to have a small-signal amplification mode of operation of transferred-electron devices. In such devices, the voltage–current static or d.c. bias characteristic is stable, often having a positive slope. In many early devices, a static non-uniform electric field profile was caused to assure a stable static I versus V characteristic while allowing a negative resistance to microwaves over a broad band of frequency centred at approximately the transit time value. These devices either had a low $N \times l$ product, or else they had a higher field at the positive electrode than they did at the negative electrode. This latter condition of high anode field can be caused by forcing that side of the device to have lower electron concentration, lower device area, or lower peak velocity due to higher temperature. Since there was a variation of electric field from the low value to the high value in the direction of electric flow, there could never be a uniform, optimum electric field in the interaction region of the devices. As a result the saturated output power, device efficiency and noise performance were not yet optimum. Ideally it would be best to have nearly uniform electric field, at a value between 5 kV/cm and 10 kV/cm, across the active layer. In order to accomplish this in a static situation, it is necessary to have the electron injection be limited, and that these electrons be hot, or energetic. One method[18] of accomplishing this is to have a thin, lightly doped layer at the negative electrode side and to have the usual value of doping throughout the rest of the layer. By proper design it is possible to have space charge limited current in the thin, lightly doped layer and yield a proper current density and the proper electric field at the start of the uniformly doped region. The uniform above-threshold electric field over the uniformly doped layer will then yield efficient, low noise amplification. Noise figures as low as about 10 dB in GaAs at 10 GHz and about 7 dB in InP at 30 GHz have been obtained as the best results in these amplifiers. The bandwidth is of the order of one octave and could perhaps be extended.

2.4 EPITAXIAL GROWTH AND PROCESSING OF MATERIALS

There are two established ways of epitaxially growing GaAs and InP. One is vapour phase epitaxy (VPE), using pure arsenic trichloride and gallium (or GaAs crystals) as chemical sources. The other is liquid phase epitaxy (LPE) using pure gallium saturated with GaAs as chemical sources.

Figure 2.13 shows such a VPE apparatus, the operation of which will be briefly reviewed. If gallium is used in the source region, at 800 °C, it must be allowed to form a crust over the entire surface before epitaxial growth is initiated to insure As saturation. Before epitaxial growth, a back etch of the surface can be accomplished by

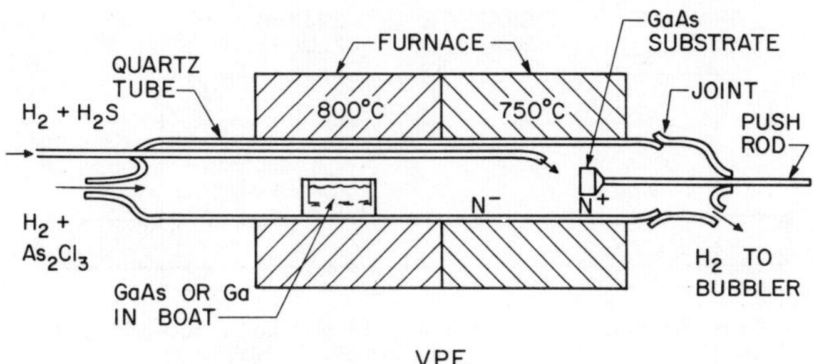

Figure 2.13

high temperature operation at 850—900 °C. The epitaxial layer is deposited at 710—750 °C, and may be moved upstream to grow lightly doped material and downstream in the H_2S to grow material that is heavily doped. By balancing the amount of arsenic trichloride flow, relative to the H_2 flow, it is possible to control the ambient net donor density over a significant range, including pure material. The vapours that transport the chemicals are As_4 and GaCl. An increase in the arsenic trichloride mole fraction and a decrease in growth temperature both cause a decrease in the electron concentration.[19] Several laboratories in Europe, USA and Japan have regular success with the epitaxial growth with low ambient donor density and with ambient acceptor densities as low as one-third the ambient donor density.

Many laboratories in Europe, USA and Japan also have succeeded in using LPE to achieve similar good results for GaAs. LPE will be emphasized in this chapter on account of the author's choice and experience.

The apparatus used for LPE is shown in Figure 2.14, involving melts in a graphite boat in a quartz tube with palladium-diffused hydrogen flowing through. A cross-section of the boat and melts are shown in Figure 2.15 with a GaAs source crystal on the bottom and top of the

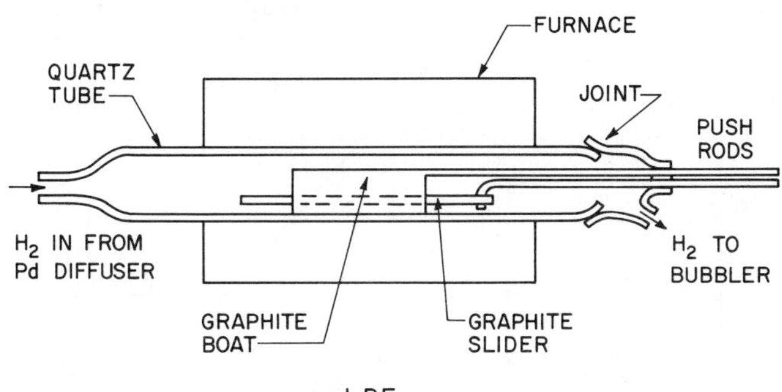

Figure 2.14

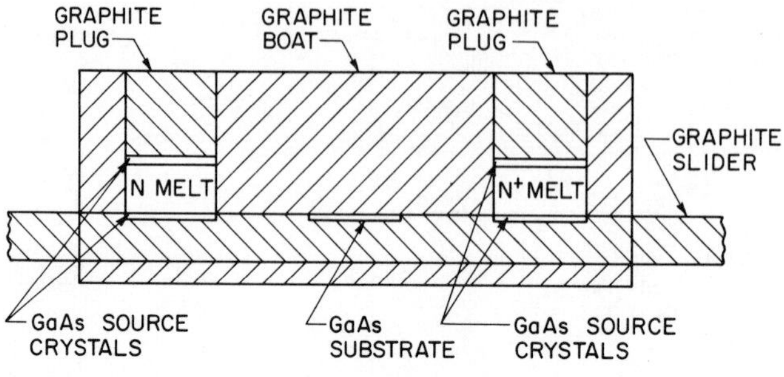

Figure 2.15

melts, in a pre-growth saturation condition. The bottom piece is a slider that is used to slide the bottom source crystals or the substrate into position under the melt.

Pure gallium can dissolve increasing amounts of arsenic, from GaAs source crystals, as its temperature is raised. This concentration rises exponentially with temperature, in the usual range of temperature

$$C(T) \cong C_0 \, e^{-(T'/T)} \tag{2.12}$$

where C_0 and T' are constants. The same is true of pure indium dissolving phosphorus from InP source crystals. Figure 2.16 shows the atomic percentage arsenic in gallium, and percentage phosphorus in indium, that can be included at saturation condition as a function of temperature. Figure 2.17 is another useful presentation that equivalently shows the thickness of GaAs crystal that can be melted into a 1 cm tall Ga melt, and of InP crystal that can be melted into a 1 cm tall In melt, for saturation, as a function of temperature.

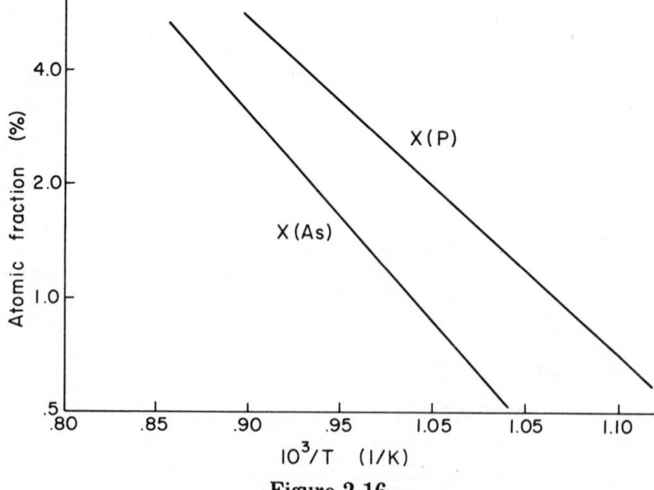

Figure 2.16

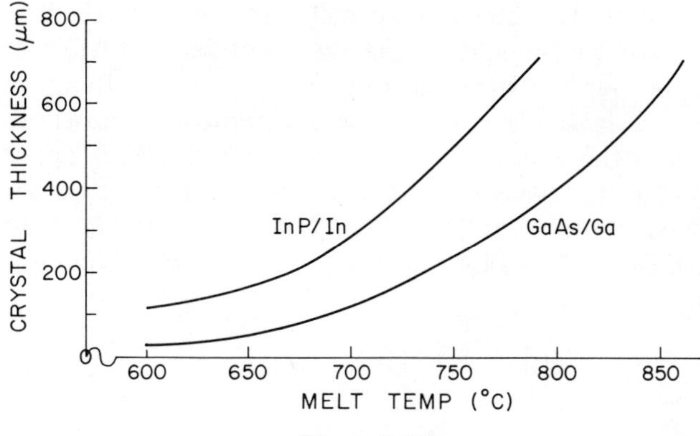

Figure 2.17

Once in the Ga melt the arsenic can be removed by cooling the melt in contact with a substrate crystal, or any other appropriate seed crystal, forcing GaAs to grow on these crystals. The removal rate of the arsenic is normally limited by the rate of diffusion of the arsenic through the melt. This diffusion constant is strongly temperature-dependent and is shown in Figure 2.18 along with the equivalent presentation for phosphorus diffusion[20] through indium.

There are three methods of causing diffusion-limited growth of GaAs in LPE. One is the simple linear reduction of temperature. Over a limited range of reducing temperature, where the diffusion constant of arsenic, D, is considered constant, and over a limited time period, Rode[21] has shown that growth thickness for $d \ll d_l$ is

$$d \cong (4d_l/3W)\sqrt{\frac{Dt}{\pi}} \tag{2.13}$$

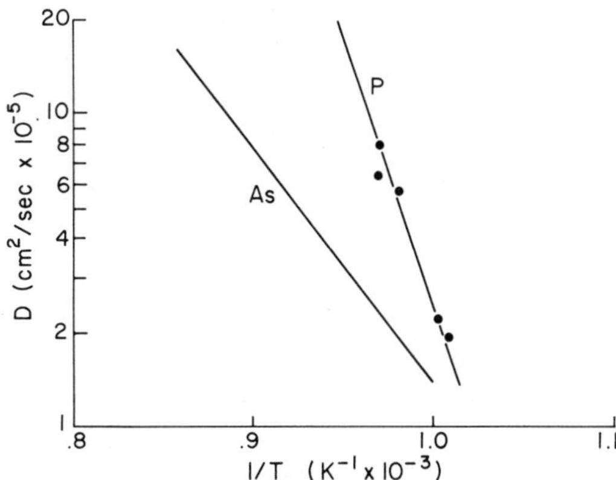

Figure 2.18

where d_l is the limiting thickness possible in long growth times, W is the melt height (or half-height if the melt has seed crystals on top and bottom), D is the diffusion constant of arsenic in gallium, and t is the growth time. d_l is slowly rising as the temperature is slowly dropped, of course. This is the method most often used for LPE growth of thin transferred-electron devices. Another method is to drop the temperature a limited amount of 10—20 °C before sliding the substrate into contact with the melt, Rode has shown that growth thickness for $d \ll d_l$ is

$$d \cong (2d_l/W)\sqrt{\frac{Dt}{\pi}} \tag{2.14}$$

where d_l is now fixed by the temperature drop used. Figure 2.19 shows the complete plot of $d(t)$ from these two methods of growth. One final diffusion-limited growth method is to cause a temperature gradient between the top, source crystal and the substrate at the bottom. This yields a growth thickness of

$$d = (D/W)(C_0 T'/T_0^2)(T_w - T_0)t/[C_0 - C_s \exp(T'/T_0)] \tag{2.15}$$

where T_0 is the seed (substrate) crystal temperature, T_w is the top source crystal temperature, C_0 and T' are the constants of Equation 2.12, and C_s is the arsenic concentration in solid GaAs. With the temperature gradients easily possible, only about 1 μm/h growth rate occurs in this steady state situation.

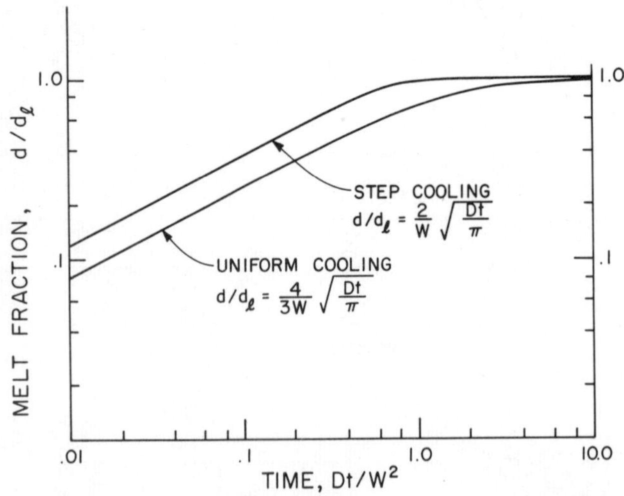

Figure 2.19

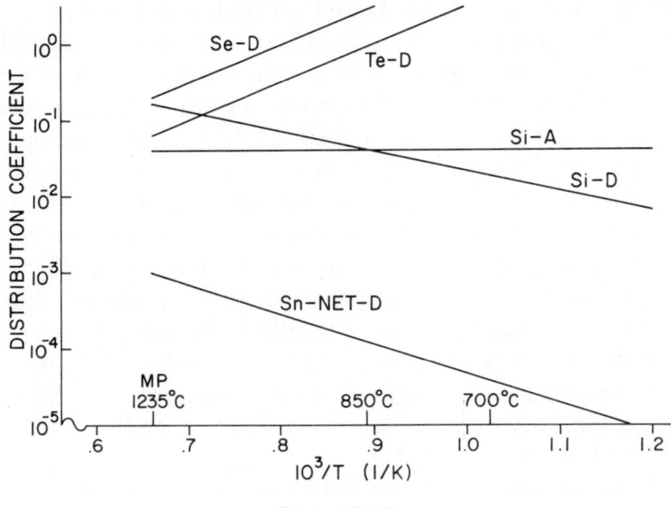

Figure 2.20

In order to cause faster growth rate, the locations of the source and substrates can be reversed. In this case melts over about 0.5 cm tall can have convection occur. Long[22] has done experimental and analytical studies of this convection. Steady state growth rates of 5—10 μm/h can be accomplished with very uniform electron concentration. Layers as thick as 200 μm, grown in this manner, have yielded peak power levels in excess of 1 kW in S, C, and X bands in pulsed LSA operation.

The inclusion of impurities in LPE GaAs has been widely studied and developed. An important parameter of these impurities is the distribution coefficient, defined as the ratio of the atomic percent of the impurity in LPE grown crystal to the atomic percent of the impurity in the melt. Figure 2.20 shows the log of the net distribution coefficients of the useful donors Sn, Te and Se, as well as the donor and acceptor distribution coefficients of Si versus the reciprocal of temperature. The exponential dependence of these impurity distribution coefficients with the reciprocal of temperature (kelvins) causes doping gradients in layers grown by linearly cooling melts during growth, especially at faster cooling rates. Tin doping is usually used, yielding a gradual decrease in donor density as the temperature is dropped during growth. Typically there is 20%—30% more donor density on the substrate side of the active layer than on the opposite side. It is possible to get controlled, lower donor density gradients in the direction of growth by slower cooling rates, by precooling the melt before growth, or by superposition of precooling and slow linear cooling during growth. In order to get a negligible donor density gradient, growth at a fixed temperature in a steady state condition is required.

The control of background impurities in LPE requires precautions.

The graphite boat should be baked by induction heating in a good vacuum to a temperature of 1200—1500 °C. The boat should also be subsequently baked in pure palladium-diffused hydrogen at a temperature of 850—900 °C. The growth system should be vacuum leak tested to prevent air leaks. Fast hydrogen flow of 0.5—1 l/min for usual quartz furnace tubes, should be used to flush air and outgassed impurities from the melt after loading the furnace for growth. The melt should be baked at or above the growth temperature for at least 8—16 h before growth is initiated. If these procedures are not used, $10^{16}/cm^3$ (or more) net donors can occur in layers grown from unintentionally doped melts. As well as oxygen, sulphur and other donors may be present and would be removed by prolonged H_2 baking. Usually $10^{14}/cm^3$ (or less) net donor concentration, with a modest amount of compensation, occurs if these procedures are used and growth occurs in the 700—750 °C temperature range.

Substrate preparation must be done carefully in order to achieve the best transferred-electron device performance. First, the substrate is carefully chemically polished on both sides to remove work damage. Even when the sawing of the substrate slices has been done carefully, 75—100 μm of work damage occurs. The substrates are usually received with 500 μm thickness and at least 100 μm must be removed from both sides. The unused side of the substrate cannot be left unpolished, because the strain due to work damage on that side can cause damage to spread throughout the substrate during the heating cycle of epitaxial growth. Etching can be used to remove work damage on the back side of the substrate, but this leaves an uneven surface. Low device yields and less reliable operation can result from this spreading work damage if it is not removed.

Next it is possible to improve device yield performance, and reliability by growing a buffer layer on the substrate. This buffer layer, first developed by the author and his associates for high performance LSA devices, is a moderately heavily doped ($1—5 \times 10^{17}/cm^3$—tin) epitaxial layer that is 20—50 μm thick. The reasons for its use are at least twofold. One reason is that the undesired ambient impurities in the buffer layer are orders of magnitude lower than those in the substrate. Compensating impurities, such as zinc, are not allowed to diffuse from the substrate into the lightly doped n-type active layers, because of the thick buffer layer between. The other reason is that etch pit densities are substantially lower on the buffer layer, compared with substrates. Significant differences in lattice constant between the substrate and the active layer may be accommodated in the buffer layer contributing some of this benefit.

Growth of the active layer on the buffer layer (or substrate) must be started in LPE by wetting the surface. If the surface is oxidized or deficient in arsenic, it is important to back-etch for 1 μm or so by melting the surface. If the surface is unoxidized and has not lost arsenic, or been otherwise degraded, wetting is quick and certain without back-melting. Indeed, to grow very thin, uniform layers with a planar interface, supersaturated melts and/or very fast cooling rates help to nucleate growth quickly over the whole surface rather than at a few isolated spots that slowly coalesce. As a result, smooth thin layers of uniform thickness can only be grown from non-degraded surfaces with little or no melting of the substrate.

Most transferred-electron device active layers have thin, heavily doped epitaxial contact layers grown on top. Sequential growth of the buffer layer, the active layer, and the contact layer can be accomplished in one growth cycle using the boat shown in Figure 2.15. Care should be taken not to supersaturate the melts by more than 10—20 °C in this operation or small crystals form in the melt and make rough surfaces.

After epitaxial growth, alloyed metal contacts are applied to both the epitaxial contact layer and the back side of the substrate. Germanium—gold eutectic mixture, melting at 356 °C, is normally used. It is evaporated on, in a vacuum, to a thickness of 0.3—0.5μm and is often covered with a layer of pure gold. In some cases a very thin nickel layer, 5% of that of Ge—Au mixture, is evaporated on the GaAs first. Alloying is done in a short, controlled transient heating cycle where the wafer is in the furnace about 1 minute and reaches a maximum temperature of 450 °C or so. The outer pure gold layer does not melt and can be used to allow either direct thermocompression bonding or electroplating of gold for better heat sinking.

The wafers are diced by wire saw or cleaving, or may be masked and etched into mesas before the devices are separated for mounting. For high performance devices it is essential to minimize the work damage on the device surface while keeping uniform device area. Whenever pressure is applied to the finished devices, such as in thermocompression bonding, damage can result in reduced electrical performance or reduced reliability. Reduction of measured device current peak to valley ratio can occur when pressures above 4,500 p.s.i. are used on small area devices and above 3,000 p.s.i. are used on larger area devices, depending on the roughness of the heat sink surface. With plated heat sinks and etched mesas it is possible to bond the heat sink without applying pressure to the device, avoiding this problem.

2.5 DEVICE ULTIMATE CAPABILITIES

The approximate Equation 2.9 can be used to estimate the highest value of pulse efficiency at high pulse power levels. Using high pulse bias voltage, near ten times the threshold value, the oscillation is nearly sinusoidal and an efficiency of 18.5% is expected and obtained from devices having a current peak to valley ratio near 2.

A nominal transit time thickness device, say 10 μm thick for 10 GHz, can thus yield 9 W peak power out if it has 1 Ω low field resistance. 35 V and 1.4 A are the resulting bias values and the optimum negative shunt resistance is about 115 Ω. The value of this peak power will scale with the square of the reciprocal of frequency, except at much higher frequencies, where the approximately 75 GHz (in GaAs, 150 Hz in InP) cut-off frequency is caused by electron relaxation time limits.

If a device is constructed to be much thicker than the nominal transit time thickness, and is operated within the limited space charge accumulation (LSA) device and circuit design criteria, then much higher pulse power can be obtained. In order to keep the same impedance level, both the device area and length are increased in proportion. Thus the active layer volume and the resulting peak power in pulsed operation are increased in proportion to the square of the ratio of active layer thickness to nominal transit time thickness. Thus a device with 1 Ω low field resistance and 100 μm thickness can be operated to 900 W at 10 GHz. Now there would be 350 V and 14 A bias values. The maximum experimental value of pulse power generated by a 200 μm thick, 2 Ω device at 10 GHz is 1.1 kW to date, while an even thicker device has yielded 6.0 kW at 1.75 GHz.

When high values of average power or continuous power are required, the device temperature rise, from average heat dissipation inside the device, causes a reduction of oscillation efficiency. This can best be seen by examining Figure 2.3 which shows the reduction of peak current, valley current and the current peak to valley ratio with increased (uniform) temperature. Since higher active layer temperature in the device reduces the current peak to valley ratio, the device microwave current swing and thus the efficiency will be reduced.

An important parameter governing the temperature rise due to power dissipation is the device thermal resistance. The thermal resistance relates the temperature rise in the device to the power dissipated as heat in the device. The temperature is not uniform in the active layer because the heat sink is not uniform since it is attached to one side. The temperature reduction from its value on the hot side, away from the heat sink, is approximately parabolic with distance. The total temperature drop from the hot side to the cooler side depends on the average power dissipated, the device geometry, and the thermal conductivity of GaAs. The thermal conductivity of GaAs is approximately (150/ T_A) watts/cm/K, where T_A is the GaAs active layer average temperature in kelvins.

The active layer length divided by its area is most easily evaluated as the device low-field resistance, R_0, divided by the room-temperature resistivity ρ_0. The portion of the thermal resistance relating the total temperature reduction across the active layer from the hot side to the cooler side is thus

$$\theta_A \cong (R_0/\rho_0)(1/2)\left(\frac{T_A}{150}\right) \tag{2.15}$$

There is another temperature drop across the GaAs contact layer, which can be evaluated as

$$\theta_c \cong \theta_A(2.5l_c/l_A)(T_c/T_A) \tag{2.16}$$

where l_c is the contact layer thickness, l_A is the active layer temperature, and T_c is the contact layer average temperature.

The final temperature drop from the cool side of the GaAs contact layer to a distant point in the copper heat sink is the thermal spreading resistance in the heat sink. This portion can be approximately evaluated as

$$\theta_s \cong \frac{\sqrt{R_0/\rho_0}}{16\sqrt{l_A}} \tag{2.17}$$

There is often a parasitic thermal resistance of 2 K/W in the threaded cartridges used, half due to a short pedestal on which the device is bonded and half due to the limited area of contact of the threads on the cartridge and in the tapped hole.

One final portion of the thermal resistance is that irregular value due to the irregularity of metal to metal contact area, to high spots on the heat sink, when thermocompression bonding with limited pressure is used. If the alloyed metal contact and its gold overlay is less than 1 μm in thickness, the heat must converge through the GaAs to the patches of good metal to metal bond. Because GaAs is only about one-eighth as good a heat conductor as copper, an appreciable added thermal resistance of (5—10 K/W), a few times the thermal spreading resistance, can be present in this portion. If gold is plated on the alloyed metal contact up to a thickness equal to the active layer thickness, this gold conducts the heat parallel to the heat sink surface to those well bonded spots with only about an additional term of from one to two times the spreading resistance for this portion of the thermal resistance. If the plated heat sink is soldered down, only about 2 K/W additional parasitic resistance results.

For a given amount of average power dissipated in a device, not only the total temperature rise above ambient on the hot side of the active layer, but the temperature on the cooler side of the active layer can be calculated from the above portions of the thermal resistance. The thermal resistance of individual devices can be inferred directly from low-field resistance changes with increased dissipation, or more easily from the reduction of peak current from its pulsed, low duty cycle value as dissipation is increased as shown in Equation 2.2 for the temperature ranges indicated.

Once the temperatures of the hotter and the cooler sides of the active layer have been determined for some value of average power dissipation, the device current peak to valley ratio can be predicted.

As an example, a device with 2 Ω low-field resistance is made from a wafer 10 μm thick and $2 \times 10^{15}/\mathrm{cm}^3$ electron concentration. With uniform power dissipation, such a device would ideally have at room temperature 5 K/W, 2 K/W, 4.5 K/W and 2 K/W for the active layer, contact layer, thermal spreading resistance and cartridge parasitic thermal resistance, for a total 13.5 K/W. At 500 K this ideal device would have about 18.5 K/W with uniform power dissipation. The parasitic thermal resistance from the thermocompression bond interface would raise the 500 K value to about 26 K/W if no plated heat sink is used, and to about 21 K/W if a plated heat sink is used.

This device is thin enough to operate c.w. near its transit time frequency. The average power dissipation in an accumulation-layer transit time oscillation rises approximately as a parabola with distance from negative to the positive electrode. As a result, the heat is much more easily removed at the positive electrode. The consequence of this result is that the active layer contribution to thermal resistance is only half that of uniform dissipation when the anode of an accumulation layer device is the heat sink. If the polarity of bias is reversed on such a device, the active layer contribution to thermal resistance is tripled. Thus at 500 K where uniform dissipation yields 8.33 K/W in the active layer, only 4.17 K/W results with heat sink positive, but 12.5 K/W results with heat sink negative, in accumulation layer oscillations near the transit time frequency. The above device, with a plated heat sink, would have about 17 K/W and 25 K/W in the two polarities, because of the reversed geometry of heat dissipation relative to the heat sink. On a small area device yielding 14.5% efficiency and 0.15 W in c.w. operation, 40 K/W and 50 K/W have been measured with an infrared instrument in these two polarities.[23]

In very thick device operation, the accumulation layer only travels across a small portion of the active layer thickness, so that the average electric fields and the power dissipation are nearly uniform across the active layer.

In determining the ultimate average and c.w. power generation capability of transferred electron devices, the reduction of the current peak to valley ratio is the critical parameter. The device current peak to valley ratio is determined only by the conditions on the hot side of the active layer when the electron concentration variations have been optimized across the active layer. In order to optimize the electron concentration variations, the valley current density on the hot side of the device must exceed that on the cold side under full operating power. The peak current density on the hot side should not exceed the peak current density on the cold side, however. Thus for optimum c.w. performance with the heat sink positive, up to 10% more electron concentration is desired on the side of the active layer away from the heat sink. If the heat sink is operated negative polarity, from 15%—30% more electron concentration is desired on the substrate side. The exact shape of the electron density variations may not be important, as long as a monotonically rising density occurs.

Once the thermal resistance is known, the hot side temperature is known, for any given power dissipation, and the hot device current peak to valley ratio can be determined for the appropriate operating polarity, with its corresponding doping profile. The efficiency drop, due to the reduction of current peak to valley ratio, is not fast until 200—240 K rise in temperature above room temperature. The efficiency has dropped to nearly two-thirds of its room temperature value at this temperature. Thus a device with $N \times 1$ product of $2 \times 12/\text{cm}^2$ has its room temperature I_p/I_v ratio reduced from 1.8 to about 1.5 for the 240 K rise. This yields a reduction of c.w. maximum efficiency from 13% to 8.5% at three times threshold voltage with no harmonic tuning.

If efforts are made to get the highest possible power with the highest possible efficiency, it is expected that 1 W with 10% efficiency can be repeatably obtained at 10 GHz in the near future from single devices. Such a result would require a plated heat sink and harmonic tuning, and would result in more reliable operation, owing to lower operating temperature, if the heat sink polarity is positive. Theoretically 3 W with 15% efficiency with a perfect metal heat sink and 5 W with 15% efficiency with a perfect diamond heat sink are also predictable. These results may not be repeatable in production, however. 2.25 W with 9.3% efficiency has been obtained to date at about this frequency with a diamond heat sink without harmonic tuning and with the heat sink polarity negative, rather than the positive polarity required for best performance.

Since thermal resistance rises nearly linearly with frequency for a fixed impedance device, while input power reduces as the reciprocal of the square of frequency, it is possible to theoretically scale c.w. power as approximately the reciprocal of frequency. It is possible that a higher ratio of bias voltage to threshold voltage, as well as lower device resistance, is necessary to obtain best results at high frequency.

2.6 OTHER DEVICE PERFORMANCE CHARACTERISTICS

The transferred-electron device range of operating frequency is rather broad. Efficiency can remain high over a tuning range of an octave for c.w. devices. Pulse operation of particular devices is also successful over an octave range of frequency. The capacitive or displacement current in transferred-electron oscillators changes as expected for a fixed capacitance, as frequency is varied, while the (negative) conduction current stays nearly constant as frequency changes. The latter is especially true if the time spent above threshold is less than the transit time for the electron accumulation layer or any other space charge disturbance. In typical high efficiency operation in simple circuits the displacement current is only a few times the conduction current so that the circuit quality factor, (negative) Q is low, its magnitude being ≤ 5. This leads to high tuneability with bias voltage unless more energy storage is designed into the resonator.

As indicated by the low noise figure in its amplification made, transferred-electron oscillators have low noise. With comparable Q factors, such oscillators have noise levels that are approximately at levels normally encountered with klystron oscillators.

REFERENCES

1. B. K. Ridley and T. B. Watkins, 'The possibility of negative resistance effects in semiconductors', *Proc. Phys. Soc.*, 78, 293—304 (1961).
2. C. Hilsum, 'Transferred electron amplifiers and oscillators', *Proc. IRE*, 50, 2, 185—189 (1962).
3. J. B. Gunn, 'Microwave oscillations of current in III—V semiconductors', *Solid State Communications*, 1, 88—91 (1963).
4. H. Kroemer, 'Theory of Gunn effect', *Proc. IEEE*, 52, 12, 1736 (1964).
5. J. A. Copeland, 'A new mode of operation for bulk negative resistance oscillators', *Proc. IEEE*, 54, 1479—1480 (1966).
6. A. Rose, 'The acoustoelectric effects and the energy losses of hot electrons — Part I', *RCA Review*, 98—139 (1966).
7. A. Rose, 'The acoustoelectric effects and the energy losses of hot electrons — Part II', *RCA Review*, 600—631 (1966).
8. J. G. Ruch and W. Fawcett, 'Temperature dependence of the transport properties of GaAs determined by a Monte Carlo method', *J.A.P.*, 41, 9, 3843—3849 (1970).
9. P. A. Houston and A. G. R. Evans, 'Electron velocity in *n* GaAs at high electric fields', *Electronics Letters*, 10, 16, 332—333 (1974).
10. C. Hilsum, 'Simple empirical relationship between mobility and carrier concentration', *Electronics Letters*, 10, 12, 259—260 (1974).
11. W. O. Camp, Jr., D. W. Woodard and L. F. Eastman, 'Bias-tuneable c.w. transferred-electron oscillators', *Proc. Fourth Cornell Conference, Microwave Semiconductor Devices*, pp. 177—183, August 1973.

12. L. F. Eastman, 'Multi-axis radial circuit for transferred-electron devices', *Electronics Letters*, 8, 6, 149—151 (1972).
13. W. O. Camp, Jr., 'Computer simulation of multi-frequency LSA oscillations in GaAs', *Proc. IEEE*, 57, 6, 220—221 (1969).
14. B. Jeppsson and P. Jeppesen, 'A high power LSA relaxation oscillator', *Proc. IEEE*, 57, 6, 1218—1219 (1969).
15. K. Mause, A. Schlachetzki, E. Hesse and H. Salow, 'Monolithic integration of gallium arsenide Gunn devices for digital circuits', *Proc. Fourth Cornell Conference, Microwave Semiconductor Devices*, pp. 211—223, August 1973.
16. T. Sugeta, M. Tanimoto, T. Ikoma and H. Yanai, 'Gunn effect digital devices for sub-nanosecond pulse regenerators and logic functions', *Proc. Fourth Cornell Conference, Microwave Electron Devices*, pp. 201—210, August 1973.
17. H. Hartnagel, 'Gunn-effect logic devices', American Elsevier Publishing Co., New York, 1973.
18. J. Magarshack, private communication.
19. J. V. DiLorenzo, 'Vapor growth of epitaxial GaAs: A summary of parameters which influence purity and morphology of epitaxial layers', *Journal of Crystal Growth*, 17, 189—206 (1972).
20. V. L. Wrick, III and L. F. Eastman, 'Diffusion-limited LPE growth of InP for microwave devices, *Proc. Conference on GaAs and Related Compounds, Deauville, France*, September 1974.
21. D. L. Rode, 'Isothermal diffusion theory of LPE: GaAs, GaP, bubble garnet', *Journal of Crystal Growth*, **20**, 13—23 (1973).
22. S. I. Long, J. M. Ballantyne and L. F. Eastman, 'Steady state LPE growth of GaAs', to be published in *Journal of Crystal Growth*.
23. S. Y. Narayan, private communication.

CHAPTER 3

Avalanche and Barrier Injection Devices

PETER WEISSGLAS

3.1 GENERAL INTRODUCTION

3.1.1 Microwave Vacuum Tubes

It is instructive to begin a discussion of the basic principles used in microwave generating solid state devices by looking at the various types of vacuum tubes which are being used for the same purpose. In the triode a grid is used to control the current between a pair of electrodes to which a high voltage is applied. In this manner d.c. is converted to a.c. One of the factors limiting the high frequency performance of this type of devices is the requirement that the transit time of electrons from cathode to anode must be short compared to the period of oscillation. With typical velocities of about 10^7 m/s and distances of about a few millimeters this leads to a maximum frequency of a few GHz. This is also about the state of the art for actual triodes being made today.

In order to reach higher frequencies different basic principles have to be used. It has turned out to be a fruitful path to follow to use the comparatively long transit time in a constructive manner. In the klystron a velocity modulation is imparted to a beam of electrons at one point and the ensuing charge modulation is allowed to interact with the voltage at a downstream position where the transit time induced phase shift is about 180°. In this way the charge modulation at the downstream point delivers energy into the r.f. field rather than extracting energy.

In the travelling wave tube (TWT) another principle is used based on a distributed interaction between electrons and fields rather than a localized one. The TWT uses an electron beam which interacts with a slow electromagnetic wave which propagates in velocity synchronism with the beam. When a velocity modulation is imparted to the beam the fast electrons will tend to overtake the slow wave and deliver kinetic energy to the wave giving rise to amplification.

There also exists a variety of crossed field devices whose operation from the present point of view is similar to the klystron and the TWT although their performance and detailed mode of operating may be different.

3.1.2 Semiconductor substitutes for microwave tubes

Let us now turn to solid state devices and examine what similarities and differencies exist. The transistor, FET or bipolar, is analogous to the triode. One electrode controls the current flowing between two other electrodes to which a high voltage is applied. The same frequency limitations therefore apply although the numbers are different. Carrier velocities in semiconductor materials are low but small dimensions can be handled and, as a consequence, there exist transistors which cover a large portion of the microwave frequency region. However, at high frequencies the useful powers are falling rapidly and it has long been realized that a need exists for solid state devices operating on fundamentally different principles.

We shall see, however, that it is not possible to directly duplicate the idea behind the klystron. The main reason for this is the difference between the way carriers are transported in vacuum and in semiconductors. In vacuum the carriers are virtually free and unless they are exposed to external electric fields move at constant velocity. In semiconductors the electrons or holes interact strongly with the lattice of atoms and mean free paths are typically much less than one μm. In weak applied fields the electron velocity is proportional to the field rather than linearly increasing with it as in vacuum. In still higher electric fields, as will be discussed below, the electron and hole velocities in semiconductors saturate and become almost independent of the field. Another way of looking at these differences is to note that an electron beam in vacuum dissipates no energy whereas in semiconductors the electrons continuously give up their energy to the lattice with typical mean free paths for energy relaxation of less than one μm. To extract the kinetic energy of drifting electrons in a solid therefore becomes much more difficult than in vacuum as it has to be done in strong competition with the ohmic dissipation of energy to the lattice.

As mentioned above the klystron operates with velocity modulated electrons. In semiconductors it is possible to obtain a velocity modulation at weak applied electric fields but this velocity modulation will everywhere be in phase with the local electric field and therefore cannot give rise to a negative resistance as in the klystron.

The wave interaction principle behind the TWT, however, is conceivable in semiconductors. The main problem is that the attainable velocities of electrons in semiconductors are quite small. Therefore the slow wave must have a very short wavelength and matching to outside circuits becomes very difficult. Nevertheless the acoustic amplifier is one example of a working device.

3.1.3 Microwave solid state transit time devices

The class of solid state transit time devices which will be discussed in this chapter is based on a different concept from those discussed above. If a density modulation is imposed on charge carriers, drifting in fields high enough to cause velocity saturation will propagate almost without attenuation. It is therefore possible to obtain a transit time delay also in semiconductors. Two important observations should be made before discussing the possibility of using this transit time effect in an active device. As will be shown below it is first of all not possible to use this mechanism alone to generate a negative resistance as the maximum terminal delay between current and voltage is 90°. The reason for this is that charges in transit contribute to the external current during all the time when they are in transit. The second thing to note is that phase mixing takes place if the electric field is so weak that the carrier velocity is proportional to the field. As a result of this phase mixing the modulation will suffer an attenuation of amplitude in addition to a delay in phase.

From these general arguments we deduce that if the transit time effect in solids is to be used in active devices it is preferable to use sufficiently high applied fields to result in saturated carrier velocities. It is also necessary to create an initial delay of the injected current modulation to give a total phase shift of more than 90°. The impatt and the baritt diodes are examples of devices in which this additional delay has been created in different ways.

In the impatt diode the current modulation arises as a result of modulating the field applied to an avalanching *p—n* junction. If the field in the avalanche region is raised above the value required to maintain a steady avalanche the current will increase linearly with time which is eqivalent to a 90° phase shift of the injected current.

The baritt diode uses a punch-through *p—n—p* structure in which one *p—n* junction is forward-biased and one is reversed-biased. Injection takes place across the forward-biased junction which can be represented as a parallel combination of a capacitance and a conductance. The capacitive delay introduces a phase shift which however, must be substantially less than 90° because if only capacitive current was flowing no charge injection would take place. For this reason the baritt diode is usually less efficient than the impatt. This is true even taking into account another injection delay mechanism in the baritt which is due to the presence of a portion of drift region where the carrier velocity is not saturated.

The trapatt diode operates on an entirely different physical principle although its physical construction is very similar to that of the impatt diode. In fact one and the same device can often be operated in both the impatt and the trapatt mode. A trapatt diode is most easily viewed as a switch capable of rapidly switching from a non-conducting or high impedance state to an almost short-circuited state. The switch is used to convert an applied d.c. voltage to r.f. To activate the switch, i.e. to cause an avalanching *p—n* junction to become completely conducting, it is necessary to apply a field high enough that the carrier generation rate exceeds the rate at which carriers can be removed by transport. Under such conditions the depletion region of the diode will become filled with a conducting plasma and will remain conducting until the generated plasma has been extracted. As the electric field drops very rapidly when the plasma is formed, the carrier velocity will be low in the extraction phase. The time required for plasma extraction will therefore be much longer than the transit time for a carrier with saturated velocity. Thus the frequency of operation in the trapatt mode is usually much lower than typical impatt mode frequencies in the same diode.

3.2 BASIC HIGH FIELD PHENOMENA IN SEMICONDUCTORS

As we have seen in the general introduction the deviations from Ohm's law which take place in semiconductors in high applied electric fields are of fundamental importance for all of the devices discussed in this chapter. We will therefore begin by reviewing some of the fundamental high field phenomena in bulk semiconductors.

3.2.1 The velocity field characteristic

At weak fields the drift velocity, v, is related to the electric field strength, E, in all semiconductors through

$$v = \mu E \tag{3.1}$$

The mobility, μ, is determined by the various interaction processes between the electrons and the lattice imperfections. If we define a mean free path, Λ, between each interaction due to a specific interaction mechanism, i, we have

$$\mu = \frac{e}{m^*} \frac{1}{\sum_i \frac{v_t}{\Lambda_i}} \tag{3.2}$$

where v_t is the mean random electron velocity and the summation is over the interaction processes. When the electric field strength is increased the values of all three of the parameters m^*, v_t and Λ may be affected.

Most semiconductors have non-parabolic conduction bands and

therefore an increase in the average electron energy caused by the applied field will affect the effective mass. In some materials there will be additional energy minima to which electrons may be scattered after gaining sufficient energy from an applied field. When this occurs the effective mass will usually be drastically changed.

Often the first deviation from a constant mobility will be due to an increase in the mean random velocity v_t when the field is increased. When the increase in mean energy is proportional to the applied field and this field is high enough to make the zero field mean energy negligible $\mu \propto E^{-1/2}$ with the result that

$$v = \mu E \propto \sqrt{E} \tag{3.3}$$

This type of behaviour is actually observed in a limited range of field strengths in some semiconductor materials.

It remains only to discuss the energy dependence of the mean free path, Λ. In an ideal unperturbed lattice the mean free path would be infinite. In this discussion we will neglect imperfections in the lattice which can be avoided by using sufficiently high quality single crystals, i.e. defects like dislocations, stacking faults and grain boundaries. However, no matter how perfect a technology is used it is not possible to avoid thermally induced lattice perturbations, i.e. phonons. Also the necessary doping atoms will lead to imperfections in the form of ionized impurities. It will be shown below that the doping densities which are used in the high field regions are usually quite low or in the range 10^{15}—10^{16} cm^{-3} and therefore ionized impurity scattering is of little importance. Very high frequency (mm wave) devices will require higher doping densities and in such devices ionized impurity scattering may have some effect. In the remaining discussion we will neglect the influence of ionized impurities and assume that the same velocity-field curves apply regardless of doping. Then phonons alone determine the velocity-field characteristics.

Most semiconductor materials crystallize in lattices of sufficiently low symmetry to admit not only acoustic phonons but also an optical phonon branch. Furthermore, most of the semiconductors are highly polar leading to a strong interaction between electrons and polar optical phonons. At low fields and moderate device temperatures this interaction is weak because, in order to emit an optical phonon, an electron must have an energy greater than that of the phonon or in practice a few times 0.01 eV. Therefore the weak field interaction is often dominated by acoustic phonons. The electron—acoustic phonon interaction is not very strongly dependent on mean electron energy and therefore the first non-ohmic effect to be observed when the applied field is increased is often due to the increase in v_t as mentioned above. As the field is further increased, however, a major part of the electrons

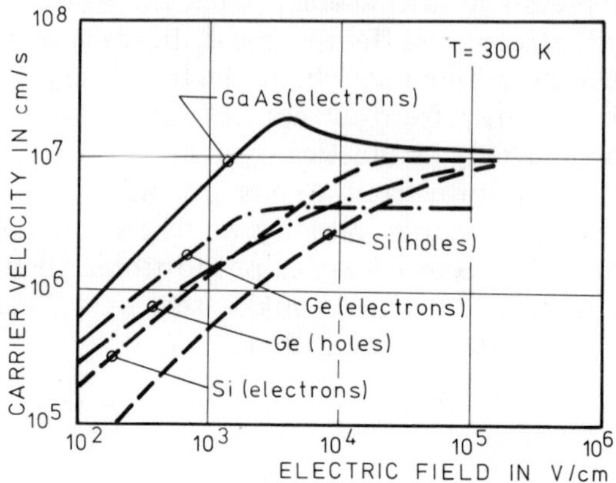

Figure 3.1 Drift velocity versus electric field of electrons and holes in Si, Ge and GaAs[1–3]

acquire enough energy to be able to emit optical phonons. The probability for such an emission is so large that effectively the electron velocity saturates.

In materials like Si and Ge the velocity field curve is dominated by this electron interaction but in several 3—5 compounds, notably GaAs and InP, the scattering of electrons to higher-lying energy minima with very low mobility is the dominant scattering mechanism, as is discussed in Chapter 2. In Figure 3.1 we give some measured values of drift velocity versus field. Accurate measurements of these velocities are difficult and even in materials which have been extensively studied a considerable uncertainty still exists. We have therefore neglected the small differences between the measured values for different crystal orientations.

This discussion of the interaction mechanisms between electrons and lattice at high field strengths would be incomplete without some reference to energy relaxation effects. At very high microwave frequencies it becomes important to note that a finite time and distance is required to accelerate the electrons up to the increased average energy that causes changes in electron phonon interaction and modifications in the mobility. Conversely a finite time is required for the electrons to give up this energy. As discussed by Rees[4] these effects are of importance for high frequency transferred electron devices. The distance required to accelerate a free electron to an energy of 0.1 eV in a field of 10^6 V/m is 0.1 μm which is beginning to constitute a significant fraction of the total device lengths for diodes made for operation above 100 GHz. It is somewhat more difficult to estimate the energy relaxation time for electrons. However, if we note that in polar optical scattering the momentum and energy relaxation frequencies are comparable[5] a first order of magnitude estimate is obtained from the measured value of mobility. From Equation (3.2) we can write the relaxation time τ as

$$\tau = 1 \Big/ \sum_i \frac{v_t}{\Lambda_i} = \frac{\mu m^*}{e} \tag{3.4}$$

With a mobility of 0.01 m^2/V s and an effective mass of 0.1 m_0 we obtain $\tau \approx 10^{-12}$ s. Crude as this estimate undoubtedly is, it nevertheless gives the correct order of magnitude and as a result we conclude that energy relaxation will begin to improve fundamental limitations only at frequencies of several hundred GHz.

No matter how outstanding the performance of an electronic device its use will be limited unless it is capable of performing over a reasonably wide range of ambient temperatures. For each of the basic physical parameters which are of importance for the device operation it is therefore of great interest to know the temperature dependence. As a general comment we recall that the carrier drift velocity is determined by interaction with phonons. As the temperature is increased the equipartition theorem tells us that the number of phonons increases linearly and it is therefore to be expected that in Equation 3.2 $\Lambda \propto T^{-1}$. In weak fields $\bar{v} \propto \sqrt{T}$ and we obtain $\mu \propto T^{-3/2}$. For the acoustic phonon interaction this is a correct result but for the more complicated optical phonon interaction a simple power law is an over-simplification. It would be carrying it too far to discuss the thermal effects in high field in any depth, but generally a decrease in the saturated drift velocity is obtained. Unfortunately, experimental results are rather scarce but in Figure 3.2 we give some published results on the saturated velocity for different temperatures. It is seen that over a 100 °C temperature range (devices are usually required to operate over such a range) more than 20% change in drift velocity may be obtained.

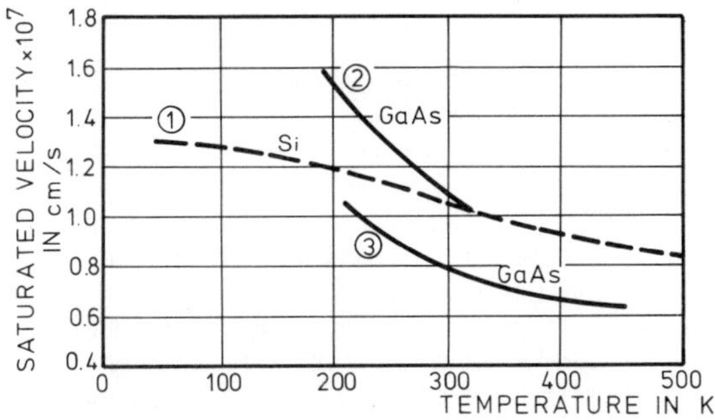

Figure 3.2 Temperature dependence of the saturated hot electron drift velocity in Si and GaAs[3,7]

3.2.2 Impact ionization

We have already discussed how a strong field will increase the average electron or hole energy and how this increase may lead to scattering to higher-lying energy minima. Carriers which are accelerated to energies larger than the band gap may also cause impact ionization (Figure 3.3).

It is customary to define an ionization rate α for each type of carrier by writing the generation rate of electron—hole pairs per unit time g as

$$g = \alpha_n n\bar{v}_n + \alpha_p p\bar{v}_p \tag{3.5}$$

Clearly α will be a strong function of the applied electric field. With no applied field there will always be some carriers in the high energy tail of the Maxwellian velocity distribution with energies in excess of the band gap energy. Owing to the detailed balance that prevails in thermal equilibrium the rate at which these carriers cause pair generation is equal to the rate of the inverse process, three-body or Auger recombination. In all the semiconductors which are of interest for transit time devices Auger recombination is almost negligible. Furthermore the total carrier lifetime is typically much longer than one period at microwave frequencies.

When an electric field is applied the rate of carrier generation will start to increase, but initially quite slowly. As a result of this increase in generation the carrier concentration will increase. However, this will be a small effect as long as the generation rate due to impact ionization is small in comparison with the total generation rate. When the field becomes so strong that the impact ionization dominates over generation

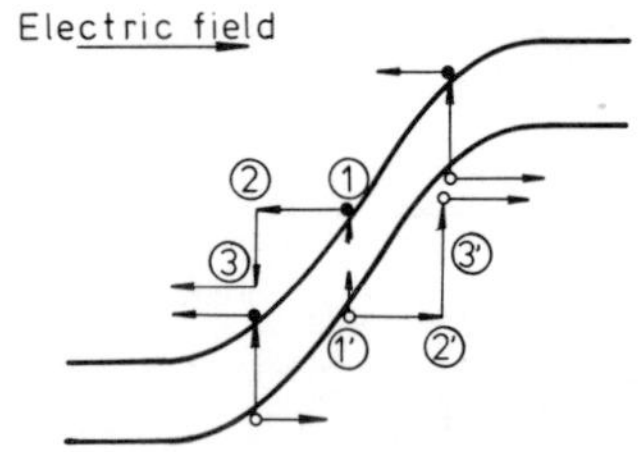

Figure 3.3 Illustration of impact ionization event. A conduction electron at 1 acquires energy from an applied field. At 2 the kinetic energy is transferred to a valence electron which is excited into the conduction band leaving a hole behind and two low energy conduction electrons at 3; similar for holes 1′—3′ [8]

due to thermally generated photons and other mechanisms, the effect on the carrier concentration will become noticeable. In reverse-biased p–n junctions the small reverse current is often (not in silicon) dominated by carriers generated outside the depletion region. In this case impact ionization may not affect the reverse current significantly, even though it is the dominant generation mechanism in the depletion region.

As the electric field is further increased in a reverse-biased p–n junction the generation rate will eventually become very high. The generation of carriers due to impact ionization will be balanced by transport of carriers out of the depletion regions and we have a stable self-sustaining avalanche. If the field is increased above the balancing value the current will increase linearly up to the value when the space charge resistance causes the field to decrease to its balancing value.

The very strong field dependence of the ionization rate on electric field is illustrated in Figure 3.4 which gives experimental values in some common semiconductors. It is also of interest to note that in materials like Si the electrons and holes have widely differing ionization rates but in GaAs the rates for both types of carriers are nearly equal.

Unfortunately, no published results are available on how the ionization rates depend on the ambient temperature. However, measurements of the dependence of breakdown voltage or reverse-biased p–n junctions indicate that in silicon the field strength required to produce a given value of α may vary by as much as 10% per 100°C[12].

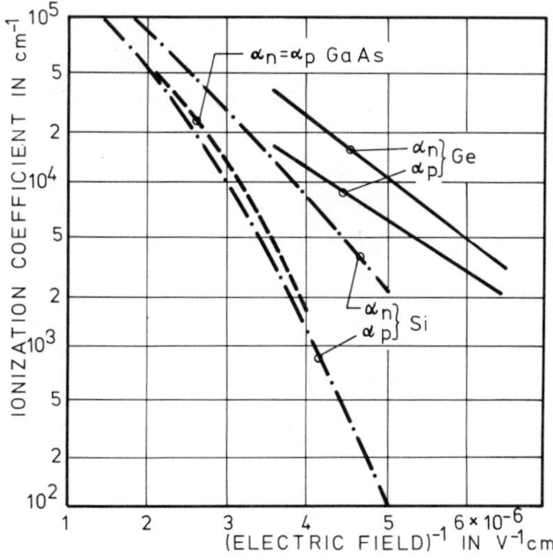

Figure 3.4 Ionization coefficient of carriers in gallium arsenide, silicon and germanium[9–11]

3.3 DYNAMICS OF A HYPOTHETICAL UNIPOLAR, VELOCITY SATURATED DIODE

3.3.1 The drift region

Before proceeding to a detailed analysis of impatt and baritt devices we shall mathematically analyse a hypothetical device in which the particle current is injected with a given phase ϕ with respect to the total current and then delayed because of transit time effects. For this purpose, consider the device illustrated in Figure 3.5.

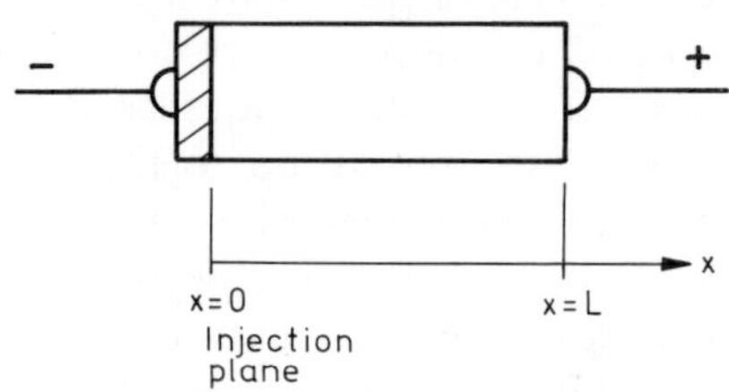

Figure 3.5 Schematic of velocity saturated hypothetical device

We assume that a d.c. voltage is applied which causes all electrons at $x > 0$ to flow with the saturated velocity v_s. At $x = 0$ the magnitude of the a.c. particle current density $\tilde{J}_p$ equals that of the total a.c. current density $\tilde{J}_t$, but the phase relation of these quantities is arbitrary.

$$\tilde{J}_p = \exp(-j\phi)\,\tilde{J}_t \tag{3.6}$$

Carrier transport in the drift region $x > 0$ is governed by the equation of continuity

$$j\omega\tilde{n} + v_s \frac{d\tilde{n}}{dx} = 0 \tag{3.7}$$

where $\tilde{n}$ stands for the oscillatory part of the electron density. The total a.c. current anywhere in the drift region is given by

$$\tilde{J}_t(x) = j\omega\epsilon\tilde{E}(x) + \tilde{J}_p(x) \tag{3.8}$$

Let us first note that the assumption of a given phase of $\tilde{J}_p$ at $x = 0$ also implies a boundary condition on the a.c. part of the electric field $\tilde{E}$ because from Equation 3.8.

$$\tilde{E}(0) = \frac{\tilde{J}_t(1 - \exp(-j\phi))}{j\omega\epsilon} \tag{3.9}$$

In particular $\phi = 0$ is equivalent to $\tilde{E}(0) = 0$. From Equation 3.8 we can easily solve for the electric field and integrate to get the a.c. impedance Z

$$Z = \frac{1}{j\omega C}\left[1 - \frac{\exp(-j\phi)(1-\exp(-j\theta))}{j\theta}\right] \tag{3.10}$$

where the geometric capacitance of the drift region, C, and the transit angle $\theta = \omega L/v_s$ are introduced. By taking the real and imaginary parts of Equation 3.10 we get

$$R = \frac{\cos\phi - \cos(\phi + \theta)}{\omega C\theta} \tag{3.11}$$

$$-X = \frac{1}{\omega C} - \frac{\sin(\theta + \phi) - \sin\phi}{\omega C\theta} \tag{3.12}$$

We are now in a position to discuss the influence of the injection phase ϕ on the value of the a.c. resistance R. Due to causality we first note that $\phi > 0$ and we further limit ourselves to the case $\phi < \pi/2$ to exclude the case when the injection mechanism is active in itself, i.e. leads to a negative R for $\theta = 0$. The sign of R is determined by the factor $\cos\phi - \cos(\phi + \theta)$. For $\phi = 0$, R is always greater than or equal to zero, i.e. no negative resistance is possible as already pointed out in the introductory physical discussion. Any non-zero ϕ will admit a negative resistance, however. The factor in the numerator of Equation 3.11 attains its maximum negative value of -1 for $\phi = \pi/2$ and $\theta = 3\pi/2$. Due to the factor θ in the denominator in Equation 3.11 the actual maximum negative resistance is achieved for somewhat lower values of ϕ and θ but the optimum value of ϕ is approximately $\pi/2$. Direct calculations for $\phi = \pi/2$ gives the optimum value of θ as the solution of the transcendental equation $\tan\theta = \theta$ or $\theta \approx 257.5°$.

The above analysis thus confirms the physical discussion in the introduction and shows the importance of an injection delay. The boundary condition Equation 3.6 is not the most general one possible but it is straightforward enough to show mathematically and also physically that, if the magnitude of $\tilde{J}_p(0)$ is smaller than that of $\tilde{J}_t$, the situation is less favourable for a large negative resistance than if these quantities are equal.

The problem of finding active transit time mechanisms has thus been reduced to finding methods of delaying the particle current injected into the drift zone with respect to the total current .

3.3.2 Methods of creating an injection delay

Before we proceed to a detailed study of impatt and baritt devices it is instructive to enter a more general discussion on how an injection delay

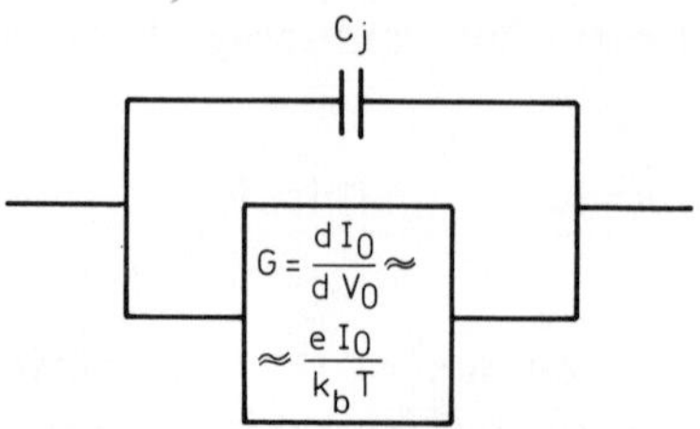

Figure 3.6 Equivalent circuit of Schottky or *p-n* junction

can be realized. The particle current injected from an ohmic contact will, as a first approximation, be in phase with the total current and therefore not useful. Non-ohmic contacts to semiconductors often consist of injection across a Schottky barrier. Let us therefore assume that a forward-biased Schottky barrier junction extends up to the plane $x = 0$ in Figure 3.5. At any given forward-bias the equivalent circuit of such a junction is a parallel combination of a conductance and a capacitive admittance as in Figure 3.6. The total current density across this junction is

$$\tilde{J}_t = j\omega C_j \tilde{V} + G\tilde{V} \tag{3.13}$$

In this case the phase of the injected current is given by

$$\tan\phi = \frac{\omega C_j}{G} \tag{3.14}$$

In order to obtain $\phi \approx \pi/2$, which is the optimum value, we should then have $\omega C_j \gg G$. However, this would mean that the magnitude of the particle current is only a very small value fraction of the total current and therefore would lead to a very small negative resistance. In actual fact the most favourable condition is to work with $\omega C_j \approx G$ thereby compromising the requirements on phase and magnitude of the injected particle current. The above arguments apply equally well to a *p—n* junction as to a Schottky junction. With some modification the capacitive delay in tunnelling junctions could also be used.

There are, however, additional considerations to be taken into account and, for instance, a forward-biased n^+—p—p^+ structure is not useful as a device. The reason for this is that in order to achieve a saturated velocity in the p or drift region a field strength in excess of about 10 kV/cm has to be reached. With the high bias voltage needed to fulfil this requirement the n^+—p junction will be very heavily forward-biased and generally we will have $G \gg \omega C_j$ leading to no injection delay. The natural way to overcome this difficulty is to replace the p—p^+ transition by a p—n^+ junction. This junction will then be reverse-biased and the field strength required for velocity saturation is easily reached without drawing an excessive d.c. current. We have in this way arrived at the baritt structure.

The basic deficiency of the baritt structure is that, as we have seen above, the RC delay cannot result in a large phase shift ($> \sim 45^\circ$) without suppressing the magnitude of the particle current. A high efficiency of a transit time device requires that the particle current is flowing in phase with the voltage only for a small fraction of a period and in order to achieve this we have to look for different mechanisms.

Avalanche breakdown is a process which is of great interest in this connection. Consider the equation governing the rate of change of particle density in an avalanching semiconductor, i.e.

$$\frac{\partial n}{\partial t} + v_s \frac{\partial n}{\partial x} = \alpha n \bar{v} \tag{3.15}$$

If we write $n = n_0 + \tilde{n}$ etc. where subscript 0 stands for steady state and tilde refers to a small a.c. perturbation we get from Equation 3.15

$$\begin{aligned} v_s \frac{\partial n_0}{\partial x} &= \alpha_0 n_0 \bar{v} \\ \frac{\partial \tilde{n}}{\partial t} + v_s \frac{\partial \tilde{n}}{\partial x} &= \alpha_0 \tilde{n} \bar{v} + \left(\frac{\mathrm{d}\alpha}{\mathrm{d}E}\right)_0 \bar{v} n_0 \tilde{E} \end{aligned} \tag{3.16}$$

Assuming that $\tilde{n}/n_0$ is independent of x we obtain

$$\frac{\partial \tilde{n}}{\partial t} = \left(\frac{\mathrm{d}\alpha}{\mathrm{d}E}\right)_0 v n_0 \tilde{E} \tag{3.17}$$

which shows that the phase difference between $\tilde{n}$ and $\tilde{E}$ will be exactly 90°. We can therefore conclude that avalanche breakdown should be a very efficient injection mechanism.

Before we proceed to a more detailed discussion of the properties of impatt and baritt diodes we shall discuss a couple of other mechanisms for creating an injection delay. These additional mechanisms may be of potential interest for the future but are not used in any practical devices today.

Let us first consider the heterodevice in Figure 3.7. Using exactly the same methods of calculation as before we obtain an r.f. impedance.

$$\begin{aligned} Z &= Z_1 + Z_2 \\ Z_2 &= \frac{1}{j\omega C_2} \left\{ 1 - \frac{\exp(-j\theta_1)[1 - \exp(-j\theta)]}{j\theta_2} \right\} \end{aligned} \tag{3.18}$$

where we for simplicity limit ourselves to the case when the injection phase $\phi = 0$. Z_1 is the impedance of the first part of the heterostructure

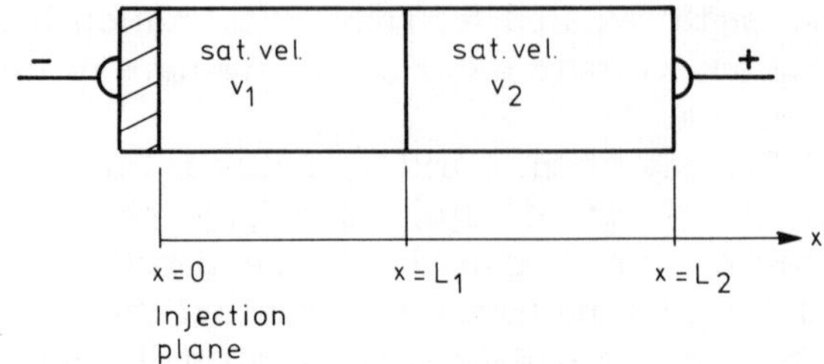

Figure 3.7 Heterodevice with two different semiconductors with different saturated drift velocities

and is given by Equation 3.10. A comparison between Equation 3.10 and Equation 3.18 shows that in Z_2 the drift delay θ_1 from region 1 in our heterostructure enters exactly as the injection delay ϕ in Equation 3.10. If we then as an example choose $\theta_1 = \pi/2$ we get

$$R = R_1 + R_2 = \mathrm{Re}(Z_1 + Z_2) = \frac{2}{\pi\omega C_1}\left(1 + \frac{C_1}{C_2}\,\frac{\pi \sin\theta_2}{2\theta_2}\right) \tag{3.19}$$

For C_1/C_2 sufficiently large and $\theta_2 > \pi$ we can clearly obtain a negative resistance without any additional injection delay. The requirement on the physical parameters of the two materials making up our heterostructure is that either $v_1 \ll v_2$ or $\epsilon_1 \gg \epsilon_2$. In both cases we are effectively achieving that the portion of the diode where the conduction current is in phase with voltage is partly short-circuited by its own geometric capacitance.

In an actual device there would also be a contribution from an injection delay at $x = 0$ and therefore a heterostructure may be capable of a performance approaching that of an avalanche diode. The technological difficulties of actually manufacturing a heterostructure of suitable materials may be very great, however.

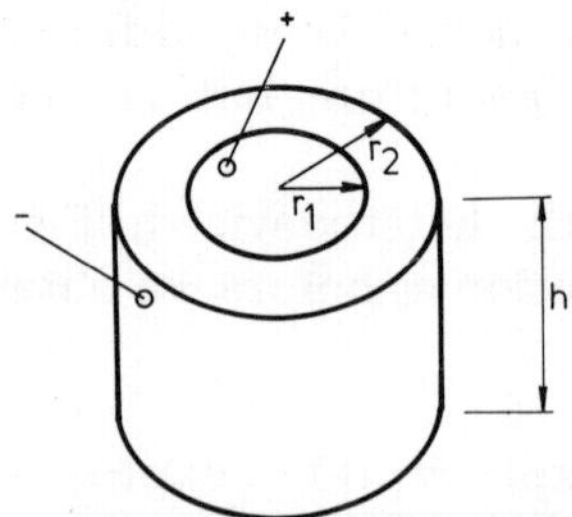

Figure 3.8 Cylindrical device. Injection of carriers takes place at r_2

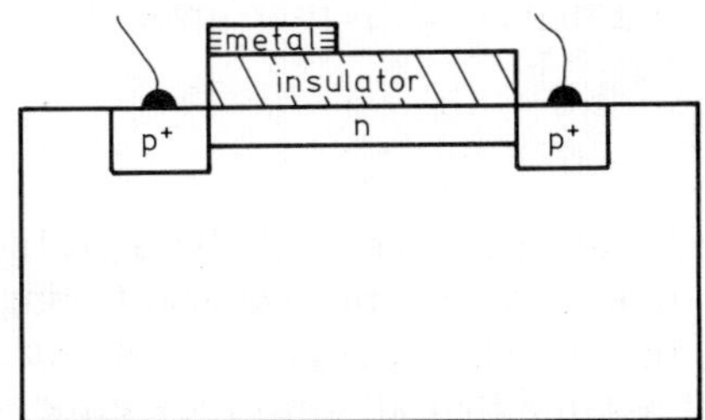

Figure 3.9 Planar *p-n-p* diode with insulated electrode for capacitance short-circuiting of in-phase current contributions

The same effect of a capacitive short-circuiting of the non-desirable parts of the sample could conceivably be achieved by abandoning the one-dimensional geometry used so far. As an example consider the cylindrical structure of Figure 3.8.

A straightforward calculation for the case when the injection phase $\phi = 0$ gives the impedance

$$Z_1 = -\frac{1}{\omega C \log(r_2/r_1)} \int_{r_1}^{r_2} \frac{\sin\left[\frac{\omega}{v_s}(r - r_2)\right]}{r} dr + \frac{1}{j\omega C}\left\{1 - \frac{1}{\log(r_2/r_1)} \int_{r_1}^{r_2} \frac{\cos\left[\frac{\omega}{v_s}(r - r_2)\right]}{r} dr\right\} \quad (3.20)$$

An examination of the real part of the impedance shows that the resistance is negative in certain frequency bands.

It máy also be possible to obtain capacitive short-circuiting in planar structures like the one illustrated in Figure 3.9.

3.4 IMPATT DIODES

3.4.1 Small signal theory, analytical treatment of an idealized diode

The basic principle underlying impatt operation were outlined in the introduction. Following Read[13] and Gilden and Hines[14] we shall proceed by developing a small signal theory for a material in which electrons and holes have identical properties, i.e. the same ionization coefficients α and saturated velocities v_s. The starting equations are those of conservation of carriers, i.e.

$$\begin{aligned} \frac{\partial n}{\partial t} + \frac{\partial}{\partial x}(n v_s) &= \alpha v_s (n + p) \\ \frac{\partial p}{\partial t} + \frac{\partial}{\partial x}(p v_s) &= \alpha v_s (n + p) \end{aligned} \quad (3.21)$$

The conduction current density J_a is then given by

$$J_a = qv_s(n + p) \tag{3.22}$$

Figure 3.10 shows the doping and electric field profile of the configuration to which we intend to apply our theory. Owing to the extremely strong variation of the ionization coefficient α with electric field (see Figure 3.4) we shall assume that all ionization effectively takes place in a thin avalanche zone as indicated in Figure 3.10. With Read we also make the assumption that the space dependence of the conduction current may be neglected in the avalanche zone. With these assumptions it is straightforward to derive Read's basic equation from Equations 3.21 and 3.22.

$$\frac{dJ_a}{dt} = \frac{2J_a}{\tau_a}\left\{\int_0^{l_a} \alpha \, dx - 1\right\} \tag{3.23}$$

where τ_a is the transit time l_a/v_s of carriers through the avalanche zone. We now want to study small periodic current fluctuations of

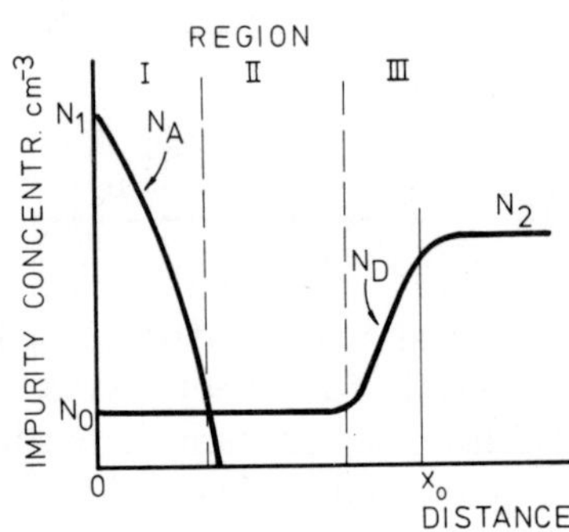

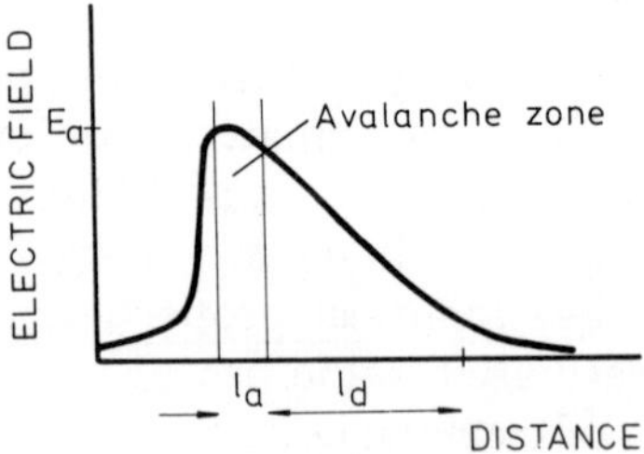

Figure 3.10 Doping and electric field profiles of impatt diode[15]
$N_A = N_1 \, \mathrm{er\chi c}(A_0 x)$
$N_D = N_0 + N_2 \, \mathrm{er\chi c}[B_0(x_0 - x)]$

frequency ω around a steady state and therefore write

$$\frac{1}{l_a}\int_0^{l_a} \alpha\,\mathrm{d}x = \bar{\alpha}_0 + \bar{\alpha}_0{}'\tilde{E}$$

$$J_a = J_0 + \tilde{J} \tag{3.24}$$

$$E_a = E_0 + \tilde{E}$$

where subscript 0 refers to the steady state and $\tilde{J}$ and $\tilde{E}$ are the small sinusoidal perturbations in conduction current density and electric field respectively. E_a is the average field in the avalanche zone. We also note that for the steady state we must have

$$\int_0^{l_a} \alpha_0\,\mathrm{d}x = 1 \tag{3.25}$$

Equation 3.23 for the small signal case then becomes

$$\tilde{J}_a = \frac{2\bar{\alpha}_0{}'l_a J_0 \tilde{E}_a}{j\omega\tau_a} \tag{3.26}$$

which again shows that the avalanche process results in a conduction current which is 90° out of phase with the electric field. To obtain the total current density through the avalanche zone, we must add to Equation 3.26 the displacement current density $\tilde{J}_d$.

$$\tilde{J}_d = j\omega\epsilon\tilde{E}_a \tag{3.27}$$

Equations 3.26 and 3.27 may be represented by the equivalent circuit for the avalanche zone in Figure 3.11 with

$$L_a = \frac{\tau_a}{2\bar{\alpha}_0{}'J_0 A} \tag{3.28}$$

$$C_a = \frac{\epsilon A}{l_a}$$

where A is the device area.

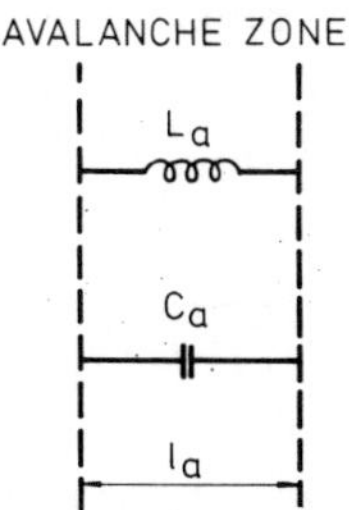

Figure 3.11 Equivalent circuit for the avalanche zone[14]

The parallel resonance frequency of the circuit in Figure 3.11 is

$$\omega_a{}^2 = \frac{2\bar{\alpha}_0{}' l_a J_0}{\epsilon \tau_a} = \frac{2\bar{\alpha}_0{}' v_s J_0}{\epsilon} \tag{3.29}$$

and it is worth noting that ω_a turns out to be independent of the avalanche zone thickness l_a and proportional to the square root of the d.c. current density J_0.

It has been shown by more accurate calculations[16] that the assumption of constant conduction current in the avalanche zone leads to a minor error in Equation 3.29. A detailed theory results in the same formula but with a number 2 in the numerator replaced by 3.

The phase factor or injection delay as defined in Section 3.3.1 takes the form

$$\exp(-j\phi) = \frac{\tilde{J}_a}{\tilde{J}_a + \tilde{J}_d} = \frac{1}{1 - \dfrac{\omega^2}{\omega_a{}^2}} \tag{3.30}$$

which allows us to obtain the total impedance of the diode directly from Equation 3.10. For completeness the series resistance, R_s, from the contact regions, external leads etc. should be added to Equation 3.10. We then have

$$Z = R_s + \frac{l_d{}^2}{v_s \epsilon A} \left\{ \frac{1}{1 - \left(\dfrac{\omega}{\omega_a}\right)^2} \right\} \frac{1 - \cos\theta}{\theta^2} + \frac{1}{j\omega C_d} \left\{ 1 - \frac{\sin\theta}{\theta} + \frac{\dfrac{\sin\theta}{\theta} + \dfrac{l_a}{l_d}}{1 - \left(\dfrac{\omega_a}{\omega}\right)^2} \right\} \tag{3.31}$$

where l_d is the length of the drift region and $\theta = \omega l_d / v_s$.

From Equation 3.30 we note that in order to get a real phase shift and a negative resistance we need $\omega > \omega_a$ which also agrees with Equation 3.31.

For small transit angles Equation 3.31 simplifies to

$$Z = R_s + \frac{l_d{}^2}{2 v_s \epsilon A} \frac{1}{1 - \left(\dfrac{\omega}{\omega_a}\right)^2} + \frac{1}{j\omega C_d} \frac{(1 + l_a/l_d)}{1 - \left(\dfrac{\omega_a}{\omega}\right)^2} \tag{3.32}$$

The impedance Z is plotted as a function of frequency in Figure 3.12.

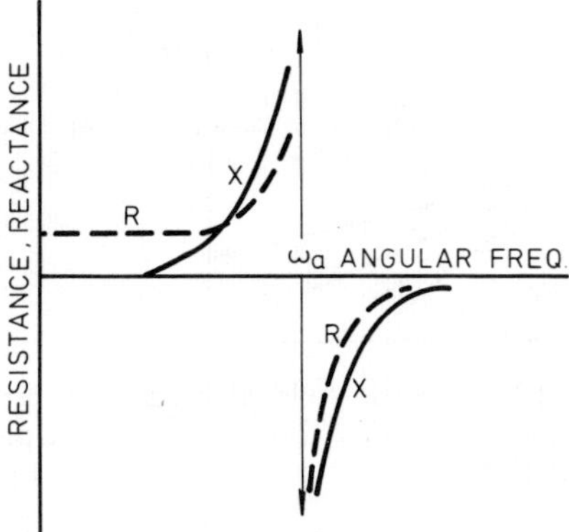

Figure 3.12 Frequency dependence of the real and imaginary parts of the diode impedance[14]

A sharp resonance occurs at $\omega = \omega_a$ at which frequency the diode develops a negative resistance and the reactance changes from capacitive to inductive. Many of the electronic properties of impatt diodes find their explanation in the resonant behaviour shown in Figure 3.12. Operation at a given frequency ω requires that the current exceeds that value which corresponds to $\omega = \omega_a$. Above this current and with a given circuit the diode will tune in frequency with current owing to the change of reactance with bias current.

At very low frequencies we obtain from Equation 3.32

$$Z(\omega = 0) = R_s + \frac{l_d^{\,2}}{2v_s\epsilon A} \tag{3.33}$$

where the second term is the space charge resistance.

3.4.2 Detailed theory of silicon diodes

Owing to the various simplifying assumptions in the Read, Gilden—Hines theory the above results are modified in several ways for practical diodes. To gain some insight into these effects we will now consider the results of some detailed theoretical calculations in which the simplifying assumptions have been removed at the price of no longer obtaining simple analytic expressions. Starting with the low frequency static I—V characteristic Bowers[15] has performed calculations on p^+—n—n^+ silicon diodes with constant doping in the n-region. One of his major conclusions concerns the behaviour of the I—V characteristic as the doping of the n-region is varied for constant diode length (see Figure 3.13). Comparatively highly doped diodes will have an essentially triangular field profile for all voltages up to breakdown. As the doping is lowered the slope of the field decreases until the condition of a diode

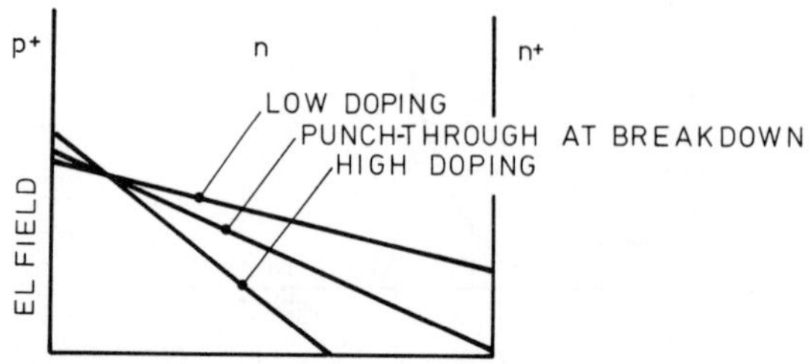

Figure 3.13 Electric field at breakdown for differently doped diodes. Note decrease in peaked field with doping

which is exactly punch-through at breakdown is reached, i.e. until the depletion layer at the breakdown voltage exactly fills the n-region. At lower dopings the diode will be punch-through even for lower voltages than breakdown.

In Figure 3.14 we reproduce some of Bowers's calculations of the I—V characteristics for diodes with constant length and different dopings (see Table 3.1 and Figure 3.10 for diode parameters). If we first concentrate on moderate currents it is seen that a positive space charge resistance like that calculated in Equation 3.33 is obtained only at high doping and low current. For highly punch-through diodes (diode No. 3) the space charge resistance is negative even at low current levels. Bowers's

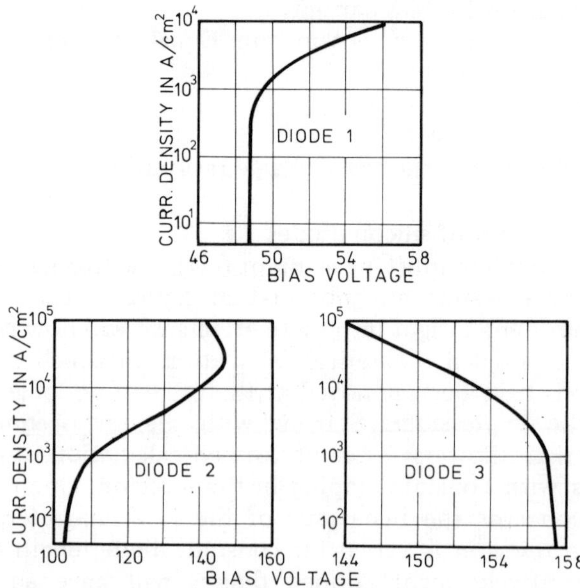

Figure 3.14 I—V characteristics of diodes Nos. 1—3 of Table 3.1[15]

Table 3.1
Diode parameters

Diode number	$N_0(\text{cm}^{-3})$	$N_1(\text{cm}^{-3})$	$N_2(\text{cm}^{-3})$	$A_0(\text{cm}^{-1})$	$B_0(\text{cm}^{-1})$	$x_0(\text{cm})$
1	1.8×10^{16}	2.0×10^{19}	5.0×10^{17}	1.17×10^4	2.28×10^4	8.0×10^{-4}
2	5.0×10^{15}	2.0×10^{20}	5.0×10^{17}	1.81×10^4	2.28×10^4	7.5×10^{-4}
3	1.0×10^{15}	2.0×10^{20}	5.0×10^{17}	1.81×10^4	2.28×10^4	7.5×10^{-4}

calculations show that unless the doping is very high, space charge effects lead to a rapidly decreasing voltage with current at high current levels as seen from Figure 3.14. This effect of the injected carriers at high current levels is further illustrated in Figure 3.15 which shows the electric field profile for various current densities. These results may be understood by noting that for diodes which are not punch-through at low current levels the net space charge is decreased by the mobile carriers. Diode No. 1 is far from punch-through and will have its voltage increased for all practical current levels. For low currents diode No. 2 will behave in a similar manner but as the current is increased it will become punch-through and eventually an almost rectangular field profile is obtained with a highly distributed avalanche. For such a distributed avalanche the space charge of both electrons and holes must be taken into account. Diode No. 3 finally shows these effects even at low currents.

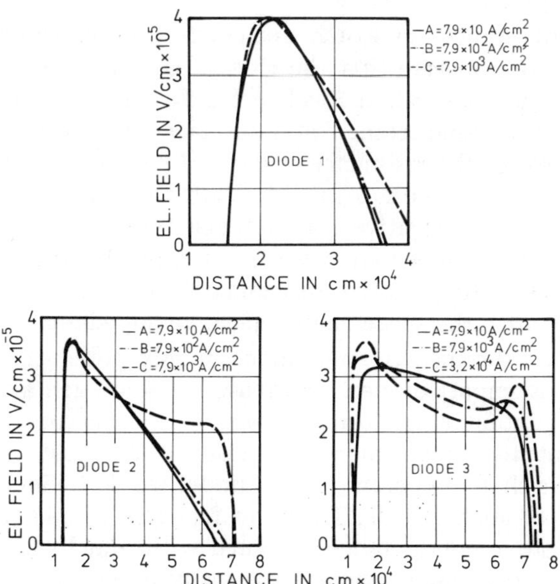

Figure 3.15 Electric field profiles for diodes Nos. 1—3 for different current densities[15]

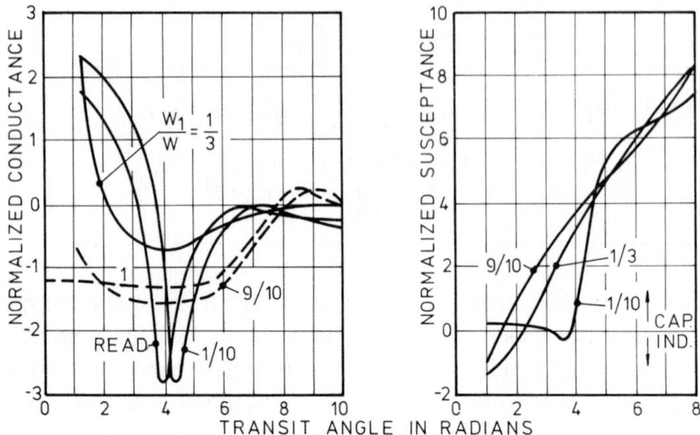

Figure 3.16 Admittance of six impatt diodes with identical total depletion layer width but different avalanche region width[17]

The small signal r.f. impedance of more realistic silicon diodes has been computed by Misawa[17] among others. The results for a fixed current and diode length w but with a variable avalanche zone width w_1 are shown in Figure 3.16. When comparing these results with those for real diodes one should observe that a wide avalanche zone corresponds to a punch-through diode whereas a very thin avalanche zone may require a non-uniform *n*-region.

A number of points are worth noting. In comparison with Figure 3.12 we first of all observe that the closest agreement is obtained with the thin avalanche zone called Read in Figure 3.16. However, even this diode has a finite conductance at resonance. As the diode becomes more punch-through the resonance at $\omega = \omega_a$ widens and the maximum negative conductance becomes smaller. With very low dopings or wide avalanche regions the negative conductance extends all the way down to zero frequency in agreement with the d.c. results of Bowers. In fact for *p—i—n* diodes the negative resistance due to transit time effects becomes mixed with the space charge negative impedance.

Owing to their d.c. negative resistance, highly punch-through diodes are unsuitable as impatt diodes. The reason is that such a diode will in general not maintain a uniform current density across its area but develop current filament which may lead to overheating and diode burn-out. As we shall see below *p—i—n* diodes would be inefficient even in the absence of such effects. This is in fact also to be expected from the smaller negative conductance obtained with punch-through diodes. The decrease in this conductance is basically due to the fact that the avalanche is distributed across the drift region so that no well defined transit time will exist. Some charge carriers will have a long transit

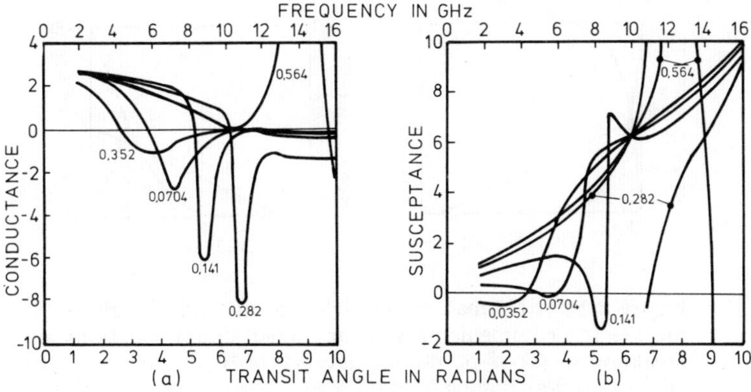

Figure 3.17 Real (a) and imaginary (b) parts of the admittance as a function of frequency at various bias currents. Current normalized to 3700 A/cm^2 [17]

delay and others a very short one and the net effect is a very small negative conductance. This discussion underlines one important aspect of impatt diode design, namely that the avalanche region should be kept as thin as possible. Another reason for keeping the avalanche region thin is that it dissipates power. The highest field strength is found there and the current is at least partly in phase with the voltage.

In Figure 3.17 we show the admittance calculated by Misawa[17] for a given diode when the bias current is varied. It is seen that in accordance with Equation 3.29 the avalanche resonance frequency increases with bias but also that the avalanche resonance becomes more pronounced as the current is increased. At this point it should be pointed out that unavoidable series resistances are neglected in Misawa's theory and if these are taken into account the conductance will be positive up to a threshold current above which microwave oscillations will start if a suitable circuit is provided. From Figure 3.17 it would appear that increasing the current would lead to an improved negative resistance without any theoretical limits. It is clear from the discussion above of Bowers's results that at sufficiently high currents the mobile space charge will begin to widen the avalanche region and lead to less efficient operation and decreasing negative conductance. The current density at which these effects will become important is given by $J = eN_d v_s$ when the mobile charge equals the background dopant space charge.

3.4.3 Impatt diode structures

It was pointed out above that a thin avalanche region is beneficial and although all the discussion so far has been concentrated on $p^+\text{—}n\text{—}n^+$ diodes with constant n-doping it is of interest to inquire into what the optimum doping configuration might be. Schroeder and Haddad[18] have

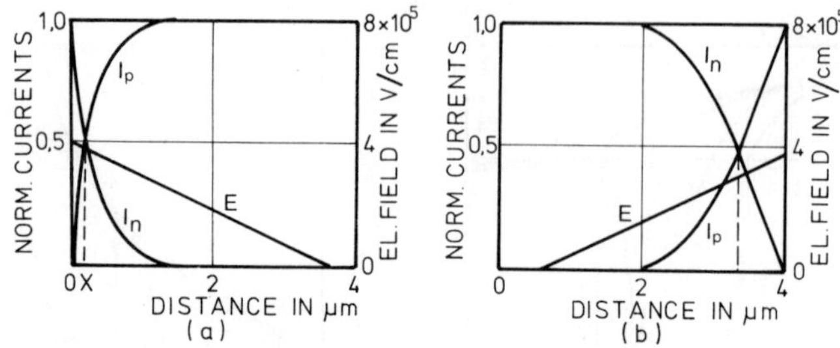

Figure 3.18 Normalized particle current and electric field profiles for complementary Si one-sided abrupt junctions at breakdown. (a) n^+-p, doping level = 7×10^{15} cm^{-3}, V_B = 76.2 V; (b) n-p^+, doping level = 7×10^{15} cm^{-3}, V_a = 71.4 V[18]

calculated the current and field profiles for p^+—n—n^+ and n^+—p—p^+ diodes with constant doping (Figure 3.18). It is clear that according to their results the latter diode should be preferred. In actual fact most practical diodes are p^+—n—n$^+$ and the reason why the p-type material is not as good as might be expected is not quite clear. Large signal calculations by Sellberg[19] do indicate, however, that the large signal impedance of p-type diodes is very small.

The original structure proposed by Read[13] was p^+—n^+—ν—p^+ which would lead to a very thin avalanche region indeed. Misawa and coworkers[20] have made a study of such diodes and find that quite good efficiencies may be obtained but the peak power capability of the Read structure in silicon is less than that of the p^+—n—n^+ structure.

A discussion of impatt diode design would be incomplete without mentioning materials other than silicon. In fact to date the highest powers and efficiencies are obtained not with silicon but with gallium arsenide and notably with the so-called lo-hi-lo structure (Figure 3.19).[7] This structure is very similar to the one proposed by Read and has the same advantages. In addition it results in a somewhat lower peak field in the avalanche region which may improve the diode

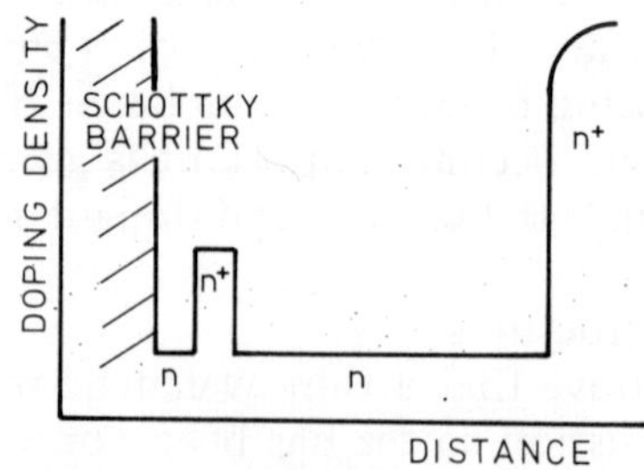

Figure 3.19 Doping profile of lo-hi-lo impatt diode

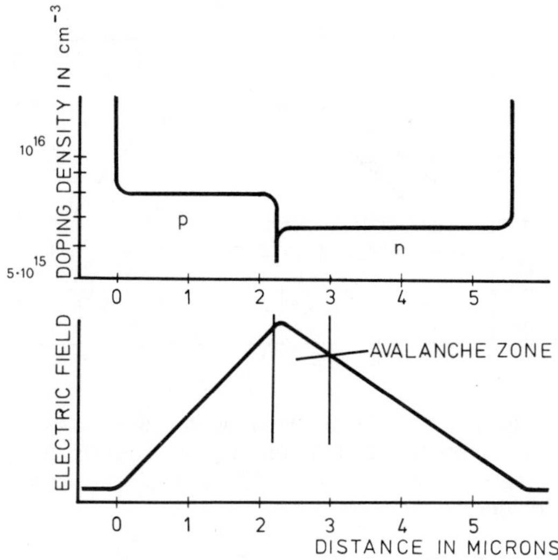

Figure 3.20 Doping and electric field profiles of double drift region impatt diode[23]

reliability. The superiority of ordinary p^+—n—n^+ gallium arsenide diodes to silicon ones seems to be due not primarily to differences in small signal properties but to the fact that gallium arsenide diodes can sustain larger r.f. voltage swings than silicon ones.[21] The reason for this difference is not completely understood at present although the higher electron velocity at low field strengths in gallium arsenide is probably of importance.

The most recently proposed impatt structure is the double drift region (DDR) diode proposed by Scharfetter and coworkers.[22] The doping and field profiles of this diode are shown in Figure 3.20.

It is seen in Figure 3.20 that the avalanche zone does not extend very far into the p-region. As a result the net avalanche zone width will be essentially unaffected by the addition of a p-region to a single drift device. It will be shown below that the output power and efficiency of an impatt diode is proportional to $l_d^2/(l_a + l_d)^2$ where l_d and l_a are the widths of the drift and avalanche regions respectively. We therefore gain in efficiency and output power by using a double drift structure. Furthermore the impedance level per unit area will be higher and the internal negative Q—value lower for the double drift structure due to the lower geometric capacitance. In addition to the higher output power we therefore also obtain a wider bandwidth as seen in Figure 3.21.

The double drift impatt diode has one further advantage which may not be so evident from theoretical considerations. As shown in Figure

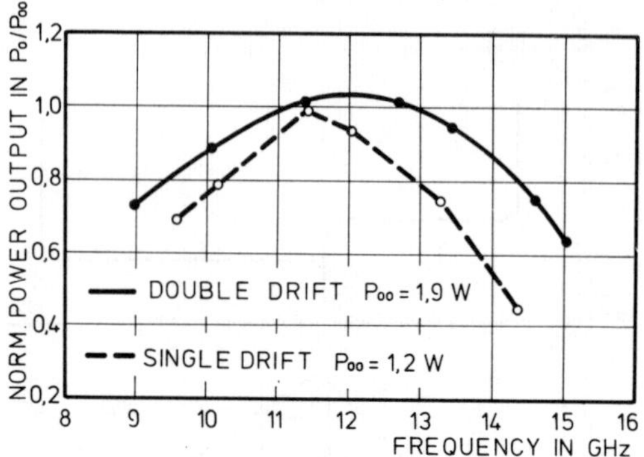

Figure 3.21 Normalized power output versus frequency at maximum power output for single mesa double drift and single drift devices[24]

3.22 the noise measure degrades much less rapidly with signal level than is the case for single drift region diodes.

The one disadvantage of the double drift structure is the higher thermal impedance per unit area. At high frequencies where the total active region width is small this is unimportant but it does impose restrictions for low frequency devices.

We shall conclude our discussion of impatt diode design by discussing how the frequency of operation influences the design and performance. The introductory discussion of this chapter suggests that the optimum transit angle θ_{opt} of the drift region should be about 180° and more detailed calculations confirm that this is at least of the correct order of

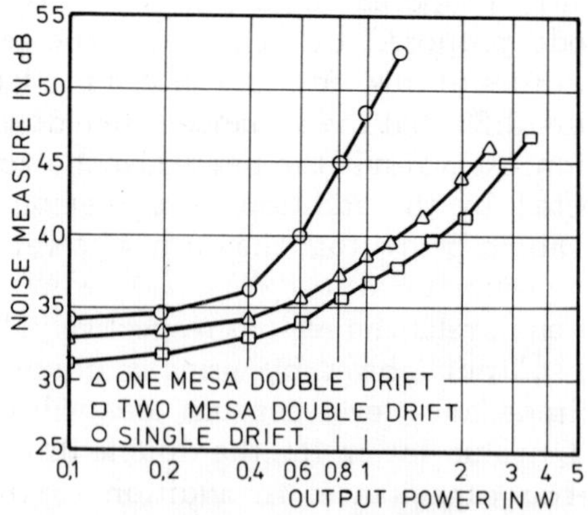

Figure 3.22 Noise measure 100 kHz from the carrier versus output power for one mesa double drift, for two mesa double drift and for single drift devices[24]

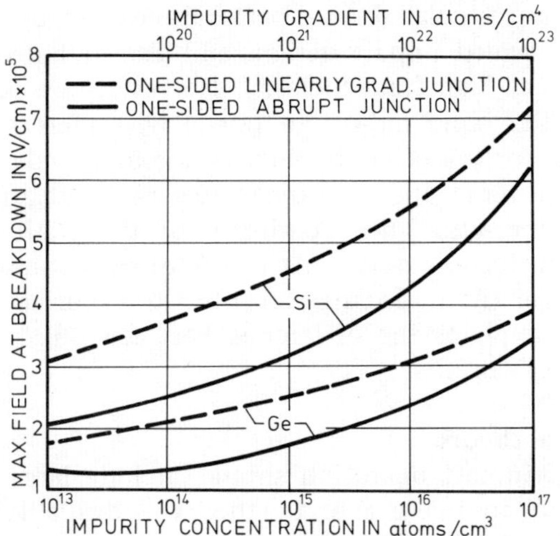

Figure 3.23 Maximum field at breakdown condition for one-sided abrupt and one-sided linearly graded Ge and Si *p-n* junction[25]

magnitude. The width of the drift region l_d is then obtained as $v_s\theta_{opt}/\omega$ and is inversely proportional to frequency.

The doping concentration for a p^+—n—n^+ diode should be chosen so that it is approximately punch-through at breakdown. As a first approximation the avalanche breakdown field strength E_a is independent of doping leading to

$$N_d = \frac{\epsilon E_a}{e l_d} = \frac{\epsilon E_a \omega}{e \theta_{opt} v_s} \tag{3.34}$$

E_a is actually somewhat dependent on doping as shown in Figure 3.23[25].

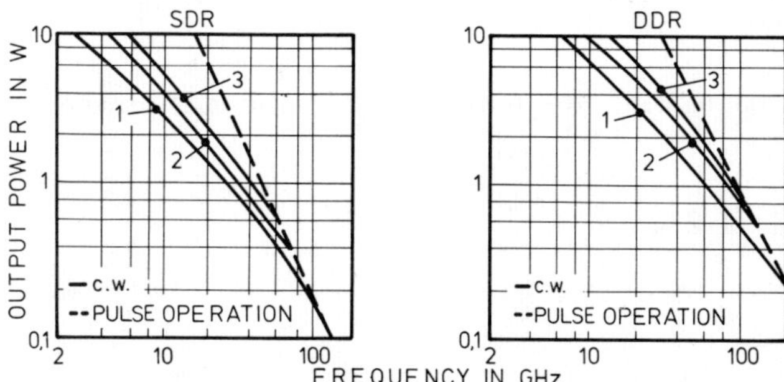

Figure 3.24 Output power versus frequency for one, two and three single and double drift region silicon impatt diodes on copper heat sinks with efficiencies of 8% and 10% respectively[26]

Scharfetter has calculated the performance versus frequency for various properly scaled impatt designs and also included the effect of skin effect losses at high frequencies.[26] His results are shown in Figure 3.24 and show the rapid fall-off of power with increasing frequency. The main reason for this effect is the decrease in impedance as the drift region becomes thinner and its capacitance per unit area larger when the frequency increases. The superiority of the DDR design is also evident from Scharfetter's results. At low frequencies the major limiting factor is no longer the impedance level but the thermal impedance of the diode. This explains the difference between pulse operation and c.w. in Scharfetter's results.

3.4.4 Large signal theory

A description of impatt operation should include large signal effects. Unfortunately no complete analytic theory is available so one has a choice of highly simplified analytic theory or computer simulation. Both approaches have been used but here we shall only take up some of the simplest analytic theories. For the results of detailed computer simulations the reader is referred, for instance, to the paper by Matsumura.[27]

The first published large signal theory is found in the original paper by Read[13] and is based on the so-called sharp pulse approximation. Read assumes a sinusoidal voltage and an avalanche current I_a which, because of the very rapid field dependence of the ionization coefficient, can be approximated as a sharp pulse delayed 90° with respect to the peak r.f. voltage. The external current will then consist of a rectangular pulse with a width equal to the transit time through the drift region (see Figure 3.25). The optimum transit angle in this approximation is π.

A Fourier analysis of the external current of Figure 3.25 leads to

$$I_{ext} = \frac{I_0}{2} + \frac{2I_0}{\pi} \sin \omega t + \text{higher order harmonics} \tag{3.35}$$

From Equation 3.35 we may calculate the r.f. power P_{rf} and the efficiency η

$$\begin{aligned} P_{rf} &= \frac{1}{2} V_a \frac{2I_0}{\pi} \\ \eta &= \frac{P_{rf}}{P_{DC}} = \frac{2}{\pi} \frac{V_A}{V_B} \end{aligned} \tag{3.36}$$

where V_A is the r.f. voltage swing as defined in Figure 3.25. Assuming $V_A/V_B = 0.5$ we obtain an efficiency of more than 30% which is in fact

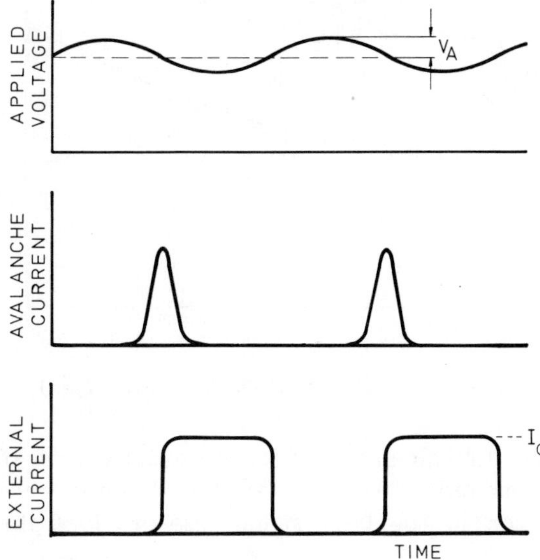

Figure 3.25 Applied voltage, avalanche current and external current for the sharp pulse approximation

very close to the best efficiency obtained in practice with gallium arsenide lo-hi-lo diodes.

A completely different approach to the large signal effects has been taken by Delagebeaudeuf[28] and Tager.[29] The starting point is the small signal theory with first order non-linear effects included. Delagebeaudeuf starts from Read's Equation 3.23 corrected for non-uniform currents in the avalanche zone. From Equations 3.23—3.25 we obtain

$$\frac{\tau_a}{3}\frac{d}{dt}\log J_a = \bar{\alpha}_0{}'\tilde{V}_1 \tag{3.37}$$

where $\tilde{V}_1$ is the r.f. voltage across the avalanche zone, assumed to be sinusoidal or

$$\tilde{V}_1 = V_1 \sin \omega t \tag{3.38}$$

From Equations 3.37 and 3.38 we obtain

$$J_a = A \exp\left(-\frac{3\bar{\alpha}_0{}'}{\omega\tau_a} V_1 \cos(\omega t)\right) = A \exp(-x \cos(\omega t)) \tag{3.39}$$

Fourier expansions gives

$$J_a = AI_0(x) - 2AI_n(x)(\cos(\omega t)) + \text{higher order terms} \tag{3.40}$$

where $I_n(x)$ are modified Bessel functions.

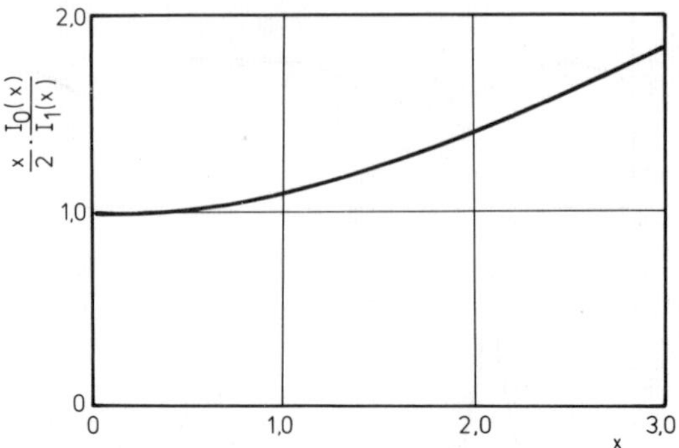

Figure 3.26 The function $(x/2)[I_0(x)/I_1(x)]$

From here on the calculations proceed exactly as in the small signal case and Delagebeaudeuf shows that the r.f. impedance in Equation 3.31 is valid even in the large signal case provided that $(\omega/\omega_a)^2$ is replaced by

$$\left(\frac{\omega}{\omega_a}\right)^2 \frac{x}{2} \frac{I_0(x)}{I_1(x)}.$$

The function

$$\frac{x}{2} \frac{I_0(x)}{I_1(x)}$$

is plotted in Figure 3.26.

The first thing we note is that as the r.f. voltage amplitude V_1 is increased the effective avalanche resonance frequency ω_a is decreased. For frequencies lower than ω_a the r.f. resistance is therefore positive for low drive levels (small V_1) but becomes negative for a sufficiently high drive. At a given frequency the negative resistance will decrease as illustrated in Figure 3.27.

For the maximum obtainable power, Delagebeaudeuf derives the expression

$$P_{max} = \frac{1}{2} J_0{}^2 \left(\frac{1 - \cos\theta}{\theta}\right)^2 \left(\frac{l_d}{l_d + l_a}\right)^2 \frac{1}{\omega^2 R_s C} \tag{3.41}$$

The first bracket in this expression gives the optimum transit angle which turns out to be somewhat smaller than 180°. The second bracket shows the importance of a thin avalanche region. The last factor finally shows that the series resistance, R_s, should be minimized and that the diode capacitance should be as small as possible. The basic features of

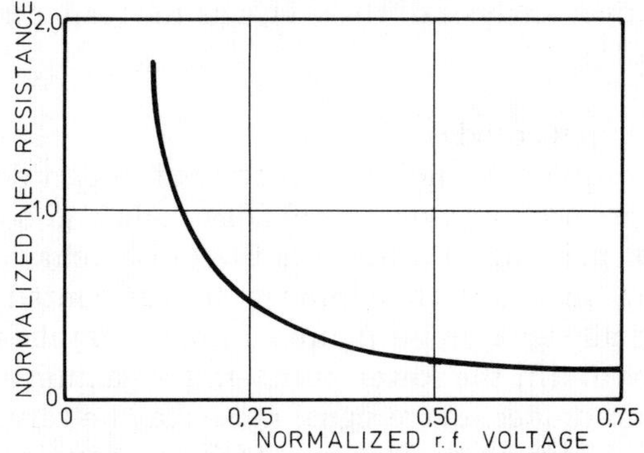

Figure 3.27 Normalized negative resistance versus r.f. drive[25]

Equation 3.41 have also been experimentally verified by van Iperen and coworkers[30] who show that as the diode is made punch-through and the avalanche zone width increases the maximum power decreases rapidly (see Figure 3.28).

The analysis by Delagebeaudeuf neglects all kinds of harmonic and parametric effects and thereby also one of the most important and difficult aspects of impatt oscillator design. A diode oscillating at $\omega > \omega_a$ may develop a negative resistance for frequencies lower than ω_a. This may lead to unwanted parametric oscillations which must be suppressed by the circuit. Particularly, the degenerate parametric oscillation at $\omega/2$ is often troublesome in actual circuit design. For a

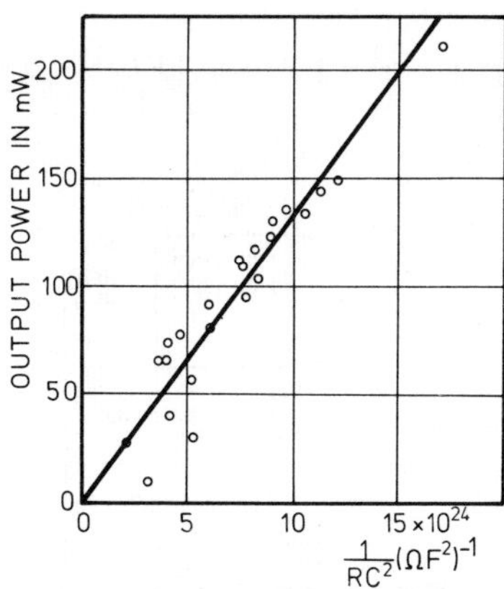

Figure 3.28 Output power of impatt diodes with varying thickness of the epitaxial layer as a function of the quantity $1/(RC^2)$[30]

detailed theoretical analysis of these effects the reader is referred to the paper by Hines.[31]

3.4.5 Noise in impatt diodes

We shall now consider the noise properties of impatt devices. Basically the avalanche process is a noisy mechanism and a good understanding of how the avalanche noise influences the noise behaviour is therefore often important in order to minimize the unwanted noise. Experimentally the small signal noise figure of impatt amplifiers ranges from about 20 dB to 40 dB, the lower numbers for gallium arsenide and the higher for silicon diodes. Large signal noise may be very much higher in hard-driven amplifiers or oscillators which are coupled to give a large r.f. voltage swing.

We shall first present Hines's theory for the small-signal noise[32] and then proceed to discuss the theories by Sjölund[33] and Kuvås[34] to explain the origin of the increased noise at large-signal level.

3.4.6 Small-signal noise

Hines starts from the representation of the diode given in Figure 3.29. The noise conduction current in the avalanche zone can be written as the sum of the short-circuit noise current $\tilde{i}_{no}$ and the induced correlated a.c. current $\tilde{i}_{ne}$. In addition the noise current in the avalanche zone will generate a displacement current. We thus have the total noise current

$$\tilde{I}_n = \tilde{i}_{no} + \tilde{i}_{ne} + j\omega\epsilon A\tilde{E}_a \tag{3.42}$$

In the drift region we have

$$\tilde{I}_n = (\tilde{i}_{no} + \tilde{i}_{ne})\exp\left(-\frac{j\omega x}{v_s}\right) + j\omega\epsilon A\tilde{E}(x) \tag{3.43}$$

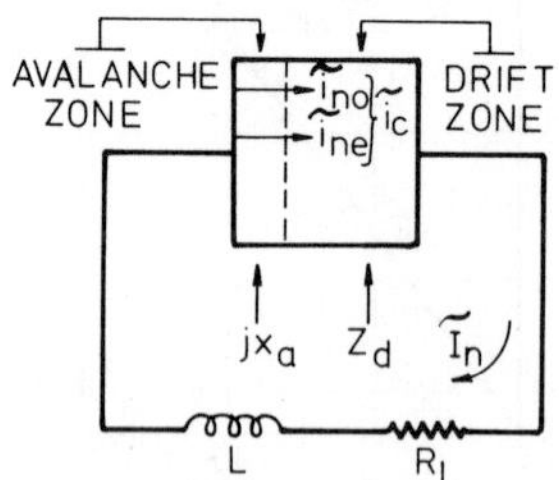

Figure 3.29 Circuit used for analysis giving the noise current $\tilde{I}_n$ as a function of the short-circuit avalanche fluctuation current $\tilde{i}_{no}$[32]

Solving for $\tilde{E}(x)$ and integrating yields the voltage drop in the drift region

$$\tilde{V}_d = \tilde{I}_n \frac{l_d}{\omega \epsilon A}\left(-j + \frac{\tilde{i}_{no} + \tilde{i}_{ne}}{\tilde{I}_n} \frac{1 - \cos\theta + j\sin\theta}{\theta}\right) \tag{3.44}$$

With $\tilde{i}_{ne}$ given by $A\tilde{J}_a$ in Equation 3.26 a relation is obtained between $\tilde{i}_{ne}$ and $\tilde{E}_a$. By applying Kirchhoff's law that the total voltage drop around the loop of Figure 3.29 must be zero we find another relation between the noise currents $\tilde{i}_{ne}$, $\tilde{i}_{no}$ and the field $\tilde{E}_a$. Eliminating the induced quantities $\tilde{i}_{ne}$ and $\tilde{E}_a$ we may calculate the total noise current $\tilde{I}_n$ in terms of $\tilde{i}_{no}$ as

$$\tilde{I}_n = i_{no} \frac{\dfrac{l_d}{\omega\epsilon A}\left(\dfrac{-1}{1-(\omega_a/\omega)^2}\right)\left(j\dfrac{l_a}{l_d} + \dfrac{1-\cos\theta + j\sin\theta}{\theta}\right)}{j\omega L + R_L + \dfrac{l_a}{j\omega\epsilon A(1-(\omega_a/\omega)^2)} + \dfrac{l_d}{\omega\epsilon A}\left(-j + \dfrac{1-\cos\theta + j\sin\theta}{\theta(1-(\omega/\omega_a)^2)}\right)} \tag{3.45}$$

where the denominator is the total impedance of the loop of Figure 3.29. Assuming that $l_a l_d$ is a small number we may simplify Equation 3.45 and calculate the mean square noise current as

$$\overline{I_n^{\,2}} = \overline{i_{no}^2}\left|\frac{R_d}{Z_t^{\,2}} \frac{1}{1-\left(\dfrac{\omega_a}{\omega}\right)^2} \frac{\tau_a}{\tau_d} \frac{l_d}{l_a} \frac{1}{\overline{\alpha_a'} I_0}\right| \tag{3.46}$$

where R_d is the real part of the diode impedance in Equation 3.31.

It remains now to calculate $\tilde{i}_{no}$. The dominant contribution to the noise is due to the statistical nature of the avalanche. In a steady avalanche the average pair of carriers leaving the avalanche region will have given rise to one new pair but often it may also have generated 0, 2 or 3 etc. pairs. This means that the current will be made up by a random sequence of step functions of equal magnitude $\Delta i = e/\tau_x$ where τ_x is the average time spent by a pair in the avalanche zone. In a steady avalanche τ_x is approximately half the avalanche zone transit time τ_a.

This situation is different from the process which gives rise to ordinary shot noise. The latter mechanism is due to the discrete nature of the electronic charge. The current will consist of short pulses of a duration equal to the transit time, τ, and a pulse height $2e/\tau$. For frequencies lower than $1/\tau$ the Fourier spectrum of such a pulse is white and appropriate summation leads to the well known result for shot noise.

$$\overline{I^2} = 2I_0 eB \tag{3.47}$$

Such noise will also be present in the avalanche but the current steps in the avalanche will generate a much larger noise contribution. The Fourier spectrum of an individual step increase (or decrease) of the current will be inversely proportional to the frequency and, as shown by Hines's proper statistical summation, leads to the current spectrum.

$$\overline{i_{\text{no}}^2} = \frac{2I_0 eB}{\omega^2 \tau_x{}^2} \tag{3.48}$$

As $\omega\tau_x$ in general is very much smaller than unity this noise contribution far outweighs normal shot noise.

For very low frequencies it may be shown from Equations 3.46 and 3.48 that the mean square noise current is frequency independent and inversely proportional to the d.c. current I_0. This white noise below the avalanche frequency is finding practical use in noise sources in which noise levels more than 30 dB above the thermal level may be obtained.

At resonance $\omega = \omega_a$ the noise current is given by

$$\overline{I_n{}^2} = 2eI_0 B \frac{1}{\left(1 + \dfrac{R_L}{R_d}\right)^2} \frac{C_d{}^2 V_a{}^2/I_0{}^2}{m^2 \tau_x{}^2} \frac{\theta^2/2}{1 - \cos\theta} \tag{3.49}$$

where m is defined below and $V_a = l_a E_a$ is the avalanche zone voltage.

In order to be able to compare impatt diode noise to other sources we shall now calculate the noise figure of an impatt amplifier with gain G. By definition this noise figure F is given by

$$F = 1 + \frac{R_L \overline{I_n{}^2}}{G k_b TB} \tag{3.50}$$

G is given by

$$\left| \frac{R_d - R_L}{R_d + R_L} \right|^2 \tag{3.51}$$

and inserting this expression, and Equations 3.46 and 3.48 into Equation 3.50, we obtain

$$F = 1 + \frac{1}{\overline{\alpha_0}' \omega^2 \tau_x{}^2} \frac{e}{k_b T} \frac{1}{\left| 1 - \left(\dfrac{\omega_a}{\omega}\right)^2 \right|} \frac{R_d R_L}{(R_L - R_d)} \tag{3.52}$$

If we assume high gain or $R_d \approx -R_L$ and an exponential field dependence of α, i.e. $\alpha l_a \simeq (E_a/E_0)^m$ we may simplify Equation 3.52 to

$$F = 1 + \frac{1}{4m\omega^2 \tau_x{}^2} \cdot \frac{eV_a/k_b T}{1 - \left(\frac{\omega_a}{\omega}\right)^2} \tag{3.53}$$

Inserting practical numbers will lead to noise figures of the order of 40 dB with the large factor in the second term of Equation 3.53 being $1/\omega^2 \tau_x^2$. The avalanche process has indeed turned out to be very noisy.

Small-signal noise calculations for more realistic diode models have been made by Gummel and Blue[35] and their results, although different in detail, confirms both the general magnitude of the noise as well as the noise pole at the avalanche resonance frequency.

3.4.7 Large-signal noise

As the r.f. voltage swing in an oscillator or amplifier is increased actual diodes will show a dramatic increase in noise generation. This increase is basically due to two different physical effects. With a large r.f. peak voltage harmonic and parametric effects will become important and lead to an increase in noise. Furthermore, fluctuations in the avalanche process will become relatively stronger when the voltage, having been reduced very much below breakdown, swings above breakdown and multiplication starts from a very low current level. The theoretical analysis, being large-signal, becomes complicated but results have been published by, among others, Sjölund,[33] Kuvås[34] and Goedbloed and Vlaardingerbroek.[36] As an illustration we reproduce Figure 3.30 from Kuvås's paper. Kuvås uses the noise measure M defined by de Loach[37] instead of the noise figure F. For large amplifier gains and high noise

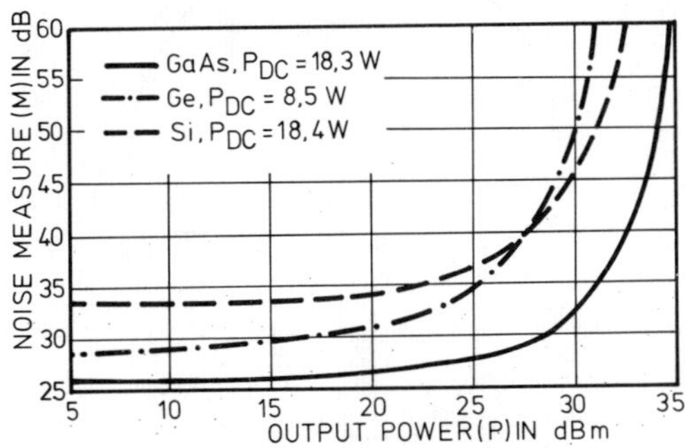

Figure 3.30 Calculated noise measure, M, versus output power, P, for 6 GHz GaAs, Ge and Si p^+-n-n^+ diodes[34]

figures the two quantities become identical, however. We have

$$M = \frac{F-1}{1-1/G} \approx F \tag{3.54}$$

Figure 3.30 illustrates the very rapid increase of noise with output power or r.f. voltage swing but also the superiority of GaAs and Ge over Si with respect to noise performance. Goedbloed has pointed out[36] that n^+—p—p^+ Si diodes may be better than GaAs diodes however.

3.4.8 Practical circuit considerations

We shall conclude this section by discussing briefly some commonly used circuit types and some practical aspects of their use.

In Figure 3.31 we show a Smith chart plot of an actual packaged Si p^+—n—n^+ X-band diode. We note that the peak negative resistance occurs very close to series resonance. Owing to the low impedance level of impatt diodes it is often advantageous to choose the package in such a way that the operating frequency is close to series resonance. Otherwise an extra tuning element with unavoidable losses may have to be introduced and the overall efficiency decreases. The primary function of the external circuit is therefore to transform the resistive diode impedance into 50 ohms.

The internal Q of the diode is typically about 10 and if no external reactive elements are supplied the frequency stability of an oscillator will be very poor. For high stability applications it is therefore necessary to supply an external resonance circuit. The required Q-value of this resonator will depend on the oscillator specification and on

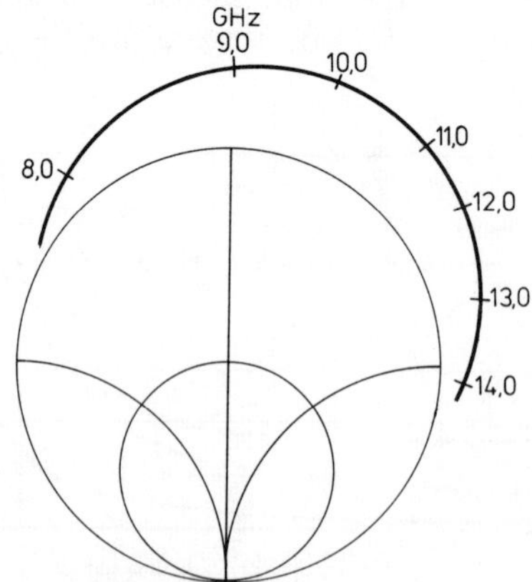

Figure 3.31 Measured impedance of packaged X-band p^+-n-n^+ impatt diode

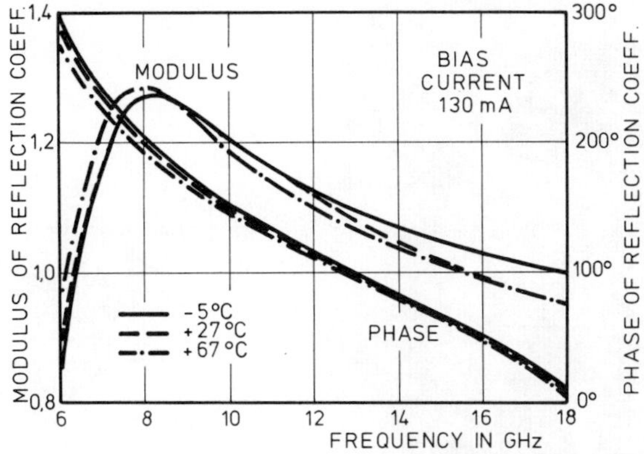

Figure 3.32 Impatt diode reflection coefficient versus frequency with heat sink temperature as parameter[38]

diode properties such as impedance variation with ambient temperature. To illustrate the latter point we show in Figure 3.32 the measured phase and magnitude of the diode impedance.

The diode properties are seen to be remarkably independent of temperature although to give an example loaded Qs of about 1000 are required to maintain a stability of a few MHz over a range of 100 °C in an X-band oscillator. As illustrations of actual circuits we show in Figure 3.33 a low Q microstrip and in Figure 3.34 a high Q circuit using direct coupling to a TM_{010} cavity.

Figure 3.33 Low Q microstrip impatt oscillator

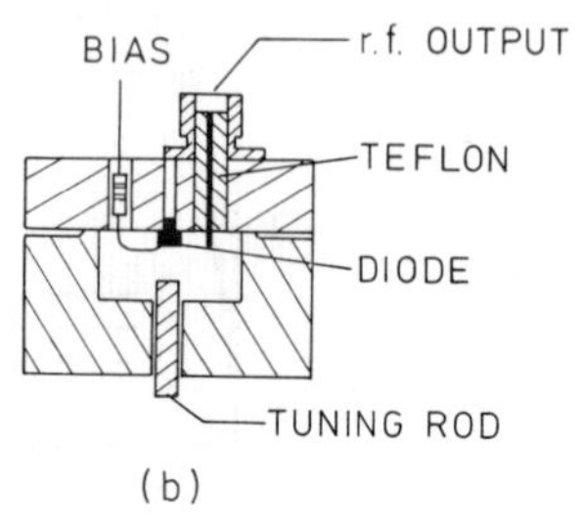

Figure 3.34 High Q impatt oscillator

At very high frequencies other waveguide circuits must be used and the cap circuit illustrated in Figure 3.35 is often used and is very convenient. The resonance frequency in this circuit is determined by the area and height of the cap and the coupling is conveniently adjusted by the sliding short.

A very important aspect of impatt oscillator design is the bias circuit which should be designed with much the same considerations as with tunnel diode oscillators and amplifiers. The reason for the special concern is the very wide band nature of the device. In tunnel diodes this is a small signal property at all frequencies but in impatt diodes it is due to large signal effects and limited to certain frequency ranges. An example of a suitable bias filter is therefore one which stops only frequencies very close to the operating frequency f_0 but transmits and resistively loads all other frequencies particularly those below f_0.

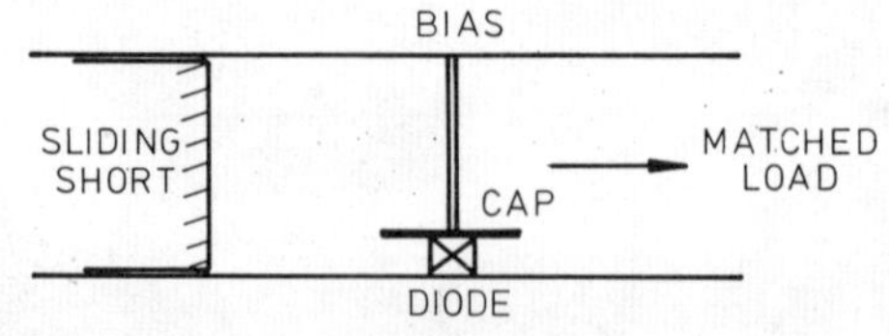

Figure 3.35 High frequency impatt oscillator

3.5 BARITT DIODES

3.5.1 D.C. properties

As discussed already in the introduction to this chapter the basic baritt structure is a p^+—n—p^+ diode with abrupt p^+-regions and a constant doping in the n-region as shown in Figure 3.36.

We shall begin by discussing the d.c. properties of this structure. When a voltage is applied between the p^+-regions one of the p—n junctions will be reverse-biased and oppose any current flow. For low applied voltages the electric field profile will be as shown by curve 1 in Figure 3.36. As the voltage is increased the depletion region of the reverse-biased junction will widen until punch-through is reached, i.e. the situation illustrated by curve 2. If no current is to flow for voltages below this punch-through voltage V_p the peak electric field must be so small that avalanche multiplication may be neglected. This condition imposes a condition on the maximum allowable doping in the n-region.

When the voltage is increased above V_p the forward-biased junction will obtain a positive bias and current will start to flow as is seen from curve 3 in Figure 3.36. The differential resistance above V_p will be given as the sum of the space charge resistance of the part of the diode where the carrier velocity is saturated, the ohmic resistance of the non-saturated region and the resistance of the forward-biased junction with its associated resistance from the reverse-biased junction. The voltage drop due to the latter resistance is given by the area between curve 2 and 3 in Figure 3.36. As a net result we will obtain an I—V characteristic as shown by the solid curve in Figure 3.37. The current below V_p is the normal reverse saturation current.

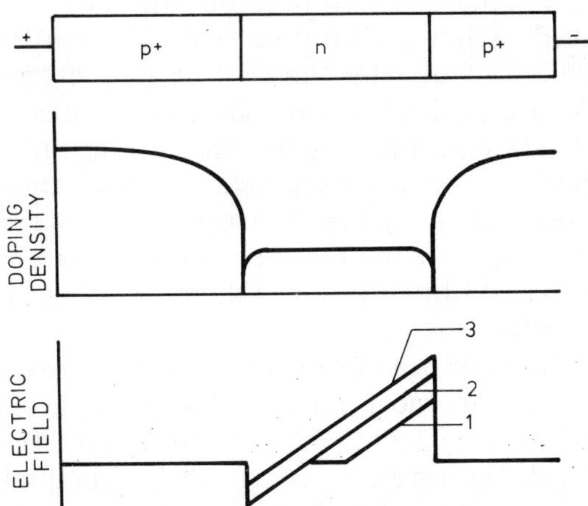

Figure 3.36 Structure, doping profile and electric field profiles of p^+-n-p^+ baritt

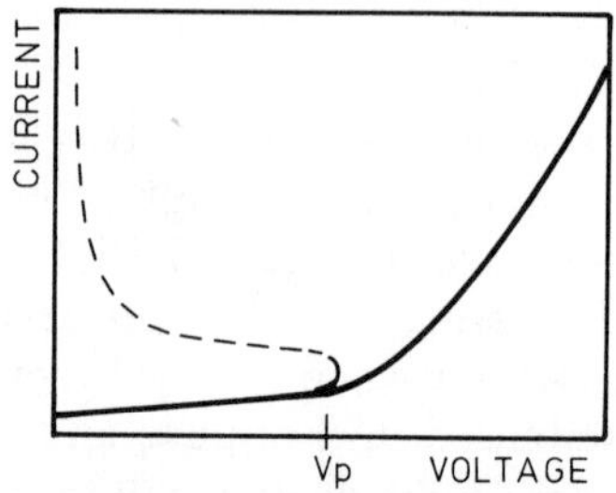

Figure 3.37 Current—voltage characteristic of baritt structure. The dashed characteristic results when the peak field is too high

If the doping in the n-region is too high avalanche multiplication will take place as pointed out above. This effect will show up at much lower peak fields than in impatt diodes, however. The reason is to be found in the voltage minimum which is present at the plane where the electric field goes through zero. Majority carriers (electrons) which are generated in the peak field region in the right-hand part of the diode will travel to the left and become trapped in the voltage minimum, where they are neutralized by minority carriers injected from the left. In this way the diode becomes filled with a plasma and loses its voltage sustaining capability. As all electrons are effectively collected in the potential minimum this effect occurs for much lower multiplication rates than are required to cause normal avalanche breakdown in p—n diodes. The phenomenon is analogous to the breakdown of bipolar transistors where the current multiplication of the transistor and avalanche multiplication multiply to cause a breakdown. The dashed curve in Figure 3.37 illustrates the I—V characteristic which will result when the peak electric field is too high. In a practical silicon p^+—n—p^+ device the field strength must be less than about 20 V/μm in order to avoid this breakdown. In the reciprocal n^+—p—n^+ device the allowable peak field is very much smaller because of the higher ionization coefficients of electrons. As we shall see below, a reasonably high field is desirable from an r.f. point of view and therefore n^+—p—n^+ devices do not appear very attractive.

The original baritt device of Coleman and Sze[39] used a metal-p-metal structure, which from most points of view behaves in an identical manner as the p^+—n—p^+ device.[40] An important fundamental difference does exist, however, between p—n junctions and Schottky barriers for use in baritts. This difference shows up particularly at and below room temperatures. In Figure 3.38 we show the measured I—V characteristic of an MSM-structure at two temperatures from Sze and coworkers.[41]

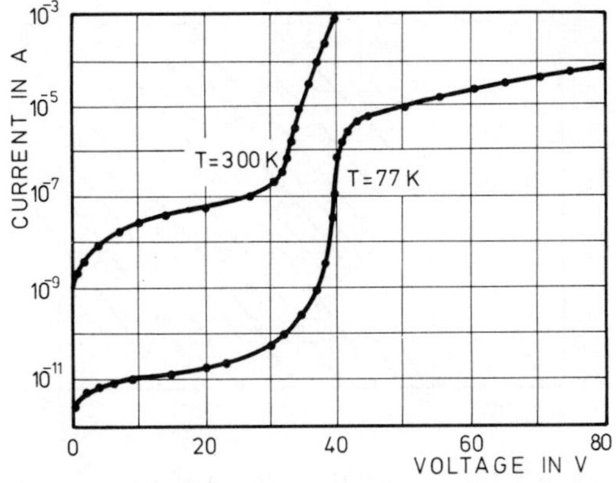

Figure 3.38 Current versus applied voltage in MSM-structure at 77 K and 300 K[41]

The difference between the behaviour in Figure 3.37 is clear and the saturation of current which takes place in forward-biased Schottky junctions will render them useless in baritts at low temperatures. In fact, many published results on metal-p-metal structures have been made possible only by the self heating of the device. When the current saturation of a metal-p-metal structure sets in, further increase in applied voltage will eventually lead to avalanche breakdown. At low temperatures only a small current will therefore flow until the bias voltage is increased far above the punch-through voltage as illustrated in Figure 3.39.

In the following parts of this section we shall therefore limit ourselves to structures with doped contact regions and holes as active carriers. As a guideline to the design of silicon $p^+—n—p^+$ structures we

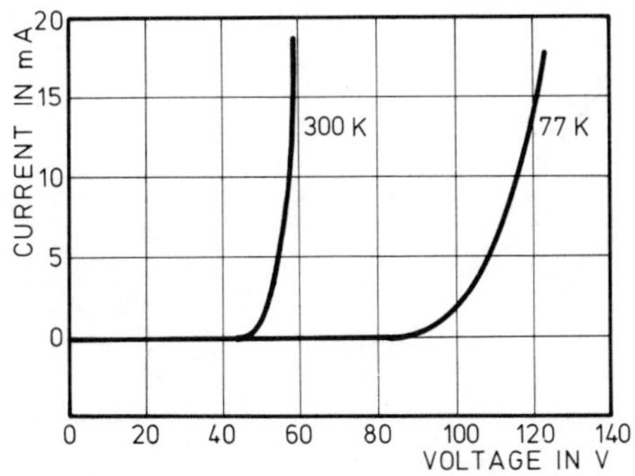

Figure 3.39 Current—voltage characteristics at 300 K and 77 K of Schottky barrier M-n-p^+ device[42]

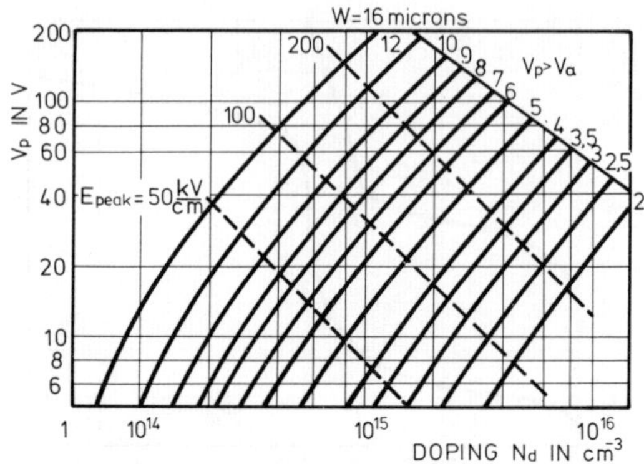

Figure 3.40 Punch-through voltage versus *n*-region doping with *n*-region width, *W*, as a parameter for p^+-*n*-p^+ structure

present Figure 3.40 which illustrates the relationship between punch-through voltage, length and doping.

3.5.2 Small-signal r.f. properties

In Figure 3.41 we show the measured r.f. chip impedance of a baritt diode with the bias current as a parameter.[42] The particular p^+—n—p^+ diode to which these measurements refer has an *n*-region length of about 7 μm and assuming a saturated hole velocity of the order of 7×10^4 m/s we find that the peak negative resistance at about 6.5 GHz corresponds to a transit angle of about 270°. This is indeed also the value predicted in some early theories.[43,44]

Before we proceed to discuss an appropriate theoretical mo del of the

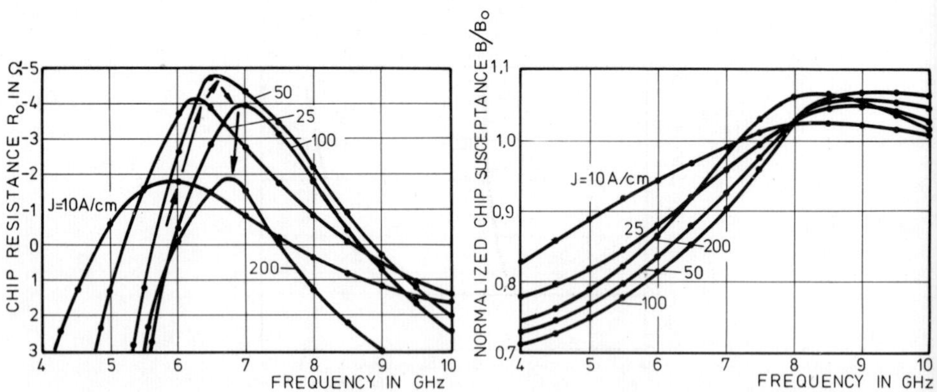

Figure 3.41 Small-signal chip resistance and normalized chip susceptance of p^+-*n*-p^+ device with current density as a parameter. Device area = 3×10^4 cm². Arrows indicate peaks in R_0. $B_0 = \omega \times 0.405$ pF[42]

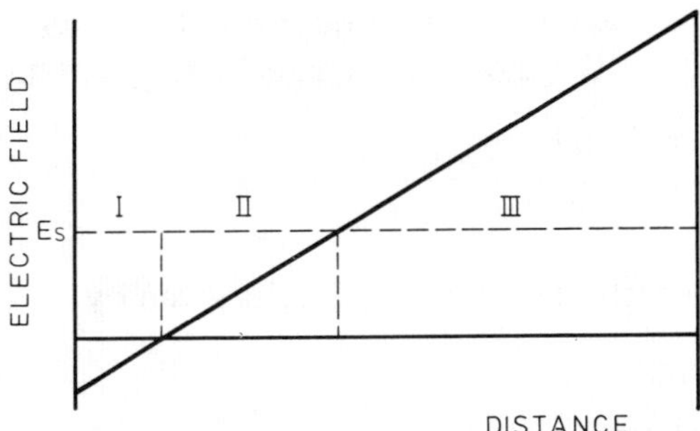

Figure 3.42 Regions in p^+-n-p^+ baritt. I Depletion region of forward-biased p-n junction. II Ohmic region. III Drift region, $v = v_s$

baritt diode we shall discuss some of the more important features of the experimental results. For low currents the diode develops a negative resistance with a certain frequency band. As the current is increased the peak negative resistance moves to higher frequencies and increases. Further increase in current leads to a decrease in the peak negative resistance and eventually also to a decrease in the frequency at which this peak occurs. The latter effect, i.e. the decrease in the optimum frequency can be shown experimentally to be a thermal effect and due to the increase in chip temperature with dissipated power. A theory in which a constant chip temperature is assumed is therefore not expected to predict this effect.

With reference to Figure 3.42 we shall now discuss a theoretical model of the p^+—n—p^+ baritt diode. The figure shows three distinct regions. Region 1 is a forward-biased p—n junction which may be represented as a parallel RC network with $C = C_1$ given by its geometrical capacitance and R from the usual diode current voltage relationship as $k_b T/eI$. In region 2 the electric field is low enough so that to a first approximation we may assume that Ohm's law applies. In region III the hole velocity is saturated and approximately constant.

The most complicated region to treat theoretically is region II and initially we shall neglect it and assume that the carrier velocity is saturated in region II as well as in region I. From Equation 3.11 in Section 3.3.1 we can calculate the small signal impedance of the diode provided that we know the injection phase ϕ. In a parallel RC network the total current is given by

$$I = j\omega C_1 V + V/R \tag{3.55}$$

The injection phase ϕ is then given by

$$\exp(-j\phi) = \frac{V}{RI} = \frac{1}{1 + j\omega RC_1} \tag{3.56}$$

Inserting this expression into Equation 3.10 we obtain a first approximation Z_1 to the diode small signal r.f. impedance.

$$Z_1 = \frac{1}{j\omega C_1}\left[1 - \frac{(1-\exp(-j\theta))}{j\theta(1+j\omega RC_1)}\right] \tag{3.57}$$

The maximum negative resistance $-R_{opt}$ is given by

$$-R_{opt} = \frac{0.048}{\omega C_1} \tag{3.58}$$

and occurs for $\phi = 62°$ and $\theta = 292°$.

Appropriate data for the diode used in the measurements in Figure 3.41 are diode area 3.10^{-4} cm^2 and n-region width 7 μm which gives an optimum frequency of 6.8 GHz and $R_{opt} = -2.5\ \Omega$ for $\tau = RC_1 = 45$ ps. The experimental value of R_{opt} is thus larger than predicted even without taking the substrate series resistance into account. This discrepancy is due to the neglect of the ohmic region II in Figure 3.42.

In region II carriers have non-saturated velocities and will therefore give an additional contribution to the phase shift which increases the negative resistance. Owing to ohmic losses region II will also cause a decrease of the magnitude of the injection phase factor $\exp(-j\phi)$ just as is the case for the RC network of region I (see Equation 3.56).

Theories which take both of these effects of region II into account have been worked out[4 5] but become quite complicated. We shall neglect the ohmic losses and represent region II as a drift region with a constant velocity v_0 (the average velocity) lower than that of region III. With this assumption we may use the basic theory for a drift region in Section 3.3.1. The calculation is straightforward and leads to the impedance

$$Z = \frac{1}{j\omega C_1}\left[1 - \frac{1-\exp(-j\theta_1)}{j\theta_1(1+j\omega RC_1)}\right] + $$
$$+ \frac{1}{j\omega C_2}\left[1 - \frac{\exp(-j\theta_1[1-\exp(-j\theta_2)]}{j\theta_2(1+j\omega RC_1)}\right] \tag{3.59}$$

where θ_1 is the transit angle of the ohmic region $\dfrac{\omega l_{II}}{v_0}$

and $\quad C_2 = \dfrac{\epsilon A}{l_{II}}$

and $\quad \theta_2 = \dfrac{\omega l_{II}}{v_s}$

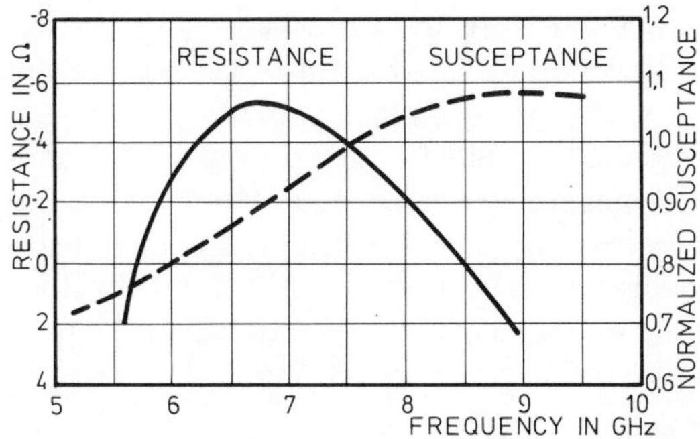

Figure 3.43 Resistance and normalized susceptance as a function of frequency for a double drift region structure with $W_1 = 3.5 \times 10^{-4}$ cm, $W_2 = 4.5 \times 10^{-4}$ cm, $v_{s1} = 5.6 \times 10^6$ cm/s, $v_s = 7.2 \times 10^6$ cm/s and $\tau = 20$ ps (corresponding to the optimum current density of the device shown in figure 3.41)

In Figure 3.43 the impedance from Equation 3.59 is plotted for the diode of Figure 3.41. It is seen that, in spite of the simplifications a reasonable fit to the experimental curve is obtained.

Having established a theoretical model of the basic $p^+—n—p^+$ baritt structure we shall now proceed to discuss more general structures. It is particularly in two respects that an improved structure would be desirable. The negative resistance begins to decrease at quite low current levels and this leads to a serious limitation in available output power. Furthermore, the experimentally observed efficiencies of $p^+—n—p^+$ diodes are quite low or of the order of 2%. What structures could possibly lead to improvements in performance on these two points?

The reason for the impedance decrease at high current is that when ωRC_1 becomes small the forward-biased $p—n$ junction becomes capacitively short-circuited. As R is inversely proportional to the current but independent of the diode structure the only way to push the optimum $\omega RC_1 \approx 1$ to higher currents is to increase C_1 which is achieved by increasing the n-doping in region I. In region III however, an increase in the electric field above the value E_s required for velocity saturation will have no effect on the small signal r.f. impedance but will lead to an increase in dissipated d.c. power and to a decrease in efficiency. In order to obtain a constant field E_s, the drift region should be slightly p-type with an acceptor concentration large enough to cancel the space charge for the current that results in $\omega RC_1 \approx 1$. Below we shall present results of a large signal calculation for a diode with the resulting $p^+n^+—p—p^+$ structure and the result is indeed a dramatic improvement in output power and efficiency. The impedance level will turn out to be very low, however. This is also to be expected as the n^+-region covers both the forward-biased junction and the ohmic region. The field will therefore reach saturation in a very short distance leading to a very small value of θ in Equation 3.59 and therefore to a small value of the negative resistance. Further elaboration of the diode structure to $p^+—n^+—n—p—p^+$ with the n-region coinciding with region II would be expected to lead to an improvement in this respect. Owing to the complexity of these structures experimental results are lacking, but measurements on $p^+—n^+—\nu—p^+$ diodes[42] show a higher power capability.

3.5.3 Large signal effects

At large signal levels the impedance of the baritt diode will differ from its small signal value. In Figure 3.44 the variation of bias voltage and r.f. conductance with r.f. voltage is shown.[42] The reduction in operating voltage is substantial and, in order to avoid bias oscillations due to an r.f. induced negative resistance, a certain amount of series resistance must be included in the bias circuit. The r.f. conductance increases with voltage amplitude. At frequencies around the lower limit for negative resistance it may also happen that the conductance decreases with drive for small signal levels and begins to decrease only for quite high r.f. voltages.

Very strong parametric effects like these occurring in impatt diodes are usually not observed in baritt diodes. A comparison of the two diodes with respect to the amplitude dependence of the conductance shows that the conductance decrease is usually more rapid in baritt diodes.[38] This will lead to a more pronounced gain saturation in baritt amplifiers.

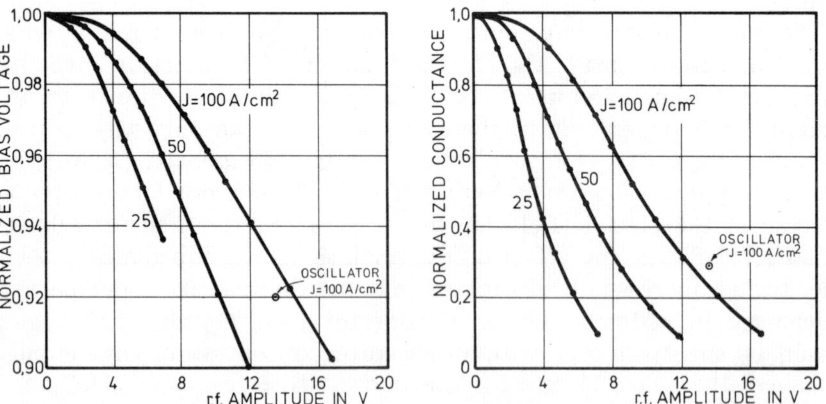

Figure 3.44 Large-signal voltage rectification and chip conductance versus r.f. voltage amplitude for p^+-n-p^+ device operated as a tuned amplifier near 6.6 GHz[42]

Comparatively few attempts to develop large signal theories for baritt diodes have been published. Lacombe[46] and Delagebeaudeuf and Lacombe[47] have developed a theory using the same approach as used by Delagebeaudeuf for impatt diodes. The dominant non-linearity in impatts is the dependence of the ionization coefficient on field strength. The corresponding non-linearity in the baritt diodes is the exponential current—voltage relation of the forward-biased p—n junction. Lacombe and Delagebeaudeuf assume that peak performance is obtained when the r.f. voltage swing across the forward-biased region equals 50% of the built-in voltage and obtains a good agreement with experiment for the predicted output power and efficiency.[49] Sellberg and Sjölund[48] have used the basic concepts in Lacombe's theory to study the p^+—n^+—p—p^+ structure discussed above. Their results are worth noting. In one case they assume the following diode parameters

n^+-region	Doping 10^{17} cm^{-3}; width 0.14 μm
p-region	Doping 3×10^{14} cm^{-3}; width 5.4 μm
Area	1.5×10^{-4} cm^2

and obtain the following results

Frequency	10 GHz
Output power	350 mW
Efficiency	7%
Large signal negative resistance	—0.55 Ω
Optimum current	60 mA

Structures of this type could easily be manufactured using ion-implantation and although the effect of a series resistance has been neglected they certainly seem to merit further study.

3.5.4 Noise in baritt diodes

The baritt diode is expected to have a better noise performance than the impatt as no avalanche process is involved and indeed experimental values for the amplifier noise figure fall in the range 10—15 dB. As it is particularly this low noise that makes the baritt interesting from an application point of view it is important to understand the basic physical processes responsible for noise generation in this device.

Ordinary shot noise will be present but will be influenced by space charge effects. Furthermore there will be an important contribution from thermal noise or noise arising mainly from velocity fluctuation in the drift region. Theoretical treatments of these mechanisms have been given by among other Haus and coworkers[49,50] and Sjölund.[51,52]

The space charge smoothed open circuit noise voltage due to fully developed shot noise in the injected current is given by

$$\overline{V^2}(\omega) = I_0 \mid u(\omega) \mid^2 /e \tag{3.60}$$

where I_0 is the d.c. current and $u(\omega)$ the voltage caused by injection of one carrier when the total a.c. current equals zero, $u(\omega)$ may be calculated using small signal theory, as developed above for instance.

The velocity fluctuation open circuit noise voltage may be obtained from the basic formula of Shockley, Copeland and James[53]

$$\overline{V^2}(\omega) = 2e^2 D_p \int \left| \frac{\partial Z_{Nx}}{\partial x} \right|^2 p_0(x)\, dx \tag{3.61}$$

where $p_0(x)$ is the steady state distribution of holes in the n-region, D_p the diffusion coefficient for holes and the impedance field vector $\partial Z_{Nx}/\partial x$ is given by

$$\frac{\partial Z_{Nx}}{\partial x} = \frac{V_N(\omega)}{I(\omega)\delta x} \tag{3.62}$$

V_N is here the voltage generated by a current $I(\omega)$ injected at x and extracted at $x + \delta x$ and may again be calculated using small signal theory.

We shall refer to the paper by Sjölund[52] for the details of these calculations and in Figure 3.45 we reproduce some of his calculated results. The diode to which these calculations refer is the same as that referred to in Figure 3.39.

Looking first at the dashed curves in Figure 3.45, we note that these include only the effect of the injected shot noise. Owing to the

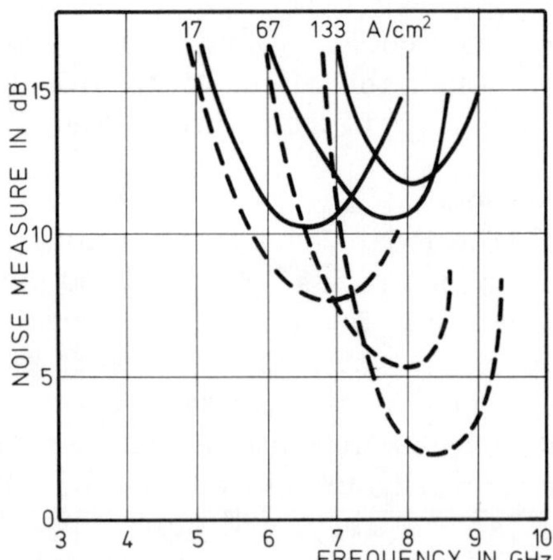

Figure 3.45 Noise measure versus frequency for the diode in Figure 3.39 for (1) injection noise only (dashed lines) and (2) injection noise and diffusion noise (solid lines); current density as parameter[52]

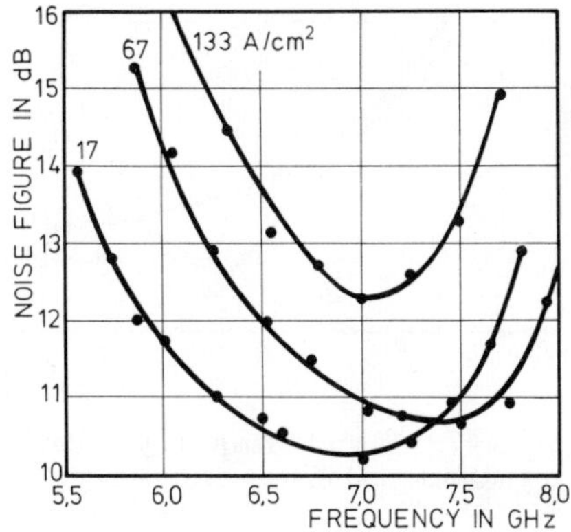

Figure 3.46 Noise figure for the diode in Figure 3.41 versus frequency, with current density as parameter[54]

increasing importance of space charge smoothing the minimum noise measure decreases with bias current. For a given current density the noise measure shows a minimum close to the frequency where the diode has its maximum negative resistance. When the diffusion or velocity fluctuation noise is also included, as in the solid curves of Figure 3.45, the optimum noise measure increases with current instead. A comparison of these predictions with the experimental values[54] of Figure 3.46 shows a reasonably good agreement and in fact a major part of the discrepancies may be explained as being due to the neglected effect of diode heating with bias current.

Just as in the small signal resistance curves this thermal effect causes the experimental optimum at high currents to move towards lower frequencies in disagreement with the isothermal theory. We therefore conclude that the present understanding of the basic high frequency noise sources in baritt diodes is adequate and that the best attainable noise measures are of the order of 10 dB.

Sometimes, however, diodes show a very much higher optimum noise figure which cannot be explained by the present theory.[55] It can be shown that this anomalous noise occurs only when the peak electric field in the drift region is above about 150 kV/cm. The reason for the excess noise generation is that avalanche generation already begins to play a role at these field strengths. In comparison, at field strengths of more than 200 kV/cm, a p^+—n—p^+ diode no longer works as a baritt diode but shows switching to low voltages as discussed in Section 3.5.1.

The noise figure of practical amplifiers may be computed from the noise measures in Figure 3.46 (using Equation 3.54). In oscillators the results in Figure 3.46 also apply at high modulation frequencies. No strong deviations due to large signal effects have been observed or

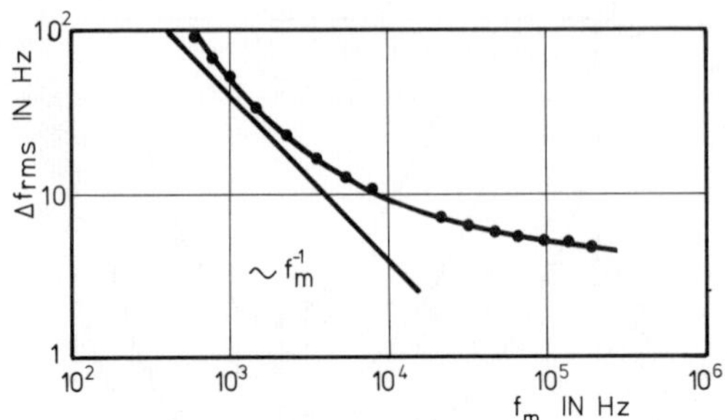

Figure 3.47 Mean frequency deviation Δf_{rms} against modulating frequency f_m[56]

predicted. At very low modulation frequencies, however, a rapid increase of FM noise shows up as illustrated in Figure 3.47.[56] A clue to the origin of this excess noise is found in Figure 3.48 which shows the equivalent noise current I_{eq} for the same diode at two different currents. This noise equivalent current is defined by

$$I_{eq} = \Delta I_{rms}^2 / 2eB \tag{3.63}$$

where ΔI_{rms}^2 is the mean squared short-circuit noise current.

In contrast to what would be expected[57] the low frequency noise exceeds shot noise. The exact origin of the excess low frequency noise is not known but most likely it is a surface effect which may or may

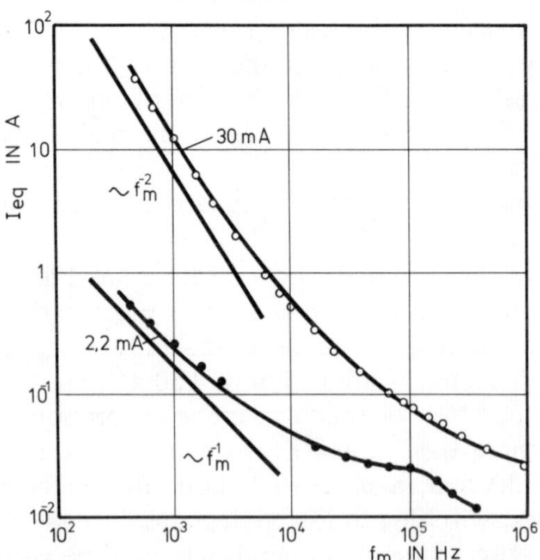

Figure 3.48 Equivalent noise current I_{eq} against frequency f_m for two values of bias current[56]

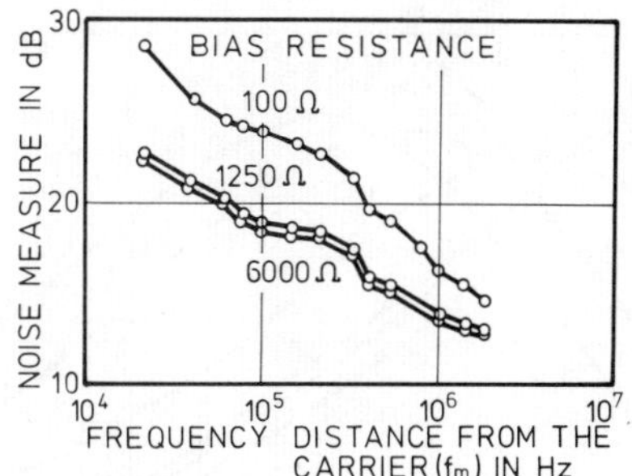

Figure 3.49 Noise measure versus frequency (f_m) with bias resistance as parameter[58]

not be possible to eliminate through improvements in technology. The low frequency noise is up-converted to FM oscillator noise through the current tunability of the oscillator. The influence of current on diode susceptance is evident from Figure 3.41. At very low frequencies thermal effects counteract the current tuning and therefore I_{eq} is expected to increase less rapidly at frequencies below that given by the reciprocal of the thermal time constant.

As the excess FM noise is due to bias current fluctuations it is not surprising that an increased bias resistance reduces the excess noise. This effect is illustrated in Figure 3.49 which also shows that the excess noise cannot be completely eliminated in this way.

3.5.5 Circuits and practical considerations

In Figure 3.50 we show a Smith chart plot of the small signal impedance of a 7 GHz baritt diode. In most respects it bears a strong

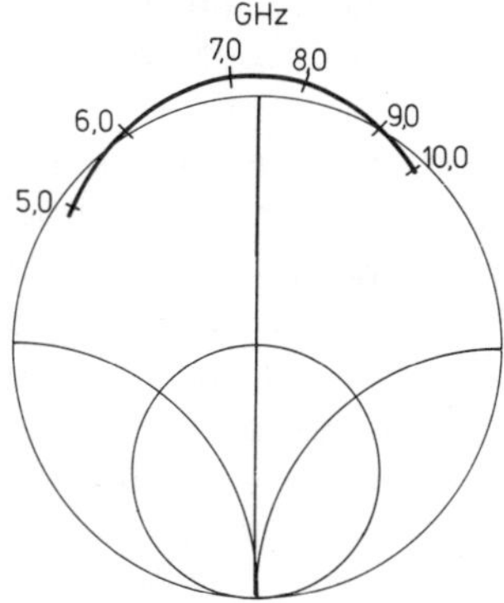

Figure 3.50 Smith chart plot of a 7 GHz baritt diode[38]

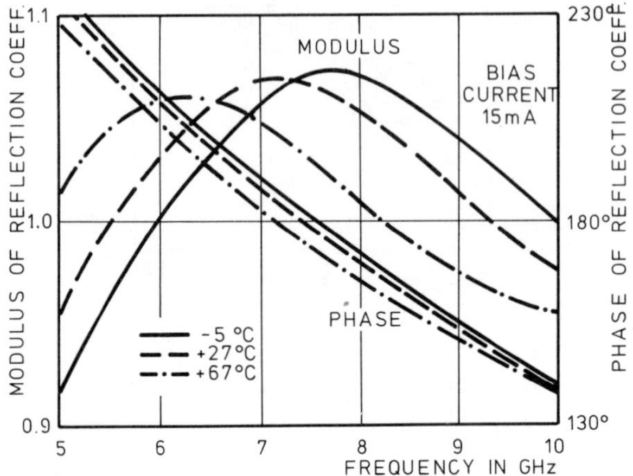

Figure 3.51 Baritt diode reflection coefficient versus frequency with heat sink temperature as parameter[38]

resemblance to the same plot for our impatt diode. In fact the only difference which is important to note is that the negative resistance is smaller. Baritt and impatt circuits therefore have much in common and we shall not separately discuss baritts but only underline the fact that unnecessary series resistance is highly undesirable.

In practical devices the dependence of diode parameters on ambient temperature is usually very important. In Figure 3.51 we illustrate this effect by plotting the 50 Ω reflection coefficient for three different ambient temperatures. In comparison with the impatt diode (Figure 3.32) the phase remains comparatively constant but the magnitude varies more rapidly. We would therefore expect that with identical circuits a baritt oscillator would have larger power variation with temperature than an impatt one but would have a better frequency stability.

A further comparison of practical problems in baritts and impatts reveals that in general impatt circuits are more critical because of the tendency of this device to develop spurious signals. Another pleasant feature of ordinary p^+—n—p^+ baritts is that at optimum current the diode is not thermally limited but can take considerable over-voltages without burnout. This advantage may well disappear if higher power structures like those discussed above are developed.

3.6 THE TRAPATT DIODE

3.6.1 Basic device physics

The trapatt mode of operation[59] of avalanche diodes differs fundamentally from the impatt mode and from the baritt diode. The

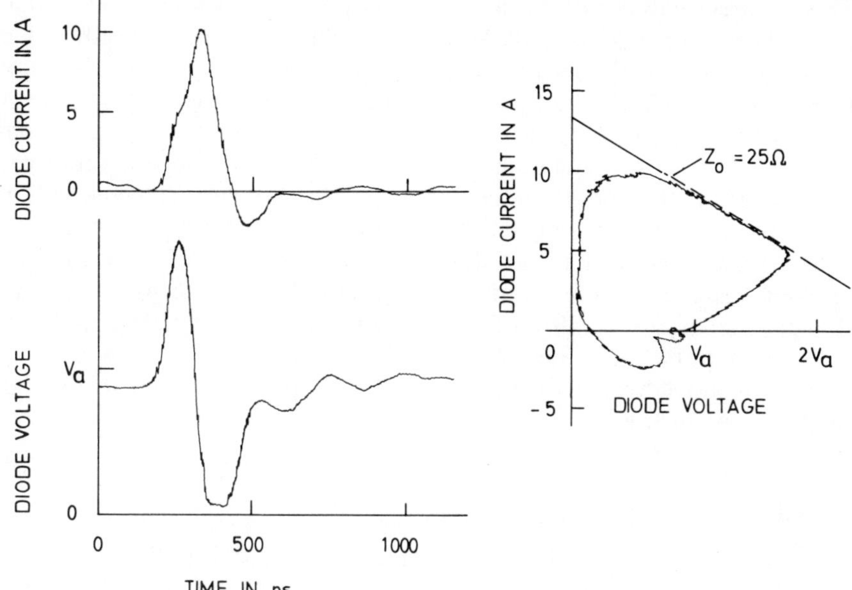

Figure 3.52 Dynamic current—voltage characteristic and waveforms[60]

fundamental mechanism no longer relies on transport and injection delays to cause a negative resistance. In fact no negative small signal resistance at all is associated with the trapatt mode.

As a starting point for our discussion of the physics of trapatt oscillations we turn to Figure 3.52 which shows the measured dynamic current—voltage loop of an oscillating trapatt diode.[60] As the r.f. voltage rises above the breakdown voltage the current initially increases in a normal manner. When the current reaches a value of, in our case, somewhat less than 5 A, the voltage sustaining capability of the diode suddenly disappears and the diode switches along the 25 Ω load line of the circuit into a high current, low voltage state. The remaining part of the cycle is characterized by a decreasing current and a recovering voltage. The key feature of the trapatt mode is exactly the voltage collapse which occurs at high currents and which is a dynamic feature which would not occur in the absence of a suitable circuit.

To gain an understanding of this voltage collapse we have to look into the concept of the avalanche shock front.[61] Let us, to this end, assume that a very sharply rising reverse current pulse is applied to a non-conducting reverse-biased n^+—p—p^+ diode. As long as the voltage remains below breakdown all the current is capacitive and the electric field is everywhere increasing at the rate

$$\frac{\partial E}{\partial t} = \frac{J(t)}{\epsilon} \tag{3.64}$$

This increase will continue until the field somewhere reaches the value E_a required for avalanching. When the field increases above this value rapid production of electrons and holes will take place. We shall now make the important assumption that the increase in current is so rapid that we may neglect the drift of the generated carriers during the part of the process now being considered. What then will happen is that very shortly after the field increases above E_a enough carriers will be generated to allow the current to be carried as conduction current with a very small residual field in all portions of the diode where E has once increased above E_a. For a punch-through diode with constant doping — the preferred trapatt structure — the situation is illustrated in Figure 3.53. The point at which $E = E_a$ is moving rapidly to the right; an avalanche shock front is propagating. The slope of the field to the right of $E = E_a$ is given by

$$\frac{dE}{dx} = \frac{e}{\epsilon} N_a \tag{3.65}$$

By dividing Equation 3.64 by Equation 3.65 we obtain the velocity of the avalanche shock front

$$\frac{dx}{dt} = \frac{J}{eN_a} \tag{3.66}$$

We are now in a position to quantify the assumption that the process is so fast that carrier drift is negligible. The maximum carrier velocity is given by the saturation velocity v_s (neglecting the difference in electron and hole velocities) and we therefore require

$$v_s \ll \frac{J}{eN_a} \tag{3.67}$$

In fact as long as the weaker condition $v_s < J/e\,N_a$ is fulfilled there will not be enough time for the generated plasma to be transported away

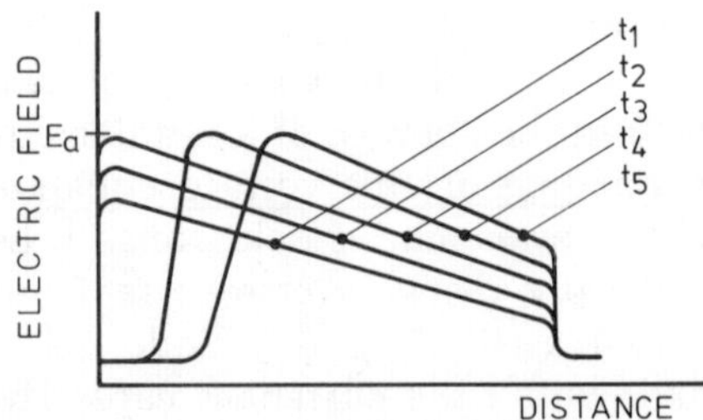

Figure 3.53 Time development of electric field before and during plasma formation

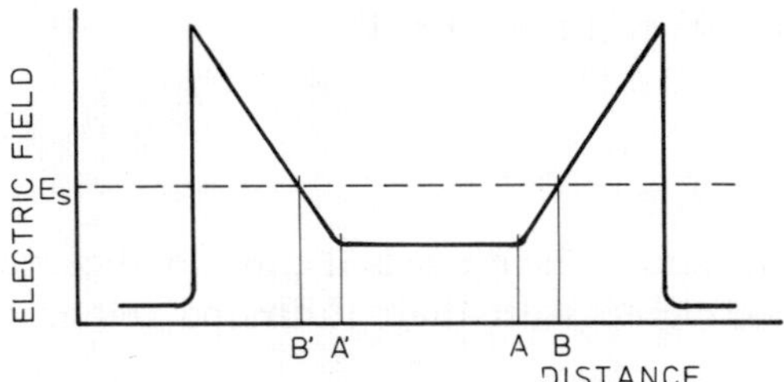

Figure 3.54 Time snapshot of entire electric field versus distance plot during recovery transient[61]

and an avalanche shock front will develop. An excellent discussion of the physics of the avalanche shock front is given by de Loach and Scharfetter[61] to which the reader is referred for details. Returning to Figure 3.52 we note that the diode dynamics is largely determined by the circuit once avalanche shock front formation occurs. From a circuit point of view the extraction of the plasma which must take place in order to reduce the current and re-initiate the cycle may be considered as a discharge of a non-linear capacitor. We have to return to the device physics and study the extraction phase to gain insight into the nature of the non-linear capacitance. We shall follow de Loach and Scharfetter in our discussion starting from Figure 3.54 which shows the electric field in the diode in the beginning of the plasma extraction period. Electrons and holes are assumed to have equal velocities. Deep in the interior a plasma with constant carrier densities and electric field E_0 exists and the current density is given by

$$J_R = e\mu E_0(n_0 + p_0) \tag{3.68}$$

where n_0 and p_0 are the electron and hole densities. Space charge neutrality requires that

$$n_0 + N_a = p_0 \tag{3.69}$$

Close to the contacts charge neutrality can no longer prevail and the electric field will increase in these end regions as in Figure 3.54. There will be small regions AB, A′B′ in which the field increases from E_0 to the value required for velocity saturation.

However, the main object of our present discussion is to estimate the time required to extract the plasma from the diode and the contribution from these intermediate regions may be neglected. We shall therefore concentrate on the ohmic central and the velocity-saturated outermost regions. The plane A and A′ will move towards the centre plane of the diode with velocities μE_0. The time to sweep through the entire diode width, W, will therefore be

$$\tau_1 = \frac{W}{2\mu E_0} \tag{3.70}$$

Once the planes B′ and B have reached the diode centre the remaining carriers will be extracted with their saturated velocity v_s leading to a delay time

$$\tau_2 = \frac{W}{2v_s} \tag{3.71}$$

The total plasma extraction time is then given by the sum of τ_1 and τ_2 with τ_1 the larger contribution. The actual value of the extraction time is then mainly determined by E_0 which in turn depends on the circuit or more accurately on the resistive impedance through which the diode is discharged. According to the analysis of de Loach and Scharfetter the discharge current density J_R is in fact independent of the circuit and given by

$$J_R = ev_s(2N_0 - N_a) \tag{3.72}$$

where N_0 is the impurity concentration for which a diode with a length equal to the one of the actual diode would be just punch-through at breakdown.

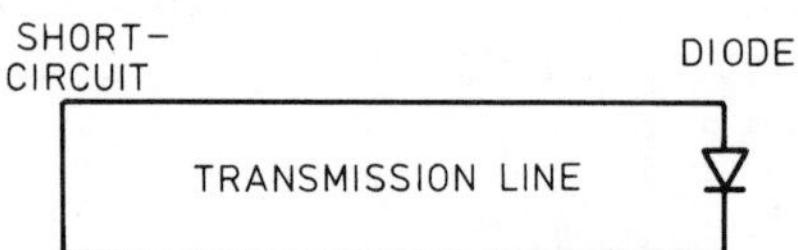

Figure 3.55 Schematic of microwave circuit for trapatt operation

According to this discussion trapatt operation starts with a short over-voltage at the beginning of the cycle. This over-voltage triggers an avalanche shock front which forces the diode to short-circuit and launch an inverted voltage pulse which will travel along a transmission line which is part of the microwave circuit (Figure 3.55).

While the plasma is being extracted from the diode this inverted pulse is reflected at the shortest end of the transmission line, inverted again and returned to the diode. Clearly the frequency of operation is mainly determined by the length of the transmission line but the returning pulse must not return before the plasma extraction phase has ended, leading to an upper frequency limit.

3.6.2 Basic circuit principles

Up to this point we have neglected to introduce any extraction of useful power into an external load and under these conditions a representative voltage waveform is that of Figure 3.56(a). If one wants to extract power at the fundamental frequency the short circuit at the end of the transmission line in Figure 3.55 should be replaced by a band pass filter partly transmitting the desired frequency to the load. Through the action of this filter the rectangular waveform of the voltage pulse travelling along the line will be distorted and the voltage waveform will acquire the shape of Figure 3.56(b) in which a sinusoidal component superimposed on the voltage of Figure 3.56(a). As the loading is increased the amplitude of this sinusiodal component increases until the situation in Figure 3.56(c) is reached. At this loading the peak of the sinusoidal component just reaches the breakdown voltage. Any further increase of the loading will result in an extra, premature, avalanching taking place, as in Figure 3.56(d), with a dramatic deterioration of output power. From this discussion it is clear that avoidance of this premature breakdown is the major limiting factor for the r.f. loading of the diode. With the simple type of circuit to which we have confined ourselves the only additional parameters are the length and impedance of the transmission line. The length directly determines the frequency and the impedance is usually given by

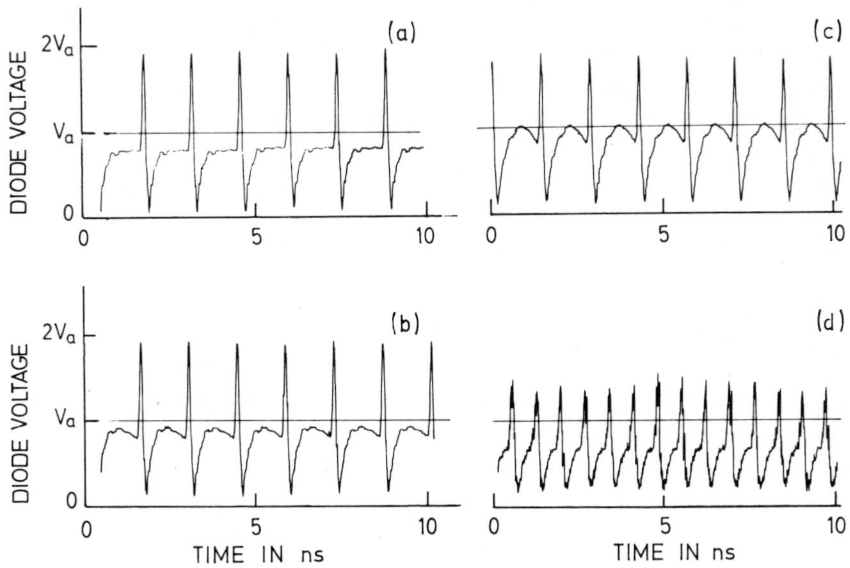

Figure 3.56 Voltage waveforms for different output coupling at the fundamental frequency[60]

practical considerations, although it does have an influence on the plasma extraction time and on the rise time of the high voltage spike.

More elaborate circuits involving loading at higher harmonics may be used but will not be considered here. It is of interest, however, to point out that the basic waveform is very rich in harmonics and therefore the output power may be extracted not only at the fundamental frequency but also at any higher harmonic. C. P. Snapp[62] has studied this possibility and concludes that the efficiency and output power at the higher harmonic frequencies vary as $1/m$ where m is the harmonic number. In view of the fact that the fundamental frequency trapatt efficiency may be above 50% this means that efficiencies comparable to those of many impatt diodes may be obtainable up to the fifth or sixth harmonic.

3.6.3 Starting mechanism

As was previously pointed out, the trapatt mode is inherently large signal and is not associated with any small signal negative resistance. We are therefore confronted with a starting problem. A d.c. bias applied to the diode might not itself start an oscillation. It is now generally accepted that the starting mechanism involves an impatt oscillation. As we shall see in more detail below, a good trapatt diode consists of a rather highly punch-through structure which will also work as an impatt diode but at a much higher frequency than the typical fundamental trapatt frequency, although it will be rather far from an optimum impatt design. The detailed manner in which the impatt oscillation acts is probably dependent on the circuit. As the impatt frequency is much higher than the trapatt one the circuit presented for impatt oscillations is very often not well controlled. As an example of how the impatt mode may act as a starting mechanism we present Figures 3.57 and 3.58 from Reference 60. Both figures show the current and voltage waveforms during start-up of a trapatt oscillator but Figure 3.57 is obtained from measurements whereas Figure 3.58 is obtained from a computer simulation.

The two results are basically the same. The interpretation of these curves is as follows. The build-up of the average voltage is given by the rise time of the d.c. bias pulse but superimposed on this rise is a ringing associated with multiple reflections of the incoming pulse at the two ends of the transmission line constituting the trapatt circuit. Once the peak voltage reaches breakdown and impatt threshold an impatt oscillation starts but is quickly quenched when the ringing causes the voltage to decrease below threshold. Synchronous with the ringing, a pulse of a few impatt cycles travels along the transmission line, is reflected and returns to the diode when the voltage is again above threshold. The impatt pulse is then reflection amplified and the cycle repeats. Eventually successive reflections against the diode amplifies the

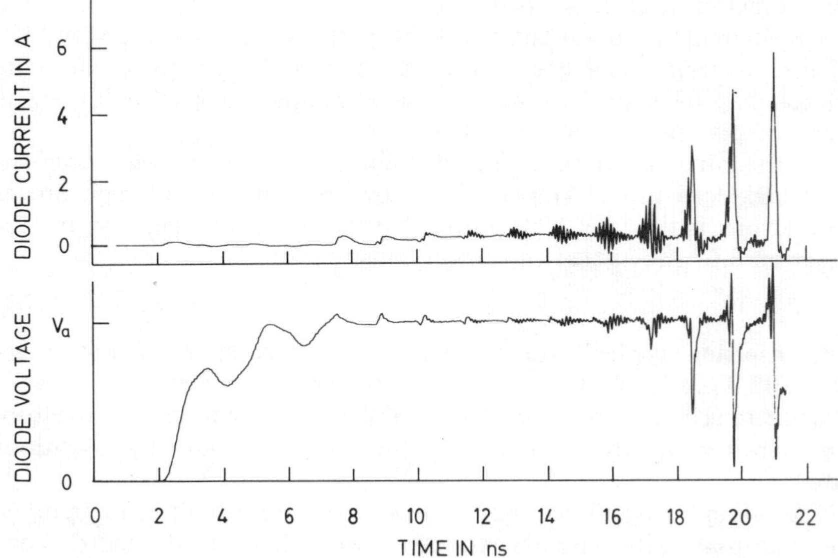

Figure 3.57 Measured current and voltage waveforms during start-up of trapatt oscillator[60]

impatt oscillation enough so that the peak current exceeds the trapatt threshold and the trapatt oscillation starts.

In circuits like the one of Evans presented below in Figure 3.60 a separate impatt circuit is provided close to the diode. The detailed starting mechanism is probably different in this type of circuit.

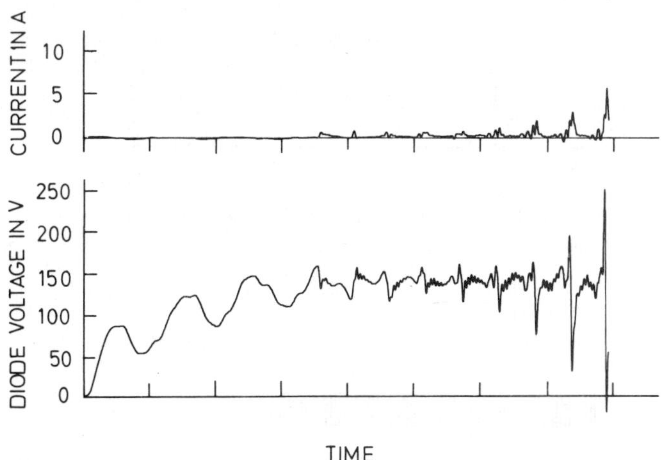

Figure 3.58 Calculated current and voltage waveform during start-up of trapatt oscillator[60]

3.6.4 Practical trapatt performance

The main reason for the interest in trapatt oscillators is the very high efficiencies which have been achieved. These in turn result because the voltage may be kept at a very low level during that part of the cycle when the plasma is extracted and current flows.

On the other hand the major difficulty which so far has prevented the widespread use of trapatt oscillators in a number of high power applications is the very high power density which is necessary. In fact Equation 3.67 shows that the power density P fulfils

$$P > V_a e v_s N_a \tag{3.73}$$

and, inserting typical numbers, we find that P is in the range 10^5-10^6 W/cm^2. With such enormous power densities it is very difficult to achieve c.w. operation with reasonable junction temperature and therefore trapatt oscillators are considered for pulsed applications only.

The rather stringent requirements add to the difficulties in practical use of trapatts. Most circuits considered are of the type discussed above and differ mainly in the way the transmission line is built up and in the design of the output filter. Clorfeine and coworkers[63] have also successfully operated a lumped-element circuit at UHF frequencies. A schematic of their circuit is reproduced in Figure 3.59. Figure 3.60 shows the more usual circuit first proposed by Evans.[64]

In pulse mode operation very impressive results have been obtained with more than 1 kW of power at L-band[65] and an efficiency of 60%.[66]

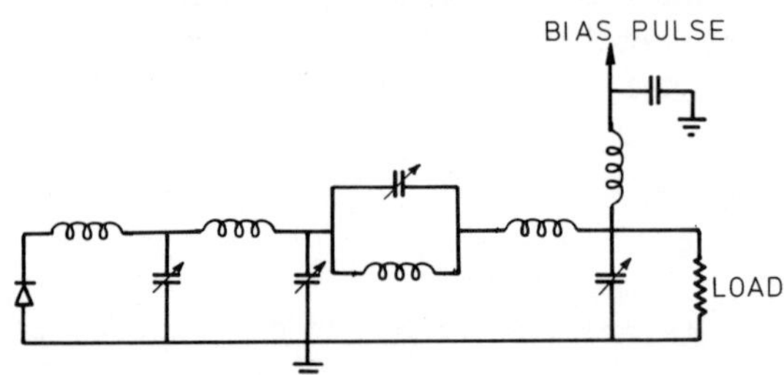

Figure 3.59 Schematic of lumped-element trapatt circuit[63]

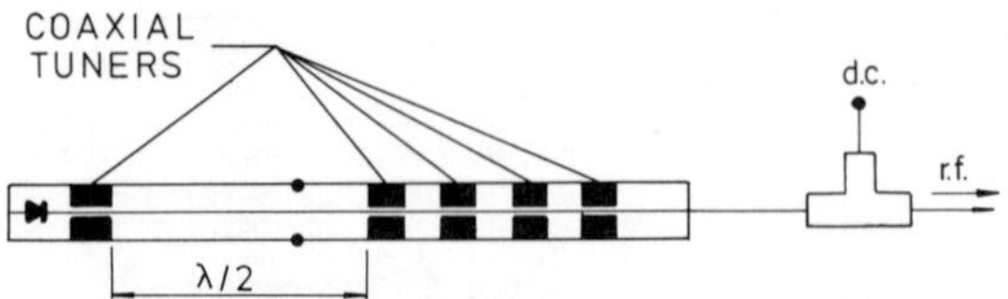

Figure 3.60 Trapatt 500 MHz oscillator circuit[64]

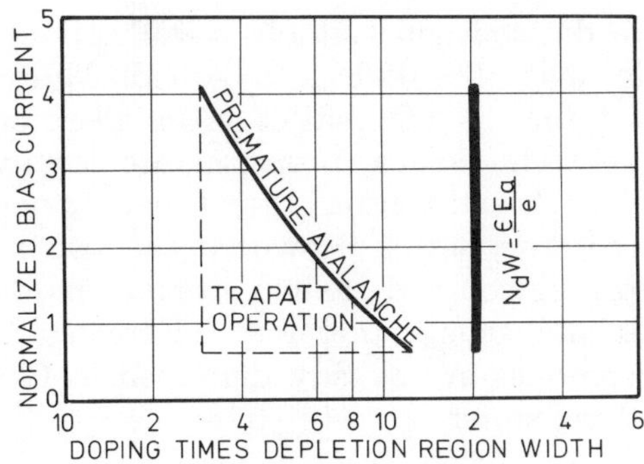

Figure 3.61 The design triangle[66] current normalized to $eN_a v_s$

At higher frequencies the power drops off rapidly and above about 10 GHz no useful trapatt power can be obtained. The major difficulty at higher frequencies is claimed to be obtaining an impatt oscillation to start the trapatt mode.[67]

The noise behaviour is poor and comparable to that of an impatt diode with a very large voltage swing. The reason for the high noise level in the trapatt mode is the same as in the overdriven impatt diode, i.e. large fluctuations will occur in the start-up of the current by avalanche multiplication.[68] When the voltage begins to exceed breakdown, very few, or even no, carriers may be present.

We shall conclude our discussion of the trapatt mode by reproducing the design criteria developed by Clorfeine.[66] Using a more complete version of the theoretical concepts discussed in Section 3.6.1 he obtains the design triangle of Figure 3.61 which shows the doping times length product for useful trapatt operation.

The low current limit is that already obtained above, the high current limit is given by the condition that premature breakdown must be avoided. Finally the low doping-length limit is less well defined but dictated by the failure of the theoretical concepts used. In practice a very small $N_d W$ product should be avoided, however, because of the d.c. negative resistance which would occur and which might lead to current filament formation and diode burn-out. The line to the extreme right in Figure 3.61 shows when punch-through and breakdown coincide. We see that all good trapatt devices should be highly punch-through.

3.7 FABRICATION TECHNOLOGY

In the present section we shall discuss some of the more important aspects of the technology used for diode fabrication. The first step in the fabrication consists of wafer processing during which the required doping profiles are defined and suitable metallizations are deposited. The next step is to divide the wafer into individual chips and to mount these in a manner that is compatible with electrical and thermal considerations. The technological processes are largely the same for different types of diodes and we shall therefore only occasionally make reference to any specific device. However, the two most commonly used semiconductors, silicon and gallium arsenide, require substantially different processes and should therefore be treated individually. As GaAs epitaxial techniques are already described in Chapter 2 we will only discuss Si in this chapter.

3.7.1 Silicon wafer processing

Let us as an example take the structure required for a p^+—n—n^+ impatt diode. The wafer must then be processed so that the three differently doped layers are established. The techniques generally used in the semiconductor industry to introduce specific impurities into the materials are alloying, diffusion, epitaxy and ion-implantation. They all have their respective advantages and disadvantages and to make a proper choice of process it is necessary to understand the relative merits of these processes and to know the specific requirements for the device which is to be manufactured.

Common to all microwave devices is that they require a very precise doping control. In our example impatt structure the transitions from the n-region to the n^+- and p^+-regions must be very abrupt and the doping level and thickness of the n-region must be controlled to within a few percent. Let us now assume that our impatt device is intended for operation at X-band frequencies. The n-region thickness should then be about 3—4 μm and the donor concentration about 10^{16} cm^{-3}. To avoid unwanted series resistance both the n^+- and p^+-regions should be as thin and highly doped as possible.

The conventional way of realizing this structure is to start with an n^+-wafer onto which an n-layer is deposited by means of epitaxy followed by a p^+ diffusion.

Let us first consider the choice of substrate n^+-material. To assure a uniform final device the substrate should be monocrystalline with a low concentration of defects. The front surface should be mechanically and chemically lapped to mirror finish. The doping concentration should be as high as possible without introducing serious defects into the lattice. In practice this means that the concentration of doping atoms must be smaller that the solid solubility in silicon. The commonly used n-type dopants in silicon are arsenic and phosphorus. As is seen from Figure 3.62 arsenic has a higher solid solubility than phosphorus[69] and should for this reason be preferred. However, arsenic also has a very high vapour pressure and therefore is difficult to handle when preparing the crystal and also during epitaxial growth. Although arsenic is sometimes used in

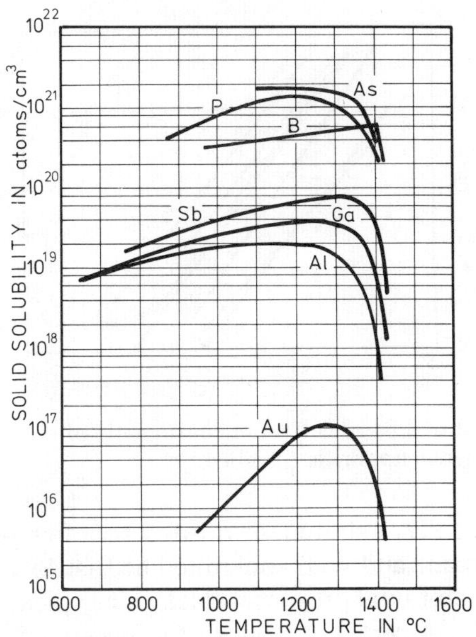

Figure 3.62 The solid solubility of impurities in silicon[69]

critical devices, phosphorus is usually preferred. A typical substrate resistivity with phosphorus is about 0.01 Ω cm.

The epitaxial processes used with silicon are all based on vapour phase reactions.[70] The starting materials are either of $SiCl_4$, $SiHCl_3$, $Si_2H_2Cl_2$, $SiHCl_3$, or SiH_4. In a hydrogen ambient they all decompose at high temperature forming Si, H_2 and HCl. In order to achieve a uniform·monocrystalline growth it is necessary to keep the substrate within a closely defined temperature range different for each of the above mentioned starting compounds. Although the most common starting materials in the semiconductor industry are $SiCl_4$ and $SiHCl_3$ these compounds are not suitable for microwave devices because of the high required growth temperature ($\gtrsim$1200 °C). $Si_2H_2Cl_2$ and SiH_4 both admit lower growth temperatures (1000—1100 °C) and therefore results in much sharper n^+—n transitions. Figure 3.63 shows some published values of the diffusion coefficients of dopant atoms in silicon[71] and it is seen that lowering the temperature by about 100 °C results in a decrease in the diffusion coefficient by as much as an order of magnitude.

The epitaxial growth rate is strongly temperature dependent[72] as seen in Figure 3.64 and good temperature control as well as uniformity across the wafer are of extreme importance. Good flow control of the reacting gases is also very important and the difficulty of controlling doping levels is well realized when one notes that the doping gas

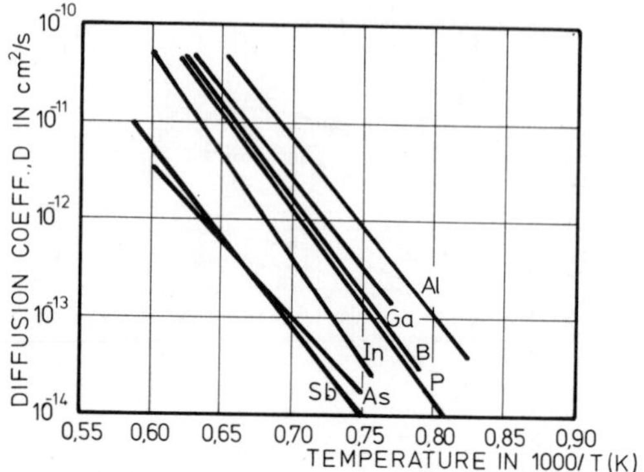

Figure 3.63 Diffusion coefficients of miscellaneous impurities in silicon[71]

(usually phosphine, PH_3) should be diluted to a few parts in 10^8 in the main hydrogen stream and well controlled at that low concentration. A very well designed epitaxial process still leaves a typical scatter in the data of the order of 10%. A certain amount of selection of wafers is therefore usually necessary in order to produce a given microwave device.

In practice there exist no alternatives to the epitaxial process for the required n—n^+ structure. For formation of the contact, however, a number of alternatives exist. The p^+ contact could also be grown by epitaxy. A shallow diffusion of boron would give almost the same result

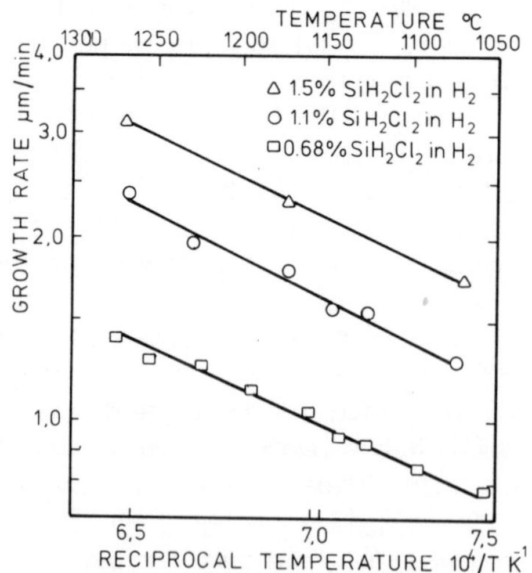

Figure 3.64 Temperature dependence of the epitaxial growth for different concentrations of dichlorosilane in hydrogen[72]

as would ion-implantation followed by a subsequent anneal at 900 °C. A slight advantage in terms of sharper n^+—n and p^+—n transitions could be noted when implantation is used. Finally, instead of a p^+ contact a Schottky barrier could be created by depositing a suitable metal onto the silicon. Some of the best Schottky barriers are made by depositing platinum or palladium and then allowing their respective silicides to form. The temperature of formation of these silicides are so low that usually the unavoidable heating during the deposition process is sufficient.[73] Which technique to use when forming the rectifying contact is largely a matter of taste although as we have seen before there may be fundamental physical differences between p—n junctions and Schottky barriers.

3.7.2 Metallization

Two major considerations are of importance for the choice of metallization. First of all the metal or metals used must result in ohmic contacts with very low series resistance. The second aspect is that the metallization must result in a reliable device. This requires that adherence between the deposited metals and the silicon and the contacting metal wire must be good and also that no ageing mechanisms will occur at the temperatures to which the device will be exposed during operation and processing.

With sufficiently highly doped n^+- and p^+-regions the first requirement is not very restrictive and a variety of metals will result in good ohmic contacts. Good adherence to silicon is a more difficult requirement but among the metals which are satisfactory in this respect one may mention titanium, chromium, tungsten, molybdenum and aluminium. In addition the silicides of palladium and platinum have excellent adherence. A metallization consisting of chromium alone, for example, would be highly unsatisfactory, however, because it would be impossible to bond a wire to a chromium surface because of the oxide layer which almost immediately forms on a free chromium surface. In fact, very few metals are suitable for attaching wires to and the most commonly used ones are gold, silver and aluminium. The merits of the inert metals gold and silver in this respect should be obvious but in spite of the fact that aluminium is normally covered by an inert oxide it is possible to make good wire bonds to this metal through either of the processes of ultrasonic or thermocompression bonding which will be discussed below.

We are therefore led to the conclusion that the only metal which appears suitable is aluminium and for many low frequency semiconductor devices it is indeed the metal which is used. For such low frequency devices heat sinking is very often not too critical and the back side of the chip may therefore be bonded to gold-plated packages by allowing the temperature during the operation to reach the gold—silicon

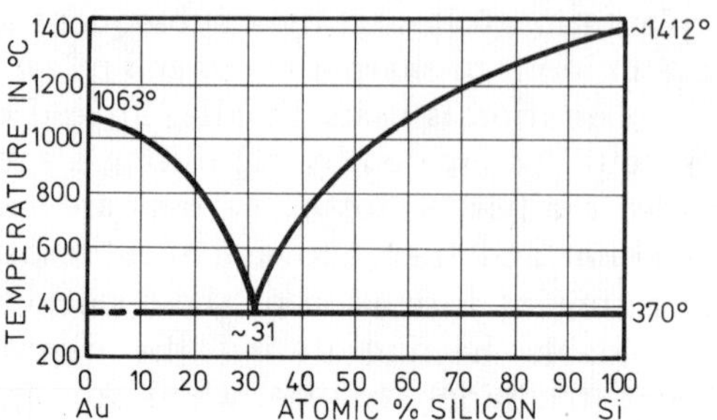

Figure 3.65 The gold—silicon system[74]

eutectic[74] (see phase diagram in Figure 3.65) and form a eutectic bond between the silicon chip and the package metallization.

This type of bonding results in a poor thermal resistance, however. One reason is that it is difficult to control the exact depth of alloying of gold into silicon and therefore the substrate thickness must be quite large. Another reason is that the eutectic in itself has a low thermal conductivity. Microwave devices often require low thermal resistance and therefore a different bonding technique must be used. Aluminium metallization is therefore very rarely used for microwave devices. Apart from thermal considerations other reasons for this omission of the simplest of all metallizations are found in the comparatively poor reliability data at higher temperatures and high series resistances to n-type material.

Leaving out aluminium one would be led to use combinations like Cr—Au or Ti—Au which are indeed very extensively used for laboratory work when requirements on high temperature operations are high reliability are moderate. It is instructive, however, to consider the ways in which these two combinations fail under more stringent requirements.

Both combinations have good adherence, are easy to bond to and result in very low series resistance. At elevated temperatures, however, material transport will take place with catastrophic results. Chromium diffuses rapidly through gold at temperatures above about 300 °C and when it reaches the gold surface will form an oxide which will make all bonding operations virtually impossible. In addition chromium—gold mixtures have a considerable electrical resistance which may be very serious if the gold surface is contacted by a gold wire. With its thin dimensions, such a gold wire must have very good electrical conductivity which will not be the case if chromium is allowed to diffuse up into the lead. Ti on the other hand admits gold to diffuse through it very rapidly, allowing gold to come in direct contact with silicon. As can be seen from Figure 3.65 gold—silicon has an eutectic at about 370 °C and even at substantially lower temperatures the rate of formation of the eutectic is considerable. Gold impurities in silicon result in catastrophic deterioration of the semiconducting properties and, indeed, eutectic formation is one of the most common failure modes of all devices in which gold is used.

One way of avoiding the problems associated with metal interdiffusion is to use a barrier metal between the actual metallic contact layer and the outer inert gold layer. Both palladium and platinum make excellent barriers separating the surrounding metals from each other.

The detailed physics of many of the processes taking place between the various possible metals and the silicon in the temperature range 200 °C to 400 °C is still unknown. As an example, gold and palladium intermix by around 200 °C but nevertheless palladium seems to inhibit the diffusion through an underlying titanium layer into silicon which would have taken place in the absence of the palladium layer. The recently developed experimental technique of α-particle backscattering[75] has provided a tool for detailed study of these processes and new information is being collected rapidly.

The metals tungsten and molybdenum appear to be non-transparent for gold up to rather high temperatures but have still found limited use, probably because of the difficulties in depositing films which are structurally stable at elevated temperatures. Similar problems are encountered when silver is used as the final layer instead of gold. Otherwise silver would be preferred on account of its much higher eutectic temperature with silicon (830 °C).

Summing up the discussion, it is found that at present the best metallizations consist of Cr or Ti followed by Pt or Pd and finally Au.

3.7.3 Heat sinking

Power densities in active microwave diodes are often enormous. In our example of an X-band impatt diode the optimum current density is in the range 500—1000 A/cm^2. The voltage is somewhat less than 100 V leading to a power density close to 10^5 W/cm^2. The temperature rise above ambient temperature should usually not exceed about 100 °C.

If we assume our device to be cylindrical, of diameter d, and neglect all internal resistance in the semiconductor we may obtain a lower limit to the thermal resistance by assuming a perfect bond to an infinite copper heat sink with thermal conductivity κ_{Cu}. The spreading resistance is then[76]

$$R_{th} = \frac{1}{2\kappa_{Cu} d} \tag{3.74}$$

The temperature rise is given by

$$\Delta T = R_{th} \frac{P\pi d^2}{4} = \frac{P\pi d}{8\kappa_{Cu}} \tag{3.75}$$

Solving for d and inserting $P = 10^5$ W/cm^2 and $\Delta T = 100$ °C we obtain $d \approx 100$ μm. In spite of our simplifying assumptions this comes close to the typical diameters of practical devices. It is also instructive to note that the additional contribution ΔR_{th} for each thickness t of silicon of diameter equal to the device area is

$$\Delta R_{th} = \frac{4t}{\pi d^2 \kappa_{Si}} \tag{3.76}$$

For each micrometre of silicon we therefore get an additional contribution ΔR_{th} of about 1.3 °C/W which underlines the importance of avoiding any substrate between the device and the heat sink.

Two principal techniques have been developed to solve the heat sinking problem in practical devices. As a starting point we note that usually the wafer has been processed in such a way that the active, weakly doped layer on the other side has a comparatively thick highly doped substrate and on one side a thin (<1 μm) also highly doped diffused or grown contact layer. It is then necessary to apply a highly conducting metallic heat sink to the thin contact side. Owing to the poor thermal conductivity of the alloys used as solders it is impossible to use soldering techniques. However, thermocompression bonding offers a technique of attaching the metallized contact layer side to the heat sink which then must have gold evaporized or plated onto it. When two gold surfaces are pressed together at elevated temperatures a good bond is formed, provided that the pressure and temperature are high enough. Figure 3.66 gives the approximate requirements which must be fulfilled to give a good bond.

One of the difficulties with using thermocompression bonding is that both the temperature and pressure which are required are so high that there is a substantial risk of damaging the chip or its metallization. Also the chip and heat sink surfaces must be extremely well aligned. Furthermore a silver, copper or gold heat sink will be quite soft at the bonding temperature and with the thin contact layer it is difficult to avoid pressing the chip so far into the metal that a short-circuit occurs, even though a clean-up etch is often applied to remove some of the semiconductor material after the bond is formed. A completed thermocompression bonded device is outlined in Figure 3.67. Owing to the various difficulties with thermocompression bonding it is mainly used with diamond heat sinks when no alternative technology is available and when no problem with pressing the chip into a soft heat sink exists. Diamond is sometimes used as a heat sink because of the extremely good thermal conductivity of type IIA diamond. At room temperature it has a thermal conductivity about 5 times better than copper although it degrades more rapidly with temperature than that of pure copper.[77,78]

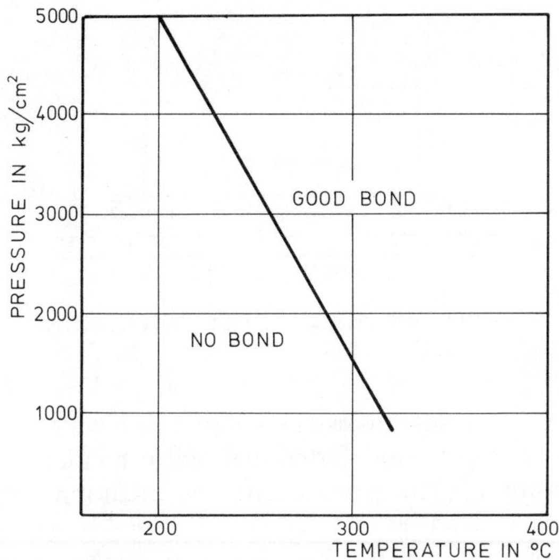

Figure 3.66 Illustration of requirements on temperature and pressure to obtain a good thermo-compression bond.

A less exotic and more commonly used heat sinking technology is the electroplated heat sink. Here the contact side of the entire wafer is plated with a layer of silver, copper or gold, typically 100—200 μm thick. The wafer is then thinned to about 20—30 μm total thickness and mesas are etched through the semiconductor from the substrate side. The major difficulty with this technique arises from the difference in thermal expansion coefficients between the semiconductor and the metal heat sink. Without taking extreme precautions to make the plated heat sink metal very soft, the entire wafer will curl up making photoresist steps difficult and possibly causing the thin semiconductor to crack. With proper handling these problems may be overcome and

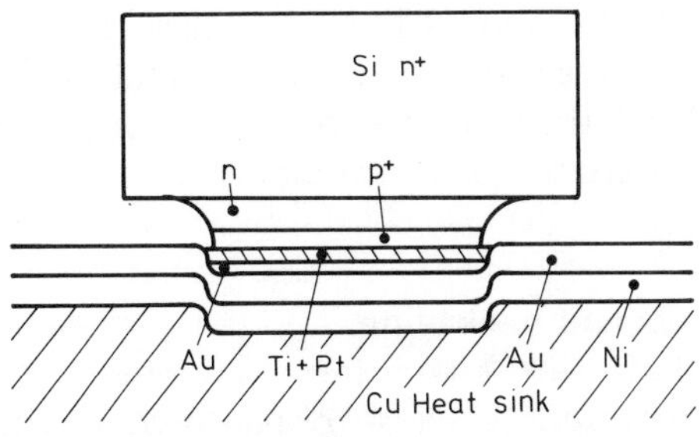

Figure 3.67 Thermocompression bonded diode

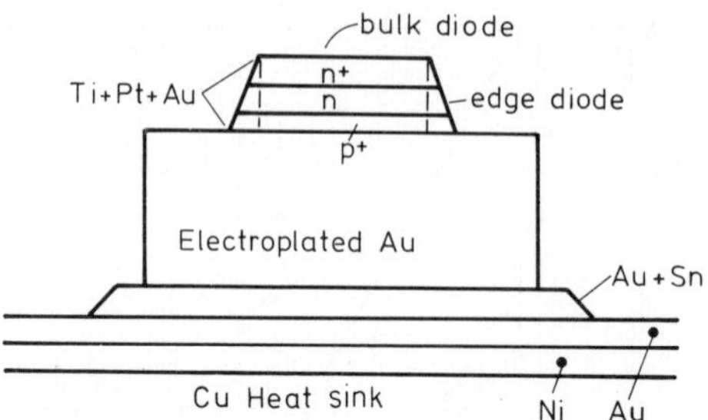

Figure 3.68 Diode with electroplated heat sink

one of the major advantages is then that the heat sink is formed in a batch process without any individual chip handling as in thermo-compression bonding. The chips cannot be diamond scribed apart like most other semiconductors but the plated metal must be sawed or cut. If the area of the metal chip is made large enough, typically 0.5—1 mm^2, it may be soldered into the package using an Au—Sn solder, for example. The heat spreading then takes place in the plated metal making the thermal quality of the soldered bond uncritical. A complete device with electroplated heat sink is illustrated in Figure 3.68.

In view of the importance of a good heat sink for impatt devices it is most fortunate that adequate techniques are available for measuring the thermal resistance of individual chips. The starting point of the various techniques is the temperature dependence of the breakdown voltage[79] as illustrated in Figure 3.69.

The only complication in using this effect is that the current—voltage characteristic in the reverse direction is affected by space charge resistance in addition to the thermally induced resistance (Figure 3.70). As a first approximation the space charge resistance can be calculated as $R_{sc} = l^2/2\epsilon v_s A$ and subtracted.

In most cases, however, the crude theory behind this expression results in insufficient accuracy and an experimental determination of R_{sc} is required. One way of doing this is by measuring the current—voltage characteristic with a short (sub-microsecond) voltage pulse during which all thermal effects may be neglected. A number of alternative and experimentally simpler techniques have also been developed.[80,81]

3.7.4 Surface effects and packaging

Avalanche devices have very high internal electric field strengths. In fact the electric field required to cause breakdown in silicon is about ten times that required to cause breakdown in air. As the minimum voltage required to cause a breakdown in air is about 300 V there is no

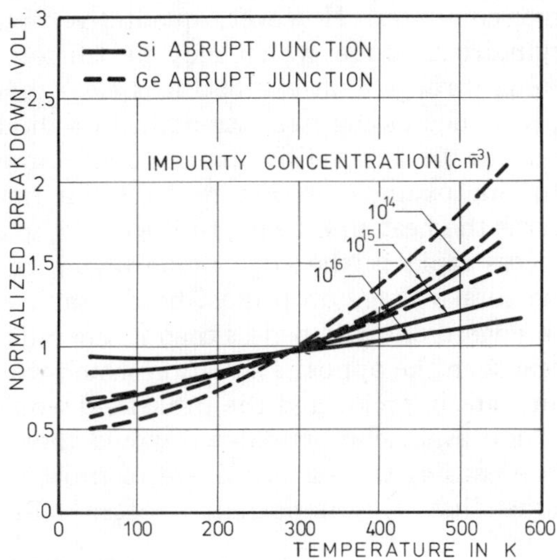

Figure 3.69 Normalized avalanche breakdown voltage versus lattice temperature[79]

immediate problem of short-circuiting the device by a gaseous discharge on the outside but, as we shall see, surface and boundary effects are nevertheless important.

There are two major reasons why the local electric field strength inside the semiconductor will be different close to the surface and deep inside at otherwise equivalent positions. One of these reasons is purely geometric and has to do with the fact that the mesa boundary is usually not aligned with the unperturbed field lines. The second reason is that extra energy levels at a free or oxidized silicon surface bend the energy bands from their level in the bulk.

To consider the geometric effect first we turn to Figure 3.68, where a mesa with its 'ideal geometry bulk diode' and 'edge diode' is shown. To calculate the complete field distribution is a formidable problem

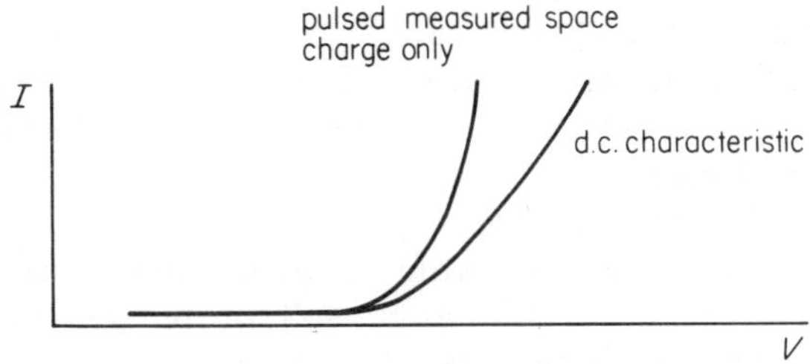

Figure 3.70 Current—voltage characteristics of avalanche diodes under pulsed and d.c. conditions

requiring a large computer. However, qualitatively the edge diode behaves as a cylindrical diode which may be treated analytically.[82] Even without going through that exercise it is clear than an expanding geometry will give a decreasing field strength. In a diode this effect is superimposed on the field variation owing to the acceptor or donor space charge. In our example in Figure 3.68 the field is increasing everywhere towards the heat sink owing to the donor space charge. This increase will be less rapid in the edge diode owing to the expanding geometry. As the total voltage drop must be the same everywhere, the peak field in the edge diode will be less than in the bulk. This situation is favourable because in the opposite situation which would occur if the p^+- and n^+-regions are interchanged the peak field would be maximum at the boundary and avalanche breakdown would therefore take place first at the surface causing the surface diode to short out the bulk. We therefore conclude that it is important to etch mesas with proper geometry.[83]

It would be taking it too far to go into a detailed discussion of all the surface effects which will contribute to band bending. We will only note that in silicon the surface usually has an excess of positive charge resulting in an edge diode with somewhat more n-type doping. In our simple diode shown in Figure 3.68 this will increase the peak field in the edge diode and must be compensated by the geometric effect in order to avoid a surface breakdown. Another effect is that some of the surface charge is due to mobile ions in the inevitable oxide layers. Sodium impurities especially, resulting from unsatisfactory processing cleanliness, give rise to such mobile surface charge. Such mobile charge may result in unwanted leakage currents and unstable microwave properties.

In order to avoid problems with surface contamination it is necessary to protect the surface throughout the life of the device. One way is to hermetically encapsulate the chip; another way, often used as a complement to encapsulation, is to protect the diode with a suitable insulating material.

Power devices made with electroplated heat sinks cannot be passivated with high temperature processes and when a surface passivation is needed a silicone rubber or a low-temperature deposited oxide or glass is used. It is difficult, however, to achieve a controlled surface state density with such passivations and often one must rely on the hermetic package alone.

Varactors and low temperature devices may have a thermal oxide and/or nitride grown onto the wafer at high temperature after etching the mesas. With such techniques a well protected stable surface is achieved although the doping profile may be changed because of the high temperature processing.

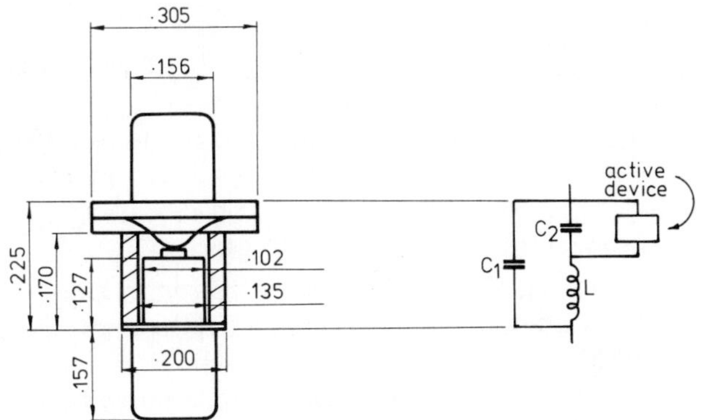

Figure 3.71 S4 package (dimensions in centimetres) and equivalent circuit[84]

In addition to offering a hermetic seal, the package must be compatible with microwave operation. A small size is imperative to avoid series inductance and parallel capacitance. Among other things the actual choice of package is dictated by the frequency at which the device is to operate. A wide variety of commercially available packages exist and as an example we only show one of the most common ones in Figure 3.71. In the same figure we also give a typical low frequency equivalent circuit of the package.[84] The equivalent circuit gives a reasonably correct description of this package up to frequencies in the range 7—8 GHz. The actual package inductance is heavily influenced by the way in which the connection to the chip is made. Typically the chip is soldered into the packages using, for example, an Au—Sn eutectic solder in a reducing atmosphere or, with greater contamination risk, with a suitable non-corrosive flux. The top of the chip is then connected to the top flange of the package by one or several gold wires or ribbons attached by thermocompression bonding. The number of such leads in parallel and their dimensions to a large part control the series inductance of the package.

The final sealing of the package is achieved by soldering or welding a cap on to the flange. In view of the above discussion it is clear that this sealing must be done in a very well controlled atmosphere. As a rule of thumb, the inert gas filling package should have an overall impurity content better than 10 ppm.

ACKNOWLEDGEMENT

The author of this chapter wishes to express his gratitude to Dr Bengt Källbäck. He has carried the major burden of collecting the background material upon which the text is based and been of invaluable help in compiling figures and references and in reading the typescript.

REFERENCES

1. C. Canali, G. Ottaviani and A. A. Quaranta, *J. Phys. Chem. Solids*, 32, 1707 (1971).
2. S. M. Sze, *Physics of Semiconductor Devices*, John Wiley, 1969.
3. J. G. Ruch and G. S. Kino, *Phys. Rev.*, 174, 921 (1968).
4. H. D. Rees, *IBM J. Res. Develop.*, 13, 537 (1969).
5. R. Stratton, *Proc. R. Soc.*, 246, 406 (1958).
6. C. Y. Duh and J. L. Moll, *Solid State Electron.*, 11, 917 (1968).
7. G. Salmer, J. Pribetich, A. Farrayre and B. Kramer, *J. Appl. Phys.*, 44, 314 (1973).
8. A. S. Grove, *Physics and Technology of Semiconductor Devices*, John Wiley, 1971.
9. R. Hall and J. H. Leck, *Int. J. Electronics*, 25, 529 (1968).
10. W. N. Grant, *Solid State Electron*, 16, 1189 (1973).
11. D. R. Decker and C. N. Dunn, *IEEE Trans. Electron Devices*, ED-17, 290 (1970).
12. P. Mars, *Int. J. Electronics*, 32, 23 (1971).
13. W. T. Read, Jr., *Bell Sys. Tech. J.*, 37, 401 (1958).
14. M. Gilden and M. E. Hines, *IEEE Trans. Electron Devices*, ED-13, 169 (1966).
15. H. C. Bowers, *IEEE Trans. Electron Devices*, ED-15, 343 (1968).
16. D. Delagebeaudeuf, *L'Onde Electrique*, 48, 722 (1968).
17. T. Misawa, *IEEE Trans. Electron Devices*, ED-14, 795 (1967).
18. W. E. Schroeder and G. I. Haddad, *Proc. IEEE*, 59, 1245 (1971).
19. F. Sellberg, *Ericsson Techn.*, 28, 103 (1972).
20. T. Misawa, R. A. Moline and A. R. Tretola, *Solid-State Electron.*, 15, 189 (1972).
21. B. Culshaw and R. A. Giblin, *Electron. Lett.*, 10, 285 (1974).
22. D. L. Scharfetter, W. J. Evans and R. L. Johnston, *Proc. IEEE*, 58, 1131 (1970).
23. A. Lekholm and J. Mayr, *Electron. Lett.*, 9, 64 (1973).
24. C. P. Snapp, G. Pfund and A. F. Podell, *Proc. European Microwave Conference, Montreux* (1974).
25. C. R. Crowell and S. M. Sze, *Appl. Phys. Lett.*, 9, 242 (1966).
26. D. L. Scharfetter, *IEEE Trans. Electron Devices*, ED-18, 536 (1971).
27. M. Matsumura, *IEEE Trans. Electron Devices*, ED-17, 514 (1970).
28. D. Delagebeaudeuf, *Rev. Techn. Thompson-CSF*, 1, 309 (1969).
29. A. S. Tager, *Soviet Phys. Usp.*, 9, 892 (1967).
30. B. B. van Iperen, H. Tjassens and J. J. Goedbloed, *Proc. IEEE*, 57, 1341 (1969).
31. M. E. Hines, *Proc. IEEE*, 60, 1534 (1972).
32. M. E. Hines, *IEEE Trans. Electron Devices*, ED-13, 158 (1966).
33. A. Sjölund, *Electron. Lett.*, 7, 161 (1972).
34. R. G. Kuvås, *IEEE Trans. Electron Devices*, ED-19, 220 (1972).

35. H. K. Gummel and J. L. Blue, *IEEE Trans. Electron Devices*, **ED-9**, 569 (1967).
36. J. J. Goedbloed and M. T. Vlaardingerbroek, *IEEE Trans, Electron Devices*, **ED-21**, 342 (1974).
37. B. C. de Loach, *IEEE Trans. Electron Devices*, **ED-9**, 366 (1962).
38. P. Weissglas and F. Sellberg, *Conf. Proc. Microwave, 73*, 285 (1973).
39. D. J. Coleman, Jr. and S. M. Sze, *Bell Syst. Techn. J.*, **50**, 1695 (1971).
40. C. P. Snapp and P. Weissglas, *Electron. Lett.*, **7**, 743 (1971).
41. S. M. Sze, D. J. Coleman. Jr. and A. Loya, *Solid-State Electronics*, **14**, 1209 (1971).
42. C. P. Snapp and P. Weissglas, *IEEE Trans. Electron Devices*, **ED-19**, 1109 (1972).
43. H. W. Rüegg, *IEEE Trans. Electron Devices*, **ED-15**, 577 (1968).
44. D. J. Coleman, Jr. and S. M. Sze, *Bell Syst. Tech.. J.*, **50**, 1695 (1971).
45. Th.G. van de Roer, *Report 74-E-46*, Eindhoven University of Technology, Netherlands (1974).
46. J. Lacombe, *Rev. Tech. Thompson-CSF*, **4**, 467 (1972).
47. D. Delagebeaudeuf and J. Lacombe, *Electron. Lett.*, **9**, 538 (1973).
48. F. Sellberg and A. Sjölund, private communication.
49. H. A. Haus, H. Statz and R. A. Pucel, *Electron. Lett.*, **7**, 667 (1971).
50. H. Statz, R. A. Pucel and H. A. Haus, *Proc. IEEE*, **60**, 644 (1972).
51. A. Sjölund, *Electron. Lett.*, **9**, 2 (1973).
52. A. Sjölund, *Solid-State Electronics*, **16**, 559 (1973).
53. W. Shockley, J. A. Copeland and R. P. James, *Quantum Theory of Atoms, Molecules and the Solid State* (edited by P. O. Löwdin), Academic Press, 1966.
54. G. Björkman and C. P. Snapp, *Electron. Lett.*, **8**, 501 (1972).
55. F. Sellberg, private communication.
56. J. Helmcke, H. Herbst, M. Claasen and W. Harth, *Electron. Lett.*, **8**, 158 (1972).
57. A. van der Ziel, *Noise. Sources, Characterization, Measurement*, Prentice—Hall, 1970.
58. A. Sjölund and F. Sellberg, *Proc. European Microwave Conference, Brussels*, 1973.
59. H. J. Prager, K. K. N. Chang and S. Weisbrod, *Proc. IEEE*, **55**, 586 (1967).
60. B. Kerzar and P. Weissglas, *Japanese J. Appl. Phys.*, **12**, 260 (1973).
61. B. C. de Loach, Jr. and D. L. Scharfetter, *IEEE Trans. Electron Devices*, **ED-17**, 9 (1970).
62. C. P. Snapp, *IEEE Trans. Electron Devices*, **ED-18**, 294 (1971).

63. A. S. Clorfeine, H. J. Prager and R. D. Hughes, *RCA Rev.*, 34, 580 (1973).
64. W. J. Evans, *IEEE Trans. Microwave Theory Tech.*, MTT-17, 1060 (1969).
65. S. G. Liu and J. J. Risko, *RCA rev.*, 31, 3 (1970).
66. A. S. Clorfeine, *IEEE Trans. Electron Devices*, ED-18, 550 (1971).
67. W. J. Evans and D. L. Scharfetter, *IEEE Trans. Electron Devices*, ED-17, 397 (1970).
68. E. F. Scherer, *Proc. European Microwave Conference, Stockholm*, 1971.
69. F. A. Trumbore, *Bell Syst. Tech. J.*, 39, 2050 (1960).
70. S. K. Ghandi, *The Theory and Practice of Microelectronics*, John Wiley, 1968.
71. R. M. Burger and R. P. Donovan, *Fundamentals of Silicon Integrated Device Technology*, Vol. 1, Prentice—Hall, 1967.
72. A. Lekholm, *J. Electrochem. Soc.*, 119, 1122 (1972).
73. J. W. Mayer and K. N. Tu, *J. Vacuum Sci. and Tech.*, 11, 86 (1974).
74. M. Hansen and A. Anderko, *Constitution of Binary Alloys*, McGraw-Hill, 1958.
75. W. K. Chu, J. W. Mayer, M-A. Nicolet, T. M. Buck, G. Amsel and F. Eisen, *Thin Solid Films*, 17, 1 (1973).
76. H. S. Carslaw and J. C. Jaeger, *Conduction of Heat in Solids*, 2nd ed., New York: Oxford University Press, 1959.
77. G. K. White, *Austr. J. Phys.*, 6, 397 (1953).
78. R. Berman, *Proc. Roy. Soc. (London) Ser. A*, 200, 171 (1953).
79. C. R. Crowell and S. M. Sze, *Appl. Phys. Lett.*, 9, 242 (1966).
80. R. H. Haitz, H. L. Stover and N. J. Tolar, *IEEE Trans, Electron Devices*, ED-16, 438 (1969).
81. B. R. McAvoy, *IEEE Trans. Electron Devices*, ED-18, 973 (1971).
82. S. M. Sze and G. Gibbons, *Solid-State Electronics*, 9, 831 (1966).
83. A. Lekholm and P. Weissglas, *IEEE Trans. Electron Devices*, ED-18, 844 (1971).
84. R. P. Owens and D. Cawsey, *IEEE Trans. Microwave Theory Tech.*, MTTT-18, 790 (1970).

CHAPTER 4

Microwave Transistors

J. S. LAMMING

4.1 INTRODUCTION

Soon after the discovery of transistor action it was realized that two different types of device would be possible. These are the bipolar transistor and the unipolar (or field-effect) transistor. In principle both of these transistors could be made with any semiconductor material with suitable electrical and physical properties. At first the choice of semiconductor was restricted to germanium, since this material had been produced with sufficient purity and in single crystal form. Later, other semiconductor materials, notably silicon and gallium arsenide, became available in a suitable state to be used for bipolar or field-effect transistors.

Transistors were at first able to operate at frequencies up to a few hundred kHz only, but theory predicted that they should be able to operate at much higher frequencies, even into the microwave frequency range, by a reduction of their overall dimensions. In particular it was recognized that the transit time of electrical charge carriers through the device, and the rate of change of electrical charge stored within the device, would limit the frequency response. In order to increase the frequency performance it would be necessary to reduce the base width of the bipolar transistor, or the gate length of the field-effect transistor, to reduce the transit time and to reduce the active device area to decrease the capacitance or stored charge. These refinements necessitated a much tighter control on all three dimensions of the device.

A better control of the dimension normal to the surface of the device has been obtained by improved methods of introducing impurities (dopants) into the material. At first dopants were added during the growth of the semiconductor crystal or were incorporated by alloying the impurity, or a metal containing the impurity, with the semiconductor. Later, methods capable of much greater dimensional accuracy, such as diffusion or ion-implantation, have been used. These methods enable dopants to be introduced with an accuracy of better than 0.1 μm in depth.

Greater control of the surface dimensions of the device resulted from the development of oxide masking and light-optical photoresist technology, which is now capable of defining regions of less than 1 μm in width, and more recently by the use of electron beam methods of defining resist, which are capable of better than 0.1 μm definition.

This improved dimensional accuracy has gradually enabled the frequency response of transistors to be extended into the microwave region of the spectrum so that bipolar transistors, capable of useful performance in C-band,[1,2] and field-effect transistors, capable of useful performance in X-band,[3,4,5] are available. This could imply that field-effect transistors are capable of higher frequency performance than bipolar transistors. This is not necessarily true since both bipolar and field-effect transistors of comparable dimensions made from the same semiconductor material have theoretically a similar high frequency performance. However, there is an important difference in the case of the C-band bipolar transistors and the X-band field-effect transistors referred to, since the former are made with silicon and the latter with gallium arsenide. Improvements in the frequency response of both types of transistor may be expected in the future by further refinement of the techniques of diffusion, ion-implantation and electron beam pattern definition.

The physical properties of the semiconductor theoretically determine the ultimate electrical performance of the transistor. For example, as the device dimensions are reduced in order to achieve microwave frequency performance, the voltage is maintained at a value sufficient to give the required power output. Ultimately a further reduction of the dimension parallel to the electric field direction will not be possible otherwise the breakdown field in the semiconductor would be exceeded. Johnson has shown[6] that the maximum frequency will be proportional to $E_B v_s$, where E_B is the breakdown field and v_s is the scattering limited drift velocity of the charge carriers. The drift velocity saturates in Ge, Si and GaAs at about 0.6, 0.8 and 2×10^7 cm s^{-1} but the breakdown field in Si and GaAs is about four times higher than in Ge.[7] In practice the maximum frequency limit determined by the $E_B v_s$ product has not been reached but for technological reasons Si is preferred to Ge and GaAs for microwave bipolar transistors.

The technological superiority of silicon is due mainly to the suitability of thermally grown silicon dioxide to act as a diffusion mask against *n*- and *p*-type impurities and the ability to etch very fine patterns in this oxide. The native oxides of germanium and gallium arsenide are not so stable as silicon dioxide and for these semiconductors chemical vapour deposited silicon dioxide or silicon nitride has often been used as a diffusion mask, or as an insulating dielectric material, but with inferior results to thermally grown silicon dioxide on silicon. Recently, however, good masking properties against zinc diffusion has been reported[8] for an anodically grown native oxide of gallium arsenide.

Other physical properties of the semiconductor material which affect the electrical properties of the transistor are the electron and hole mobilities, the dielectric constant and the thermal conductivity. For given doping levels the electron and hole mobilities will determine the base transit time, the base resistance and the collector spreading resistance of the bipolar transistor and the source, drain and channel resistance of the field-effect transistor. Low values for all of these parameters will improve the microwave gain and noise performance of the transistor. It is for this reason that *n*-type gallium arsenide is a preferred material for microwave transistors, since the low-field electron mobility in gallium arsenide is about four times as great as that for silicon. Thus, for example, the input power for a field-effect transistor should be a factor of four lower for gallium arsenide than silicon and this will give an improved intrinsic gain compared with a similar silicon device. For given doping levels and applied voltages, the dielectric constant will determine the capacitance which should be low for efficient microwave performance and in this respect both silicon and gallium arsenide have a lower dielectric constant than germanium,[7] giving them a slight advantage. It is important to have good thermal conductivity to remove heat generated within the transistor, especially for power devices, and silicon is a factor of two better than gallium arsenide[7] in this respect. It is for this reason that silicon is generally used for power transistors. If the transistor is allowed to become too hot its microwave performance will deteriorate and its long-term reliability may be impaired.

A most important property of the semiconductor material is that the device manufacturing technology should be sufficiently developed to enable the microwave transistor to be made with an economic yield and with high reliability. It is primarily for this reason that silicon has dominated the microwave bipolar transistor field. The diffusion properties of arsenic, phosphorus and boron in silicon have all proved to be controllable to within limits of 0.1 μm, using planar technology, and the attainable doping levels are especially suitable for *n-p-n* transistors. These narrow base width *n-p-n* transistors have good microwave properties because of the short transit time for electrons across the base region. Gallium arsenide, on the other hand, is not so amenable to planar processing technology and the *p*-type dopants diffuse appreciably faster than the *n*-type ones,[9] so that it is difficult to make an *n-p-n* transistor with a very narrow base width. Gallium arsenide bipolar transistors have been made[10,11] but they have not operated at microwave frequencies. The situation is different for the recently developed Schottky barrier gate field-effect transistor where, because of the simpler processing technology required, the intrinsically better microwave properties of gallium arsenide have been used to great advantage.

4.2 MICROWAVE BIPOLAR TRANSISTORS

The theory of transistor operation is covered in many books concerned with the physics and design of these devices, for example those by Valdes,[12] Phillips[13] and Grove.[7] Assuming that the basic theory of transistor operation is familiar to the reader attention will be directed to those parts of transistor design and operation that are important for the microwave operation of these devices. Much of the pioneering work on the theory and design of high frequency transistors was carried out by R. L. Pritchard and J. M. Early. Pritchard[14] derived a figure of merit for a high frequency transistor

$$K = (\text{power gain})^{1/2}\,(\text{bandwidth}) = \left(\frac{f_{\mathrm{T}}}{8\pi r_{\mathrm{b}} C_{\mathrm{c}}}\right)^{1/2} \tag{4.1}$$

where f_{T} is the current gain bandwidth frequency, r_{b} is the base resistance and C_{c} is the collector capacitance. This figure of merit was expressed by Early[15] in terms of the emitter-to-collector signal delay time, τ_{ec}:

$$K = \frac{1}{4\pi(r_{\mathrm{b}} C_{\mathrm{c}} \tau_{\mathrm{ec}})^{1/2}} \tag{4.2}$$

This equation contains the three basic parameters, namely base resistance, collector capacitance and signal delay time, which must be minimized in the design of a microwave transistor.

4.2.1 Base resistance

The base resistance arises because of the small base current which flows out of the base region contact. In a planar transistor this current flow is parallel to the emitter, and collector, junction planes. Because of the resistivity of the base region the base current flow will cause a transverse voltage drop to be developed in the base region. Both the current and the voltage will be distributed in the base depending upon the dimensions of the structure and the specific resistivity in different parts of the base region. It is possible for the transverse voltage drop to affect the performance of the transistor since those parts of the emitter junction furthest away from the base contact will be operating at a lower bias than the parts nearest to the base contact. This condition arises since the whole of the emitter side of the junction is essentially at the same potential because of the low resistivity of the emitter region. In order to overcome this emitter self-biasing effect, the emitter is designed with a large periphery-to-area ratio, very often as a series of long narrow strips, so that very little transverse voltage is developed across the width of each strip. The effective emitter width of such a strip structure was analysed by Hauser[16] as a function of current density. Microwave transistors are designed so that for the narrow emitter strip width and current densities employed the current effectively flows uniformly over the whole of the emitter area. Under these conditions it can be shown[13] that the contribution to the base resistance from that part of the base region underneath the emitter is given by

$$r_{bi} = \frac{sR_{\square e}}{12\,ln} \tag{4.3}$$

where s is the emitter strip width and l is its length. n is the number of emitter strips, each of which has a base contact on either side. $R_{\square e}$ is the base sheet resistance underneath the emitter, measured in ohms per square, which is equal to the resistivity divided by the thickness.

The part of the base region between the emitter edge and the edge of the base contact has a resistance given by[13]

$$r_{bo} = \frac{tR_{\square b}}{2\,ln} \tag{4.4}$$

where t is the emitter edge to base contact edge spacing and $R_{\square b}$ is the sheet resistance in this part of the base. Generally $R_{\square e}$ is greater than $R_{\square b}$ because the base doping level and the base thickness are both smaller underneath the emitter.

A transistor structure of this type, with two emitter and three base contact strips, is shown in Figure 4.1. Because of the alternating arrangement of the emitter and base strips this transistor structure is said to be 'interdigitated' (see Section 4.5.1).

If the impurity density in the base region is not uniform then allowance must be made for the impurity distribution $N(x)$ within the

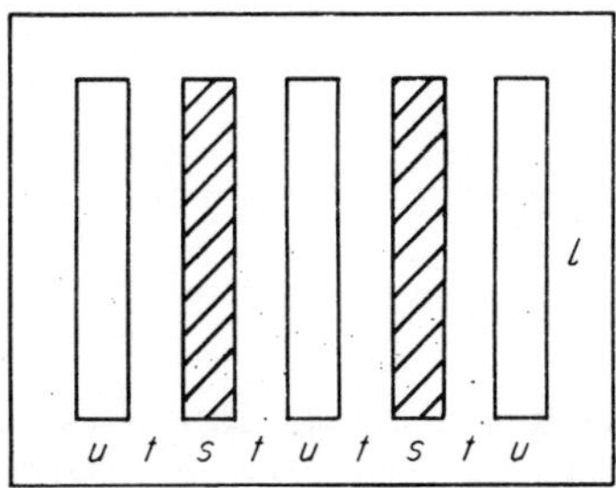

Figure 4.1 An interdigitated transistor structure with two emitter strips (n = 2)

base and the fact that the minority carrier mobility μ_{nb} in the p-type base is a function of $N(x)$, such that μ_{nb} decreases as $N(x)$ increases. Such an analysis has been carried out by Irvin[17] for silicon diffused base transistors for a range of impurity levels and for different impurity distributions.

r_{bi} is referred to as the intrinsic or internal base resistance and r_{bo} is called the extrinsic or external base resistance, there is, however, another component of the base resistance which is of importance at microwave frequencies; this is the base contact resistance, r_{bcon}. The total base resistance is therefore

$$r_b = r_{bi} + r_{bo} + r_{bcon} \tag{4.5}$$

Archer[18] gives the following expression for r_{bcon}

$$r_{bcon} = \frac{(R_{\square b}\rho_{bcon})^{1/2}}{2\,ln \tanh[(u/2)(R_{\square b}/\rho_{bcon})^{1/2}]} \tag{4.6}$$

Where ρ_{bcon} is the base contact resistivity and u is the width of the base contact strip. The following approximate expressions can be used instead of Equation 4.6

If

$$\frac{u}{2} \gg \left(\frac{\rho_{bcon}}{R_{\square b}}\right)^{1/2},$$

then

$$r_{bcon} = \frac{(R_{\square b}\rho_{bcon})^{1/2}}{2\,ln} \tag{4.7}$$

If

$$\frac{u}{2} \ll \left(\frac{\rho_{bcon}}{R_{\square b}}\right)^{1/2},$$

then

$$r_{bcon} = \frac{\rho_{bcon}}{uln} \tag{4.8}$$

A circuit model[18] representing the three base resistance components for a single emitter strip in an interdigitated transistor structure is shown in Figure 4.2(b) and the base resistance and the distributed collector capacitance is shown in Figure 4.2(c). The circuit model of Figure 4.2(c) is similar, except for the treatment of the base contact

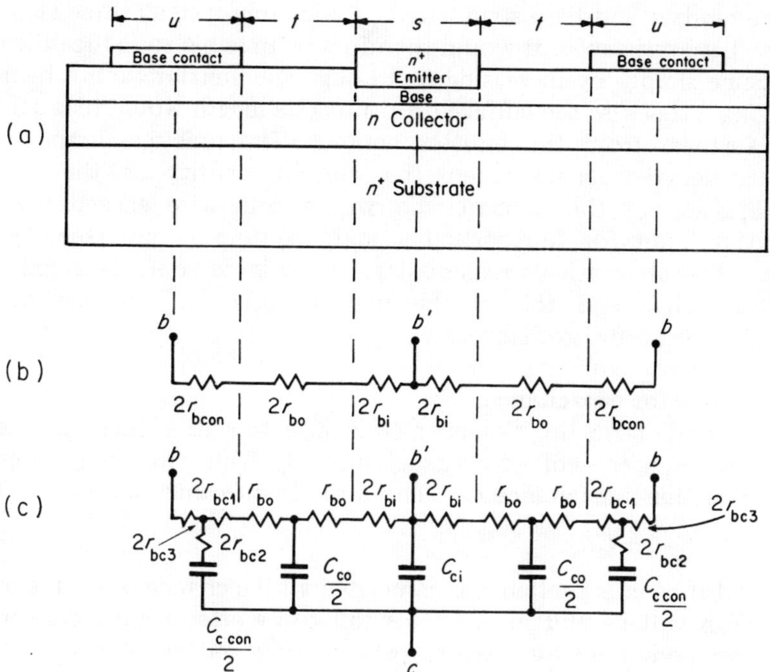

Figure 4.2 (a) Cross-section of single emitter strip in an interdigitated transistor structure. (b) Circuit model representing base resistance components. (c) Circuit model representing base resistance and distributed collector capacitance.[18] (Reproduced by permission of Pergamon Press Ltd.)

resistance, to that given by White and Thurston.[19] In Figure 4.2(c) if

$$\frac{u}{2} > \left(\frac{\rho_{bcon}}{R_{\square b}}\right)^{1/2},$$

then

$$r_{bc1} = \frac{(R_{\square b}\rho_{bcon})^{1/2}}{2l}, \; r_{bc2} = \frac{\rho_{bcon}}{ul}, \; r_{bc3} = 0 \tag{4.9}$$

If

$$\frac{u}{2} < \left(\frac{\rho_{bcon}}{R_{\square b}}\right)^{1/2},$$

then

$$r_{bc1} = 0, \; r_{bc2} = 0, \; r_{bc3} = \frac{\rho_{bcon}}{ul} \tag{4.10}$$

The emitter and base strip length, l, is common to all three elements of the base resistance, it would therefore be possible to reduce the base resistance simply by increasing l, although the emitter cannot be made too long, otherwise the voltage drop along its length would bias off that part furthest from the emitter contact. The practical limit to the emitter length occurs when the current density in the emitter metallization at the contact end would otherwise exceed the safe operating limit for the particular metal system in use (see Section 4.5.1). The alternative to increasing l is to increase n, the number of emitter strips, and this is the method used in the case of the interdigitated transistor structure.

4.2.2 Collector capacitance

The second term in Equation 4.2 affecting the high frequency performance is the collector capacitance, C_c. This capacitance consists of a transition capacitance C_{Tc} and a diffusion capacitance C_{Dc}. Thus

$$C_c = C_{Tc} + C_{Dc} \tag{4.11}$$

The diffusion capacitance is associated with a change of charge in the base with voltage and in order for this to be effective the base width must be modulated by the a.c. voltage swing on the collector. In the case of a silicon microwave transistor the base doping is greater than the epitaxial collector region doping and the depletion layer is almost entirely in the collector epitaxial region, so that the base width does not change significantly with collector voltage. Consequently for a diffused-epitaxial microwave transistor C_{Dc} is negligible.

The transition capacitance, C_{Tc}, depends upon the collector area and the width of the depletion layer. Because the base impurity profile of a microwave transistor has such a steep gradient the capacitance formula for an abrupt junction is often used

$$C_{TC} = A_c \left[\frac{e\epsilon\epsilon_0 N_C}{2(V+\phi)}\right]^{1/2} \tag{4.12}$$

or for silicon

$$C_{TC} = A_c(2.88 \times 10^{-4})\left(\frac{N_C}{V+\phi}\right)^{1/2} \quad \text{pf} \tag{4.13}$$

where A_c = collector area, ϵ = dielectric constant, ϵ_0 = permittivity of free space, N_c = doping level in epitaxial collector region, V = applied voltage and ϕ = contact potential of junction. Alternatively, use may be made of curves giving the junction capacitance and depletion layer width of graded junctions for a wide range of doping levels and junction depths, calculated by Lawrence and Warner.[20]

A good microwave transistor will be designed so that the collector epitaxial layer is fully depleted at the operating bias level, in which case

$$C_{TC} = \frac{A_c \epsilon \epsilon_0}{W_{epi}} \tag{4.14}$$

where W_{epi} is the epitaxial layer thickness. Under these conditions, in order to ensure that premature base punch-through does not occur before the collector breakdown voltage is reached, Hewlett and coworkers[21] have shown that

$$\frac{W_{epi}}{W_b} = \frac{N_b}{N_c} \tag{4.15}$$

where W_b is the base width and N_b the impurity concentration in the base.

In order to minimize the collector capacitance, thereby improving the microwave frequency performance, the collector area must be minimized. This means that the dimensions s, t and u in Figure 4.1 must be made as small as possible and that l and n must be no greater than is required for current or power handling capability.

The total collector capacitance, C_c, for the purpose of transistor modelling is subdivided into elements forming an RC transmission line with associated parts of the base resistance, as indicated in Figure 4.2(c). If C_0 is the collector capacitance per unit area, then

$$C_{ci} = C_0 sl \tag{4.16}$$

$$C_{co} = 2C_0 tl \tag{4.17}$$

$$C_{ccon} = C_0 ul \tag{4.18}$$

For the equivalent circuit of Figure 4.2(c) Equation 4.1 may be written, in modified form,

$$K = \left\{ \frac{f_T}{8\pi[r_{bi} + r_{bo} + r_{bc1} + r_{bc3})C_{ci} + (r_{bo}/2 + r_{bc1} + r_{bc3})C_{co} + r_{bc3}C_{ccon}]} \right\}^{1/2} \tag{4.19}$$

This equation shows that if the base contact strip is sufficiently wide for $u/2 > (\rho_{bcon})^{1/2}/R_{\square b}$ to be satisfied, then K is independent of the width of the base contact and, for base contact widths less than this, K decreases as the strip width is reduced. The base contact strip width should therefore be determined by the relative values of the base contact resistivity and inactive base sheet resistance available. The base contact resistivity depends upon the particular contact metal system used and the doping level at the silicon surface; data for aluminium [22,23] aluminium—silicon,[18] titanium[23] and molybdenum[23,24] have been published. The base sheet resistance underneath the base contact, and the doping level at the silicon surface, are often modified by including a p^+ diffusion underneath the base contact strip.

4.2.3 Signal delay time

The signal delay time, τ_{ec}, in Equation 4.2 is the total transit time of a charge carrier constituting part of the input signal at the emitter which is taken out of the device at the collector. The charge carrier encounters successive regions of time delay in its transit through the device, each of which are discussed in the following. These time delay elements may be summarized

$$\tau_{ec} = \tau_e + \tau_{eb} + \tau_{bc} + \tau_b + \tau_d + \tau_c \tag{4.20}$$

τ_e = emitter delay time owing to excess holes stored in the emitter
τ_{eb} = emitter—base junction capacitance charging time through the emitter
τ_{bc} = base—collector junction capacitance charging time through the emitter
τ_b = base transit time
τ_d = collector depletion layer delay time
τ_c = base—collector junction capacitance charging time through the collector.

Each of these time constants may be expressed as a characteristic frequency $\omega = 1/\tau$, thus

$$\frac{1}{\omega_T} = \frac{1}{\omega_e} + \frac{1}{\omega_{eb}} + \frac{1}{\omega_{bc}} + \frac{1}{\omega_b} + \frac{1}{\omega_d} + \frac{1}{\omega_c} \tag{4.21}$$

The frequency $\omega_T = 2\pi f_T$ is a most important parameter in a microwave transistor since it dominates both the gain and the noise figure. f_T is the frequency at which the magnitude of the grounded emitter current gain, h_{fe}, is equal to unity; it is known as the current gain bandwidth frequency.

Emitter delay time, τ_e

Until recently[25,26,27] it had been assumed that the emitter should be as heavily doped as possible, without introducing too much crystal lattice damage, in order to ensure a high emitter efficiency. However, high doping will affect the silicon band structure[25,28] because of impurity level broadening into an impurity band and because the large number of impurity atoms disturbs the periodicity of the silicon crystal lattice, leading to the formation of band edge tails. The heavy doping in the emitter therefore leads to a reduction in the silicon band gap, which depends upon the total doping level and hence the position in the emitter. DeMan introduced[25] a doping-dependent intrinsic carrier concentration, n_{ie}, to account for the effect of the heavy doping on the classical expressions for the electrical parameters. The validity of this approach was later verified theoretically.[26,28] Because n_{ie} is dependent upon the total doping level it will be greater at the surface and the sidewall of a planar diffused emitter than along the plane part of the emitter—base *p-n* junction. Since n_{ie} is more effective close to the junction the emitter sidewall region will show most markedly the effects of band bending.

Under heavy doping conditions the hole current density injected into the emitter from the base is given by

$$J_p = eD_p\left(\frac{p}{n_{ie}^2}\frac{dn_{ie}^2}{dx} - \frac{p}{N}\frac{dN}{dx} - \frac{dp}{dx}\right) \tag{4.22}$$

where e = electronic charge, D_p = hole diffusion coefficient, p = hole concentration and N = net doping concentration.

The holes experience an additional drift force proportional to the derivative of the change of the valence band edge. Therefore an additional component of the hole current flowing from the base to the emitter of an *n-p-n* transistor is introduced because the gradient of n_{ie} in this region is negative. Additional holes are stored in the emitter and have a marked effect on the predictions of f_T and h_{fe}. DeMan and coworkers[29] showed that the heavy doping effect causes an emitter delay time given by

$$\tau_e = \frac{Q_e}{I_c} = \frac{1}{\beta_0}\int_0^{x_{eb}^*}\frac{n_{ie}^2}{N}\left(\int_0^x \frac{N}{D_{pe}n_{ie}^2}\,\mathrm{d}x\right)\mathrm{d}x \tag{4.23}$$

where Q_e = excess charge storage of minority carriers in the emitter region, β_0 = d.c. current gain and x_{eb}^* = boundary between the emitter neutral region and the emitter—base space charge region.

Equation 4.23 is based on the assumption of no recombination in the emitter neutral region, which is satisfied in almost all cases.[26] Because τ_e is given by the integral of the stored charge from the surface of the emitter ($x = 0$) to x_{eb}^* it can be reduced by decreasing the emitter junction depth. Microwave transistors have always been designed with shallow junction depths both to increase the gradient of the emitter doping through the emitter—base junction, in order to reduce the emitter capacitance, and to reduce the emitter sidewall capacitance. A more important reason for reducing the emitter depth even further is now seen to be to bring a reduction in τ_e. In order to take advantage of a reduced emitter depth, to increase f_T, it will be necessary to ensure that the emitter metallization system not only makes a good low resistance ohmic contact to the emitter surface, but that it also does not diffuse or migrate to the emitter—base junction during the subsequent operating life of the transistor, otherwise the junction would be impaired.

Equation 4.23 shows that τ_e may be reduced by decreasing the doping gradient through the emitter to decrease n_{ie}^2. This is contrary to the trend in the past when it had been assumed that the surface concentration of the emitter should be increased to steepen even further the doping gradient, thereby improving both f_T and the gain. This was the reason for the use of arsenic rather than phosphorus as an emitter dopant, because of its steep doping gradient through the junction. However, arsenic is better than phosphorus for physical reasons, because it is a better fit in the silicon crystal lattice.

The emitter and base doping profiles in silicon microwave transistors often deviate considerably from the error function or Gaussian curves which should be obtained ideally. For example both arsenic and phosphorus emitter doping profiles tend to be 'flat-topped'. Henderson and Scarbrough[30] considered the special case of flat arsenic and boron concentrations in the emitter region. In this case the net doping concentration, N, and n_{ie}^2 in Equation 4.23 are both constant, since N_A and N_D are constant,therefore Equation 4.23 reduces to[30]

$$\tau_e = \frac{1}{D_{pe}\beta_0} \int_0^{x_{jeb}} x \, dx$$

or

$$\tau_e = \frac{x_{jeb}^2}{2D_{pe}\beta_0} \tag{4.24}$$

For simplicity, since the arsenic emitter profile is so steep at the emitter—base junction, the limits of integration have been taken from the surface to the emitter—base junction depth, x_{jeb}, rather than the edge of the depletion region in the emitter, x_{eb}^*.

Emitter capacitance charging time through the emitter, τ_{eb}

The emitter—base junction is forward-biased and the emitter current will be divided between the parallel combination of the emitter resistance, r_e, and the emitter transition capacitance, C_{Te}. Only the current that flows through r_e gets injected into the base and is amplified. The current flowing in C_{Te} is parasitic. This gives rise to a simple RC type of frequency cut off so that

$$\tau_{eb} = \frac{1}{\omega_{eb}} = r_e C_{Te} \tag{4.25}$$

r_e is the derivative of the emitter current, I_e, with respect to emitter voltage and is given by

$$r_e = \frac{kT}{eI_e} \tag{4.26}$$

In order to reduce τ_{eb} it is possible to reduce r_e by increasing the emitter current, I_e. Microwave transistors are generally designed to operate at emitter current levels such that τ_{eb} is not a dominant part of the overall emitter to collector delay time, τ_{ec}.

The emitter transition capacitance is a result of charge stored at the edge of the emitter depletion layer and is proportional to the emitter area, A_e. Since the emitter and base regions are generally formed by diffusion the emitter junction is graded. The transition capacitance of double-diffused junctions has recently been described[31] for a range of doping conditions, but since the emitter—base junction is normally operated in forward bias the step-junction equation is a good approximation and is simple to use

$$C_{Te} \simeq A_e \left[\frac{\epsilon\epsilon_0 e N_{BE}}{2(V+\phi)}\right]^{1/2} \tag{4.27}$$

or for silicon

$$C_{Te} \simeq 2.88 \times 10^{-4}\ A_e \left(\frac{N_{BE}}{V+\phi}\right)^{1/2} \quad \text{pF} \tag{4.28}$$

For a typical silicon microwave transistor ϕ is approximately 0.7 V. The applied voltage is negative when the device is forward-biased, so that the total voltage is reduced and C_{Te} is increased.

For design purposes, in order to increase the frequency response, it is better to reduce the emitter area to the minimum required by power handling and reliability requirements by reducing the emitter width, s. The lower limit on s will be set by the definition available from the photoresist or electron beam technology employed. N_{BE} is usually set by r_b and h_{fe} requirements but occasionally N_{BE} can be reduced to reduce C_{Te} if the emitter width is also narrowed to keep r_b from increasing.

Heavy doping effects in the emitter[29] will also affect the value of C_{Te} and hence the value of τ_{eb}. The emitter transition capacitance for a linear gradient junction is given by[32]

$$C_{Te}^3 = \frac{A_e^3 e a \epsilon^2 \epsilon_0^2}{12(V_g - V)} \tag{4.29}$$

where a is the gradient and V_g is the gradient voltage.

$$V_g = \frac{2kT}{3e} \ln \left(\frac{a^2 \epsilon \epsilon_0 kT}{8e^2 n_{ie}{}^3} \right) \tag{4.30}$$

The concentration of the impurities in the emitter—base space charge region of a microwave transistor is of the order of 10^{18} cm^{-3} in the plane part of the junction, and even higher in the sidewall region. The value of n_{ie} may be greater than the classical value of n_i by a factor from 1.5 to 4.0 depending upon position over the emitter junction. This causes a shift in V_g equal to $kT/e \ln(n_{ie}/n_i)^3$, or of the order 30—100 mV, which gives rise to capacitance values 30% to 40% higher than those predicted classically. The effect of the increased capacitance in the emitter sidewall becomes more marked as the emitter width is reduced.

The emitter series resistance, r_{se}, due to the resistivity of the silicon within the emitter region and the emitter metallization, will be in series with $r_e C_{Te}$ and the emitter capacitance will be charged through the combination of r_e and r_{se}. The time constant, τ_{eb} becomes therefore

$$\tau_{eb} = \frac{1}{\omega_{eb}} = (r_{se} + r_e) C_{Te} \tag{4.31}$$

In a small-signal low-noise microwave transistor r_{se} may be 0.5—1.0 Ω, which is an order of magnitude lower than the typical value of r_e for these transistors. Microwave power transistors often include built-in emitter resistors to promote current sharing in different parts of the emitter and since these transistors are operated at much higher current levels than their low-noise counterparts, r_e will be relatively small (see Equation 4.26) and Equation 4.31 should be used for τ_{eb}.

Collector capacitance charging time through the emitter, τ_{bc}

In a microwave transistor, account has to be taken of signal delay times which would be relatively unimportant in lower frequency transistors. Thus the signal delay due to the charging time of the collector transition capacitance through the emitter must be considered. This gives rise to an RC type of frequency cut-off

$$\tau_{bc} = \frac{1}{\omega_{bc}} = (r_{se} + r_e) C_{Tc} \tag{4.32}$$

If emitter resistors are included to promote current sharing in the emitter then they should be included in r_{se}. If emitter resistors have not been included in the metallization pattern then r_{se} will generally be quite small compared with r_e and may be neglected. In the latter case, which will be true for small-signal low-noise microwave transistors, τ_{bc}

reduces to

$$\tau_{bc} = r_e C_{Tc} \tag{4.33}$$

Base transit time, τ_b

The time for charge carriers to cross the base region is derived by solving the transport equations.[13] The solution is obtained in terms of the base transport factor, β^*, which is the ratio of the current at the collector edge of the base to that at the emitter edge. β^* is complex since the carriers undergo a phase shift as well as a reduction in amplitude as they cross the base region.

If the carriers cross the base by diffusion alone, then

$$\beta^* = \text{sech} \left[\left(\frac{W}{L} \right)^2 + j\omega t_b \right]^{1/2} \tag{4.34}$$

where the transit time of electrons across the p-type base, t_b, can be written in terms of the base width, W, and the diffusion constant of electrons in the base, D_{nb}.

$$t_b = \frac{W^2}{D_{nb}} \tag{4.35}$$

To obtain the base cut-off frequency, ω_b, which is the frequency at which $|\beta^*|$ is 3 dB below its low frequency value, Equation 4.34 is evaluated using a hyperbolic secant series to obtain $\omega_b t_b$ when $|\beta^*| = \sqrt{2}/2$. The solution is

$$\omega_b t_b = 2.43 \tag{4.36}$$

Substitution for t_b from (4.35) gives

$$\omega_b \left(\frac{W^2}{D_{nb}} \right) = 2.43 \tag{4.37}$$

or

$$\frac{1}{\omega_b} = \frac{W^2}{2.43 D_{nb}} \tag{4.38}$$

The base transit time constant of Equation 4.20 thus becomes

$$\tau_b = \frac{W^2}{2.43 D_{nb}} \tag{4.39}$$

Equation 4.39 describes the base transit time when the signal charge carriers (electrons) move through the base by diffusion only. If a drift field, E, is present in the base, owing to a concentration gradient of impurities through the base region, such that

$$E = -\frac{kT}{e}\frac{1}{N(x)}\frac{dN(x)}{dx} \tag{4.40}$$

then the transit time of electrons through the base will be reduced[33] since they will be accelerated by the electric field, as well as moving by diffusion. This effect can be taken into account by using a modified form of Equation 4.39

$$\tau_b = \frac{W^2}{nD_{nb}} \tag{4.41}$$

Values of n for a linear, error function complement, and exponentially graded profile were given by Moll and Ross.[34] The optimum profile shape for minimum τ_b has been calculated by Marshak.[35] For an exponential distribution of impurities the value of n is given by[36,37]

$$n = \frac{m^2}{m - 1 + \exp^{-m}} \tag{4.42}$$

where

$$m = \frac{eEW}{kT} \approx \ln\left(\frac{N_{BE}}{N_{BC}}\right) \tag{4.43}$$

where N_{BE} and N_{BC} are the base impurity concentrations near the emitter and collector, respectively, and provided the value of m is not too small. In the case of silicon diffused base *n-p-n* microwave transistors n will be in the range from 4 to 7, typically. Within this range, Equations 4.41 and 4.42 may be combined to give, within 2% error, the simple expression[38]

$$\tau_b \simeq \frac{(m-1)W^2}{m^2 D_{nb}} \tag{4.44}$$

Another approximation to Equations 4.41 and 4.42, within 1% error in this range, is that given by Cooke[39]

$$\tau_b \simeq \frac{W^2}{(1.6 + 0.92m)D_{nb}} \tag{4.45}$$

For simplicity the electron diffusion coefficient in the base, D_{nb}, has been assumed constant. However, D_{nb} will be reduced as the concentration of impurities in the base increases towards the emitter, because of impurity scattering, and this will nullify to a large extent the effect of the built-in drift field. Taking into account the variability of

D_{nb}, the base transit times for a Gaussian and error function complement base impurity distribution were calculated by Lo[40] and for a Gaussian and exponential impurity distribution by Gover and coworkers.[41] Under these conditions the effective value of n in Equation 4.41 is reduced to about 3.

In the analysis of charge carriers in transit through the base of a transistor it is usually assumed that the charge density is zero at the collector edge of the base region. Matz[42] pointed out that this could not actually be the case since in order to maintain continuity of current density through the device the electron mobility at the collector edge of the base would have to approach infinity, whereas the maximum electron velocity in a semiconductor is limited to the scattering limited velocity, v_s. The electron current density at the collector side of the base will be given by

$$J(W) = ev(W)n(W) \tag{4.46}$$

where $v(W)$ is the velocity and $n(W)$ the electron density at the collector side of the base. Because of the non-zero value of $n(W)$ additional charge will be stored in the base and Kirk[43] showed that the fluctuation of this stored charge in response to the input signal will introduce an additional time delay, approximately equal to W/v_s for uniform base doping, in the expression for the base transit time. The total base transit time thus becomes

$$\tau_b = \frac{1}{\omega_b} = \frac{W^2}{nD_{nb}} + \frac{W}{v_s} \tag{4.47}$$

If the impurity concentration, N, in the base is a function of distance, x, from the emitter then Equation 4.47 becomes

$$\tau_b = \frac{W^2}{nD_{nb}} + \frac{1}{v(W)} \int_0^W \frac{N(W)}{N(x)} \, dx \tag{4.48}$$

Collector depletion layer delay time, τ_d

When the injected signal carriers have traversed the base they enter the depletion layer of the reverse-biased collector junction and are swept across the depletion layer under the influence of the strong electric field existing there. The carriers attain their scattering limited velocity, v_s, at electric fields of the order of 10^4 V/cm and since the electric field in the collector depletion layer of a microwave transistor is usually appreciably greater than this value the carriers reach their limiting velocity so quickly that it is justifiable to assume that they travel through the entire depletion layer at their limiting velocity. During normal operation this velocity is nearly independent of voltage over the entire signal excursion.

The transit time, t_d, of carriers crossing the depletion layer of width x_d, regardless of whether the junction is graded or abrupt, is given by

$$t_d = \frac{x_d}{v_s} \tag{4.49}$$

The a.c. voltage across the depletion layer excites a displacement current in the depletion layer and this causes a phase shift in the total current flowing across the depletion layer such that the signal delay time corresponding to the carrier transit time is given by $t_d/2$.[44] The collector depletion layer delay time is therefore

$$\tau_d = \frac{1}{\omega_d} = \frac{x_d}{2v_s} \tag{4.50}$$

τ_d is minimized by keeping x_d small. For a given collector voltage this is accomplished by lowering the resistivity of the collector epitaxial layer. τ_d is quite sensitive to the applied collector voltage, since x_d varies between the square root and the cube root of the voltage, depending on the nature of the junction. A microwave transistor is generally designed, however, so that the collector space charge layer fully depletes the epitaxial collector region and under these conditions τ_d is given by

$$\tau_d \simeq \frac{W_{epi}}{2v_s} \tag{4.51}$$

where W_{epi} is the width of the epitaxial layer.

Collector capacitance charging time through the collector, τ_c

The collector transition region capacitance, C_{Tc}, must be charged through the combined series resistance from emitter to collector. This resistance consists of three parts, the series resistance of the emitter region, r_{se}, the a.c. resistance of the emitter junction, r_e, and the series resistance of the collector region, r_{sc}. The effect of r_{se} and r_e have been discussed previously and will not be considered here. The final term in the total signal delay time from the emitter to the collector is therefore

$$\tau_c = \frac{1}{\omega_c} = r_{sc}\, C_{Tc} \tag{4.52}$$

Silicon *n-p-n* microwave transistors are made with *n*-type epitaxial silicon on an n^+ silicon substrate and are designed so that under operating voltage conditions the collector depletion layer extends through the epitaxial layer almost to reach the substrate. Under these conditions the collector series resistance will be determined almost entirely by the resistivity, ρ_c, and thickness, l_c, of the substrate, such that

$$r_{sc} = \frac{\rho_c l_c}{A_c} \tag{4.53}$$

The collector transition capacitance, under these fully depleted conditions, will be given by

$$C_{Tc} \simeq \frac{A_c \epsilon \epsilon_0}{W_{epi}} \tag{4.54}$$

so that

$$\tau_c \simeq \frac{\rho_c l_c \epsilon \epsilon_0}{W_{epi}} \tag{4.55}$$

The collector substrate resistivity, ρ_c, will be proportional to the reciprocal of the doping level, so that the substrate should by highly doped n-type to reduce τ_c. Antimony is often used as the substrate dopant since it diffuses relatively slowly in silicon and will not, therefore, readily diffuse from the substrate into the epitaxial layer during transistor processing. The substrate thickness, l_c, should be reduced as much as is practicable both to reduce τ_c and, more importantly in power transistors, to reduce the thermal resistance from the active region of the device to the heat sink.

The collector depletion layer reach-through condition is fulfilled for combinations of collector epitaxial layer doping level, N_C, and thickness, W_{epi}, for various collector voltages, V, by the curves in Figure 4.3. A step-junction approximation for the collector–base

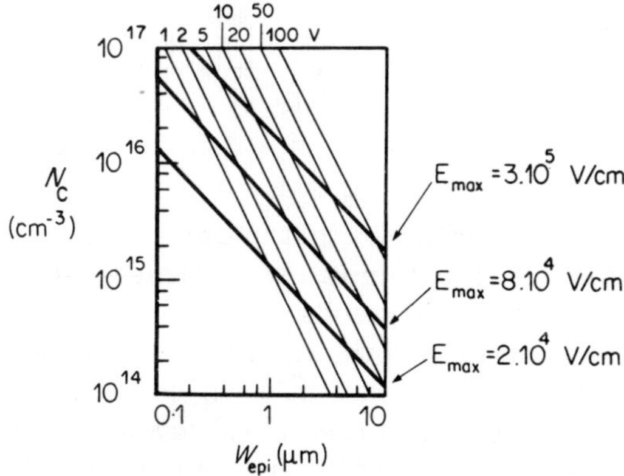

Figure 4.3 Collector design chart for transistors operated at reach-through.[18] (Reproduced by permission of Pergamon Press Ltd.)

junction was assumed so that V, N_C and W_{epi} are related by

$$V = \frac{eN_C W_{epi}^2}{2\epsilon\epsilon_0} \tag{4.56}$$

For the collector depletion layer delay time to be given by Equation 4.51 the field in the depletion region should be sufficient to maintain the carriers at their scattering limited velocity throughout the collector depletion region. Provided the carriers attain v_s at the base edge of the collector depletion region the assumption of a constant drift velocity thereafter is a good approximation. This condition is satisfied when the maximum field in the collector depletion region, $E_{max} > 2 \times 10^4$ V/cm where, at reach-through E_{max} is given by

$$E_{max} = \frac{eN_C W_{epi}}{\epsilon\epsilon_0} \tag{4.57}$$

Values of N_C and W_{epi} for $E_{max} = 2 \times 10^4$ V/cm may be found from Figure 4.3. Collector junction breakdown occurs when $E_{max} \simeq 3 \times 10^5$ V/cm and this limit is also shown in Figure 4.3. In practice[4 5] the common-emitter operating voltage cannot exceed about a quarter of the breakdown voltage, so that suitable design values for epitaxial layer doping and thickness may be found by using the line marked $E_{max} = 8 \times 10^4$ V/cm in Figure 4.3.

Summary

The results presented in the preceding parts of Section 4.2.3 for each of the time delay elements comprising the total signal delay time from the emitter to the collector of an *n-p-n* transistor may be summarized by the following equation

$$\tau_{ec} = \frac{x_{jeb}^2}{2D_{pe}\beta_0} + r_e(C_{Te} + C_{Tc}) + \frac{W^2}{nD_{nb}} + \frac{W}{v_s} + \frac{x_d}{2v_s} + r_{sc}C_{Tc} \tag{4.58}$$

Inspection of this equation shows that in order to decrease τ_{ec}, thereby increasing f_T, the transistor must be designed and constructed to achieve the following practical results.

(a) The emitter junction depth, x_{jeb}, must be shallow. For example C-band low noise transistors may have an emitter junction depth of only 0.1—0.2 micrometre. The theoretical lower limit on x_{jeb} will be set by a loss of emitter efficiency, but practically the limit may be determined by micro-alloying of the emitter contact metal through the shallow emitter, during either fabrication or subsequent operational life. For this reason a composite metal system must be used for contacting these very shallow *p-n* junctions; either aluminium with an underlayer of titanium, or an aluminium—silicon alloy has been used.

(b), The diffusion length of holes in the *n*-type emitter, D_{pe}, should be as large as possible and this means that the emitter region should not be too heavily doped.

(c) The small-signal common-emitter current gain, β_0, should be as large as possible. This implies that the emitter should be more heavily doped than the base, but not too heavily doped otherwise band gap narrowing will reduce the emitter efficiency.[26,27] Both (b) and (c) dictate that the emitter should not be too heavily doped. This is contrary to earlier transistor design theory when it was generally assumed that the emitter should be as heavily doped as possible, without introducing lattice damage in the host crystal.

(d) The emitter resistance, r_e, should be small and this can be achieved by increasing the emitter current, I_e, since r_e is inversely proportional to I_e (Equation 4.26). Any resistance, r_{se}, in series with r_e must also be kept small. Resistance in series with the emitter junction will include the resistance of the emitter region itself and the resistance of the emitter metallization, together with any contact resistance between them. Any emitter ballast resistors should be included in r_{se} and any contact resistance between the emitter metal bonding pad and the emitter lead wire.

(e) The emitter transition capacitance, C_{Te}, should be made small by reducing the emitter area as much as is compatible with current density limitations in the transistor and in the emitter metallization (see Section 4.5.1). In addition the doping in the base region part of the emitter transition region should not be too great, so that the transition region is not too narrow, and the doping in the emitter itself should not be too high otherwise a shift in V_g and increased capacitance will occur.

(f) The collector transition capacitance, C_{Tc}, should be made small by reducing the collector area as much as is compatible with power handling and thermal dissipation requirements. C_{Tc} may be reduced by increasing the collector resistivity and the collector voltage unless the collector depletion layer is limited by reach-through to the collector substrate, in which case the epitaxial layer width would have to be increased. However, except for area adjustment, these methods all increase the collector depletion layer width, x_d, which will adversely effect τ_{ec}.

(g) The base width, W, must be as small as possible. Before the advent of planar transistor technology this used to be the limiting factor to the achievement of high frequency operation. Nowadays, with improved diffusion technology, base widths of only 0.1—0.2 of a micrometre are possible, but care must be taken that the emitter-dip effect (see Section 4.2.5) does not limit the attainment of narrow base width. Very narrow, and reproducible, base regions for microwave transistors can also be achieved by ion-implantation.[1]

(h) The drift-field or excess phase factor, n, should be optimized by an appropriate doping density and gradient in the base.

(i) The diffusion coefficient of electrons in the base, D_{nb}, should be made as large as possible by ensuring that the base region is not excessively doped. In this respect it is an advantage to have the intrinsic base region underneath the emitter less heavily doped than the extrinsic base region between the emitter and the base contact; this state of affairs was realized in the low-noise transistor of Archer.[1]

(j) The collector depletion layer width, x_d, should be made as small as the operating voltage requirement, coupled with the collector doping level, will allow.

(k) The collector series resistance, r_{sc}, is made small by the use of epitaxial silicon. The transistor is designed so that the epitaxial layer is fully depleted at the operating voltage. Under these conditions r_{sc} is determined by the properties of the substrate, which consists of low resistivity silicon made as thin as practicable. Both of these factors will make r_{sc} small (see Equation 4.53) and all that remains is to ensure a low resistance contact between the back of the substrate and the header.

To give an idea of the relative magnitude of the time constants involved in Equation 4.58 they have been evaluated for values, given in Table 4.1, appropriate for a small-signal, low-noise, C-band transistor. Equation 4.58 has been written in the form of Equation 4.20.

$$\tau_{ec} = \tau_e + \tau_{eb} + \tau_{bc} + \tau_b + \tau_d + \tau_c$$

$$\tau_{ec} = [1.2 + 3.9 + 0.9 + (4.2 + 1.2) + 7.5 + 0.8] \text{ ps}$$

$$\tau = 19.7 \text{ ps}$$

therefore,

$$f_T = \frac{1}{2\pi\tau_{ec}} = 8.0 \text{ GHz}$$

It is apparent that for this transistor the base transit time and the collector depletion layer delay time, τ_b and τ_d, are quite important in determining the value of f_T.

Part of the collector capacitance is charged through part of the base resistance, as shown in Figures 4.11 and 4.12. The time constant associated with this RC product has not been included in Equation 4.58 and would have added about another 1 ps to τ_{ec}. The base—collector bonding pad capacitance, $C_{bc(pad)}$ in Figure 4.11, is in parallel with C_{Tc} and is charged through r_e, $r_{bb'}$, and $r_{bc(pad)}$. Another RC time

Table 4.1
Device design variables used to calculate τ_{ec} for small-signal, low-noise, C-band transistor

Variable	Symbol	Value
Emitter junction depth	x_{jeb}	1.2×10^{-5} cm
Base width	W	1.0×10^{-5} cm
Collector depletion layer width	x_d	1.2×10^{-4} cm
Diffusion coefficient of holes in emitter	D_{pe}	$1.5\ cm^2\ s^{-1}$
Diffusion coefficient of electrons in base	D_{nb}	$8.0\ cm^2\ s^{-1}$
Scattering limited velocity	v_s	$8.0 \times 10^6\ cm\ s^{-1}$
Field dependent constant	n	3
Emitter depletion capacitance	C_{Te}	0.3 pF
Collector depletion capacitance	C_{Tc}	0.07 pF
Current gain	β_0	40
Emitter current	I_e	2×10^{-3} A
Emitter area	A_e	$1.26 \times 10^{-6}\ cm^2$
Collector area	A_c	$8.6 \times 10^{-6}\ cm^2$
Epitaxial substrate resistivity	ρ_c	$0.01\ \Omega$ cm
Epitaxial substrate thickness	l_c	0.01 cm
Boltzmann's constant	k	$1.38 \times 10^{-23}\ J\ K^{-1}$
Absolute temperature	T	300 K
Electronic charge	e	1.6×10^{-19} coulombs

constant, equal to $(r_e + r_{bb'} + r_{bc(pad)})C_{bc(pad)}$, should therefore be added to Equation 4.58. For this small-signal, low-noise, C-band transistor $C_{bc(pad)} = 0.04$ pF, $r_e = 13\ \Omega$, $r_{bb'} = 15\ \Omega$ and $r_{bc(pad)} = 16\ \Omega$; this will add 1.76 ps time delay to τ_{ec}. Because of these additional time delays f_T will be reduced to about 7 GHz.

4.2.4 High current limitations on f_T

The d.c. current gain, β_0, in the first term on the right-hand side of Equation 4.58 is proportional to the emitter efficiency, γ. When the current level reaches values such that the concentration of majority carriers in the base increases from the equilibrium value, the emitter efficiency drops. The decrease in emitter efficiency takes place when the integral of the injected minority charge per unit area in the base becomes comparable to the majority carrier concentration per unit area. Since the injection efficiency for electrons into the base decreases, more of the emitter current will consist of holes injected into the emitter from the base and hole storage in the emitter will be increased. This will cause τ_e in Equation 4.20 to increase which, in turn, will cause a reduction in f_T, since

$$f_T = \frac{1}{2\pi\tau_{ec}} \tag{4.59}$$

It is assumed that D_{pe} will not alter much at high injection levels because the injected hole concentration in the emitter will still be much less than the majority electron concentration.[29]

Under high current conditions the emitter resistance, r_e, will be reduced, since it is inversely proportional to emitter current (Equation 4.26), and this will reduce the second term of Equation 4.58.

The denominator of the third term of Equation 4.58 is the product of n and D_{nb}. When the injected minority carrier concentration in the base approaches that of the ionized base impurities then a deterioration of the built-in field of a graded base transistor will occur leading to a reduction of n. However, under these conditions D_{nb}, which is a suitably averaged diffusivity of minority carriers in the base, will be effectively doubled owing to the fact that the electric field gradient in the base will be determined by the injected carriers rather than the fixed ionized impurities. The critical current for the transition of the base transit time from the low to the high injection value is a complication function of the built-in field. It has been shown[38] for an exponentially graded base impurity profile that the overall effect of the transition from low to high injection level conditions should be that the $n\, D_{nb}$ product remains essentially the same. Thus in Equation 4.58 the same values of n and D_{nb} can be used both at moderate and high injection levels.

By far the most important change of f_T at high current levels is the reduction brought about by the modification of the emitter and collector depletion layer boundaries by the charge in transit through them. This effect produces changes in C_{Te}, C_{Tc}, W, x_d and, possibly, r_{sc}. The greatest effect on f_T is brought about by an increase of the effective base width, W. This is often referred to as the Kirk effect.[43]

In the case of an *n-p-n* transistor at high current levels the electrons in transit in the epitaxial part of the collector depletion layer will reduce the overall charge density, ρ, according to

$$\rho = e(N_{DC} - n) \tag{4.60}$$

where N_{DC} is the donor impurity concentration in the collector epitaxial layer (assumed constant). The electrons move across the space charge region primarily by drift and give rise to a current density

$$J_c = ev(x)n(x) \tag{4.61}$$

From (4.60) and (4.61) Poisson's equation may be written in the form

$$\frac{dE}{dx} = \frac{e}{\epsilon\epsilon_0}\left[N_{DC} - \frac{J_C}{e\,v(x)}\right] \tag{4.62}$$

This equation may be integrated, subject to the boundary condition that $E = E(0)$ at $x = 0$ and $v(x)$ has the constant value v_s, to give

$$E(x) = E(0) + \frac{e}{\epsilon\epsilon_0}\left[N_{DC} - \frac{J_C}{e\,v_s}\right]x \tag{4.63}$$

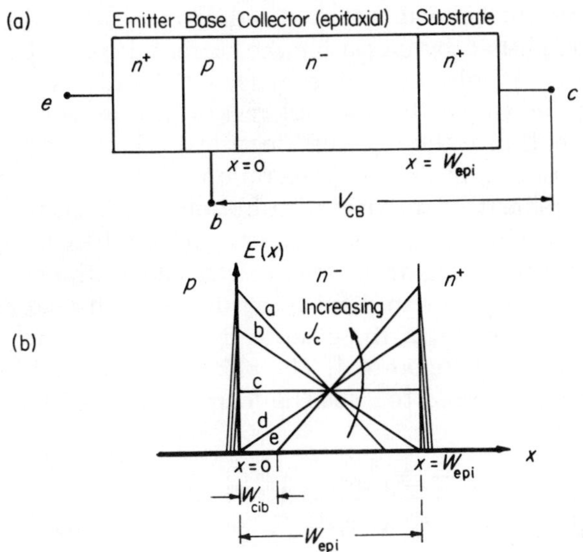

Figure 4.4 (a) *n-p-n* transistor model. (b) Field distribution in the collector epitaxial region as a function of collector current density for a fixed value of $|V_{CB}|$.[46] (Reproduced by permission of the Institute of Electrical and Electronic Engineers, Inc.)

The field distribution in the epitaxial region of the collector is linear, with a slope depending upon the current density, J_c. Figure 4.4 shows the field distribution as a function of current density for a fixed value of V_{CB}.[46] Curves (a) to (e) represent increasing values of J_c. Curve (a) is for low to moderate current levels. Curve (b) represents the situation when, with increasing current density, the collector edge of the depletion layer has just reached the substrate. Further increases in the current cause a reduction of the field at $x = 0$ until that field reaches a value approximately equal to zero (curve (d)). The critical current density required to create the conditions represented by curve (d) is given by

$$J_0 = ev_s \left[\frac{2\epsilon\epsilon_0 V_{CB}}{eW_{epi}^2} + N_{DC} \right] \tag{4.64}$$

Two models have been proposed to describe the base—collector space charge region at currents exceeding J_0, these are the one-dimensional model[43,47] and the two-dimensional model.[48]

(a) One-dimensional model
This model is based on the assumption that the current flow is strictly one-dimensional so that, because the area cannot be increased, an

increase of current beyond that depicted by curve (d) of Figure 4.4 can only be accomplished by increasing the current density. To achieve this large numbers of holes enter the collector in order to eliminate the space charge due to electrons in this region (just as holes are stored in the base to neutralize the injected electrons). These extra charges must be taken into account in the integration of Poisson's equation and increase the gradient of the field distribution; this leads to a decrease in the length over which majority carriers travel by drift in the collector. Part of the collector region is thus electrically equivalent to the base region, since the carriers in it move by diffusion instead of drift. The reduction in the length of the electric field region is shown schematically by curve (e) of Figure 4.4. The electrically neutral region of the epitaxial layer is referred to[46] as the 'current induced base', of width, W_{cib}, given by

$$W_{cib} = \left[1 - \left(\frac{I_0 - eN_{DC}v_sA_e}{I_c - eN_{DC}v_sA_e}\right)^{1/2}\right] W_{epi} \tag{4.65}$$

where $I_0 = J_0 A_e$ and A_e is the emitter area.

(b) Two-dimensional model

The basic assumption of this model is that the current density cannot exceed J_0 anywhere in the collector. This is equivalent to the assumption that holes cannot be stored in the collector and that in order to overcome the space charge limitation existing at J_0 it is necessary to allow two-dimensional effects to take place. That is, as the current increases the injected minority electrons in the base spread out laterally and are collected over a larger area of the collector, such that

$$A_c(\text{effective}) = A_e \frac{I_c}{I_0} \tag{4.66}$$

According to this model the limit on I_c would be reached when A_c(effective) is equal to the total collector area A_c. In practice this current can easily be exceeded and the real situation can probably be expressed by a combination of these two extreme models.

At high current density the charge in transit through the emitter junction depletion layer will cause the depletion layer to contract towards the emitter in a similar manner to the way in which the collector junction depletion layer moves towards the collector. However, in the case of the emitter, the movement of the depletion layer boundaries is relatively small because the doping levels in the emitter junction region are much greater than those in the collector. Archer[1] has given an expression for the forward-biased linearly-graded junction emitter capacitance which takes into account the free carriers in transit in the emitter junction transition region. This is called the emitter 'neutral' capacitance

$$C_{Ne} = \frac{2eV_l Q_b I_c}{a\,kT\,D_{nb}} \tag{4.67}$$

where V_l is the normalized voltage at the edge of the transition region, Q_b is the integrated net base doping and a is the doping gradient in the junction. At sufficiently high current levels C_{Ne} can dominate the space charge capacitance, C_{Te}. The associated time constant will be

$$\tau_{Ne} = r_e C_{Ne} = \frac{kT}{eI_e} C_{Ne} \simeq \frac{2V_l Q_b}{a\,D_{nb}} \tag{4.68}$$

The net effect of the movement of the depletion layer boundaries at high current levels is to produce changes in C_{Te}, C_{Tc}, W and x_d. Also, if the collector epitaxial layer was not fully depleted at low current levels, the movement of the collector depletion layer boundary to the interface between the epitaxial layer and the substrate will reduce r_{sc}.

Assuming,[46] for the one-dimensional model, that the total base transit time comprises the transit time across the metallurgical base W_{bo}, plus the transit time across the current induced base W_{cib}, then Equation 4.58 becomes, at high current level,

$$\tau_{ec} = \frac{x_{jeb}{}^2}{2D_{pe}\beta_0{}'} + r_e{}'(C_{Ne} + C_{Te}{}' + C_{Tc}{}') + \frac{W_{bo}{}^2}{n\,D_{nb}} + \frac{W_{cib}{}^2}{4D_{nc}}$$

$$+ \frac{W_{bo} + W_{cib}}{v_s} + \frac{W_{epi} - W_{cib}}{2v_s} + r_{sc}{}'\,C_{Tc}{}' \tag{4.69}$$

where the primed quantities refer to their high current values. D_{nc} is the diffusion coefficient of electrons in the induced base part of the collector epitaxial layer and the factor 4 is the effective value of n in this region.[46] Saturation of f_T with current should occur when the current induced base region occupies the entire epitaxial layer, so that further base widening cannot take place. The predicted saturation of f_T for thin epitaxial devices, at a value corresponding to an effective base width of $W_{bo} + W_{epi}$, was not observed for transistors measured by Whittier and Tremere.[46] However, Bowler and Lindholm[49] have suggested that this result may have been due to the operating conditions, since if V_{cb} is lower than a critical value, V_{crit}, f_T fall-off will occur because the transistor enters a saturation or quasi-saturation mode of operation,[50] whereas if V_{cb} is greater than V_{crit} fall-off occurs because the transistor enters a mode of operation associated with space charge limited flow.[43,47,48]

In the case of the two-dimensional model[48] Equation 4.58 becomes, at high current level

$$\tau_{ec} = \frac{x_{jeb}^2}{2D_{pe}\beta_0} + r_e'(C_{Ne} + C_{Te}' + C_{Tc}') + \frac{W_{bo}^2}{n\,D_{nb}} + \frac{(I_c/I_0 - 1)^2 s^2}{4n\,D_{nb}} + \frac{[W_{bo}^2 + \tfrac{1}{4}(s_c - s)^2]^{1/2}}{v_s} + \frac{W_{epi}}{2v_s} + r_{sc}'\,C_{Tc}' \qquad (4.70)$$

where s is the emitter width and s_c is the effective width of the collector over which the emitter current is collected. When $I_c = I_0$ then $s_c = s$ and Equation 4.70 reduces to the low current value given by Equation 4.58.

4.2.5 Practical limitations on frequency response

Semiconductor material

The maximum frequency response of bipolar transistors is determined practically by the technological limitations inherent in their construction. The first choice to be made is that of the semiconductor material itself. Eckton has shown[51] that at base widths less than 0.15 μm a silicon *n-p-n* transistor has a lower τ_{ec} than a germanium *p-n-p* transistor of identical geometry because of the narrower collector depletion layer width, x_d, required to support the same sustaining voltage. This is because the breakdown field in silicon is higher than that in germanium. The sustaining voltage, $V_{CE(sus)}$, is defined as the voltage at which the output impedance becomes zero when the base of the transistor is open-circuited. Silicon is therefore to be preferred to germanium on design theory grounds as well as having great practical advantages such as a native oxide eminently suitable for planar technology, better developed epitaxial growth techniques, a higher scattering limited electron velocity, lower dielectric constant and a higher thermal conductivity.

Gallium arsenide microwave bipolar transistors have yet to be made, because of technological difficulties associated with achieving the narrow base width and fine emitter dimensions required, and will not be considered further.

Silicon is the preferred semiconductor for a microwave bipolar transistor, furthermore, epitaxial *n*-type silicon of closely controlled doping level (±10%) is required on an n^+-type substrate. Since a microwave transistor is designed to operate under conditions of collector depletion layer reach-through to the substrate, the thickness of the epitaxial layer must be accurately controlled to within 0.5 μm. Suitable values of epitaxial layer doping level and thickness may be found from Figure 4.3. This thickness and doping level control must be maintained during the subsequent high temperature processing of the transistor structure, thus the dopant in the epitaxial substrate must be a relatively slow diffusant to maintain the integrity of the epitaxial layer and the abruptness of the doping gradient at the substrate layer interface. Because this interface forms a boundary, or is close to, the electrically active part of the transistor the crystal perfection at the interface must be of a high standard, otherwise localized crystal imperfections or impurity density fluctuations could cause electric field inhomogeneities, giving rise to low breakdown voltage microplasma sites. For this reason, in high reliability microwave transistors, it may be an advantage to grow an n^+ epitaxial 'buffer' layer in between the active n-type epitaxial layer and the n^+ substrate. This technique is employed in gallium arsenide field-effect transistors (see Section 4.6.2) and Gunn diodes (see Chapter 2). After fabrication of the transistor structure in the epitaxial layer the substrate should be thinned by lapping or chemically etching the back surface to reduce r_{sc} (Equation 4.53) and to reduce the thermal resistance.

Dielectric material

Apart from the silicon itself another electrically active substance used in the construction of a transistor is the dielectric material. This will generally be silicon dioxide, which is intrinsic to the planar method of transistor manufacture. The silicon dioxide is grown *in situ* by the oxidation of the silicon surface in an oxidizing atmosphere at a high temperature. Some of the silicon is incorporated in the SiO_2 and the thermal growth of an oxide will therefore cause a movement of the silicon surface. The thickness of silicon to the thickness of the oxide is in the ratio 0.45:1 therefore, since the oxide may be approximately 0.5 μm thick, about 0.2 μm of silicon will be consumed. This is about the same as some of the diffusion depths and would have to be allowed for in some microwave transistor designs. The dielectric material over the collector field area must be thick enough to minimize the capacitance between the bonding pad metal and the collector body, since this capacitance is a parasitic element in parallel with the active device. A bonding pad area of 600 μm^2 results in a feed through capacitance of 0.04 pF with 5,000 Å of SiO_2. The relatively small value of the bonding pad capacitance allows the pad area to be made sufficiently large for contact to be made to the device with wires that have a large diameter compared with the dimensions of the active part of the transistor, this enables contact to the device to be realized physically and the relatively large wire diameter helps to reduce the lead wire inductance which will improve the gain at microwave frequencies.

A quantitative treatment of the kinetics of silicon dioxide growth, the redistribution of impurities in silicon and the oxide during growth, and the use of silicon dioxide as a diffusion mask may be found in the book by Grove.[7] A review of the current understanding of charges existing in thermally oxidized silicon has recently been published.[52] Silicon dioxide plays a key part in silicon planar technology in its role as a diffusion mask and is electrically important in microwave transistors as a dielectric medium and to stabilize and passify the silicon surface.

Pattern definition

In a microwave transistor it is important that the emitter width, the emitter-to-base contact spacing and the collector area, i.e. the transistor 'pattern', should be defined with the utmost precision. Early[15] defined a figure of merit for a high frequency transistor in terms of the minimum width attainable in this pattern. Early's figure of merit, K, for an interdigital transistor pattern, was expressed in terms of the emitter strip width, s

$$K \simeq \frac{40}{s} \text{ GHz} \tag{4.71}$$

where K, the (power gain)$^{1/2}$ (bandwidth) product, is equivalent to f_{max}, the maximum frequency of oscillation of the transistor, and s is in micrometres. This calculated figure of merit for a silicon transistor, which assumes the emitter-to-base contact spacing, t, to be equal to $s/2$, was rather optimistic and a more practical value of f_{max} was deduced by Lathrop[53] from measurements made on several silicon microwave transistor types with different values of s and t (s and t are defined in Figure 4.1). This[53] figure of merit is

$$K = f_{max} \simeq \frac{40}{s + 2t} \text{ GHz} \tag{4.72}$$

where s and t are in micrometres. Recent improvements in transistor technology probably allow this figure of merit to be increased by a factor of about three.

Using light-optical definition of the transistor pattern in photoresist a linewidth and spacing of about one micrometre may be achieved. Figure 4.5 is a photograph of a transistor with one micrometre wide emitters. This represents the maximum resolution at present obtainable using light-optical methods. To define dimensions of less than one micrometre resort must be made to electron-optical techniques when, because of the much smaller wavelength of electrons, it is feasible to produce patterns with 0.1 μm linewidth and spacing. This very fine resolution cannot be maintained during subsequent transistor processing, but nevertheless it is possible to make transistors with lateral dimensions of less than one micrometre which should enable their

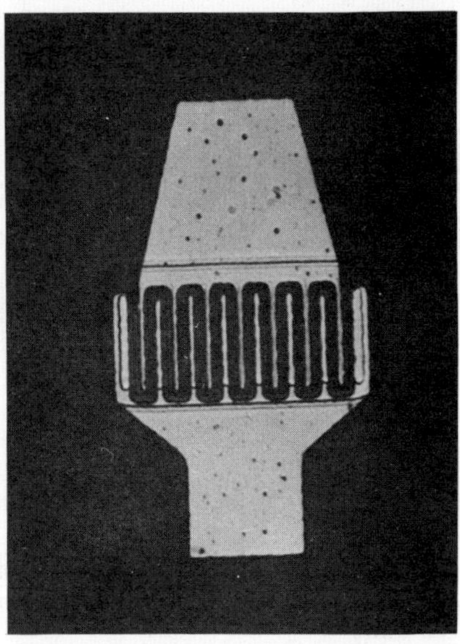

Figure 4.5 Low-noise, small-signal, microwave silicon bipolar interdigitated transistor, with 1 μm emitter width

useful electrical performance to be extended into the X-band frequency range.

Diffusion

The use of diffusion as a means of doping semiconductors with closely controlled amounts of impurity, in precise regions defined by oxide masking, first enabled the fine dimensions necessary for the microwave operation of transistors to be obtained. The subject of diffusion in semiconductors has been comprehensively treated in many textbooks, for example Boltaks[54] and Burger and Donovan.[55] In this chapter, only those aspects of diffusion that are of particular relevance to microwave transistors will be considered.

An important phenomenon in this respect is the enhanced diffusion of the base impurity underneath the emitter. This has been referred to as the emitter-dip effect, the base push-out effect, or base run-on. The experimental observations of the different aspects of this effect have been reviewed by Willoughby,[56] who concluded that a complete explanation of all of the observed effects was not available.

The emitter-dip effect is important in the fabrication of microwave transistors because it can limit the minimum base width attainable and this is a very important parameter in determining the frequency response of the transistor, see Equation 4.58. An example of the

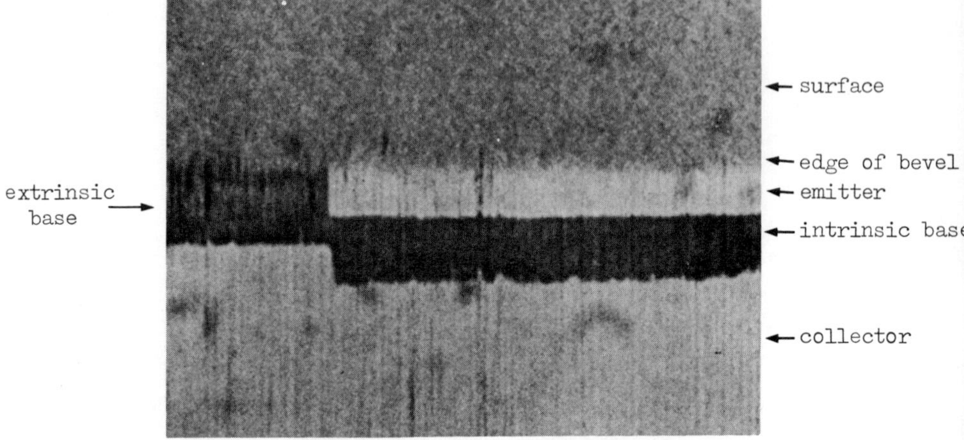

Figure 4.6 Emitter-dip effect in silicon *n-p-n* transistor. Emitter depth = 0.3 μm; base width = 0.5 μm; extrinsic base depth = 0.5 μm.

emitter-dip effect is shown in the stained and bevelled section photograph of Figure 4.6. In this case the emitter depth was 0.3 μm and the base, which would have been 0.2 μm thick under the emitter in the absence of the emitter-dip effect, had been increased to 0.5 μm. Clearly the frequency response of a transistor, made under diffusion conditions such as this, would have been curtailed.

The emitter-dip effect is most pronounced in silicon *n-p-n* transistors with phosphorus doped emitters. The enhancement of the base impurity (boron) diffusion underneath the emitter is thought to occur because the tetrahedral covalent radii of the phosphorus dopant atoms are significantly less than the radii of the silicon host crystal atoms. When a critical amount of misfit between dopant and host crystal atoms occurs a dislocation network is created in the doped layer. If the Burgers vectors of these dislocations lie in the surface plane any subsequent propagation of the dislocations into the interior of the crystal during further heat treatment must occur by a climb mechanism, which generates vacancies. The vacancy concentration in the vicinity of the emitter will be supersaturated and the diffusion of the boron atoms in this region will be enhanced. The diffusion coefficient of boron, assuming a substitutional vacancy diffusion process, will be given by

$$D = KC_{\mathrm{v}} \exp - \left(\frac{E_{\mathrm{i}} + E_{\mathrm{v}}}{kT} \right) \tag{4.73}$$

where K is a constant related to the lattice constant of silicon, C_{v} is the local vacancy concentration, E_{i} is the activation energy of migration of the diffusant atoms from a substitutional site to a vacant site, E_v is the activation energy of vacancy formation, k is Boltzmann's constant and T is the absolute temperature. Since the vacancy concentration, C_v, will be enhanced in the vicinity of the emitter because of the non-conservative motion of the diffusion induced dislocations, the boron diffusion coefficient underneath the emitter will also be enhanced.

The emitter-dip effect may be reduced by reducing the phosphorus concentration in the emitter and by reducing the emitter diffusion temperature or, more effectively, by changing the emitter dopant to arsenic. The covalent radius of arsenic is close to that of silicon, therefore no misfit dislocation networks form during arsenic diffusion. Some prismatic dislocation loops are formed in the emitter but they do not have Burgers vectors in the surface plane. These dislocations can propagate into the crystal by glide and do not, therefore, generate vacancies.

The lack of an emitter-dip effect is so complete in the case of an arsenic diffused emitter that the converse effect, the retarded base effect, can actually occur. The phenomenon has been treated quantitatively by R. B. Fair[5 7] who attributes the retarded base effect, in this case, to the formation of $[V_{Si}\ As_2]$ complexes, which cause vacancies to be removed from solution in the silicon lattice according to the equilibrium reaction

$$V_{Si} + 2\ As^{+} + 2\ e^{-1} \rightleftharpoons V_{Si}\ As_2 \qquad (4.74)$$

V_{Si} represents a vacancy in the silicon crystal lattice and the $[V_{Si}\ As_2]$ complex is a vacancy with two arsenic atoms and two silicon atoms as its tetrahedral neighbours. Vacancy undersaturation occurs in the emitter owing to the $[V_{Si}\ As_2]$ complex formation; thus the emitter acts as a sink for vacancies which diffuse in from the silicon region within a vacancy diffusion length of the emitter. The effective boron diffusivity in the vicinity of the emitter will be lowered by a factor that is equal to the degree of vacancy undersaturation, since C_v in Equation 4.73 will be reduced. The reduced boron diffusivity produces a retardation of the base region underneath the emitter.

A retardation of the base can also be produced if the sequence of the base and emitter diffusion is reversed, so that the emitter is diffused first and the base is then diffused through the emitter. In this case the retardation of the base dopant atoms is due to the built-in electric field of the ionized emitter impurities.

If it can be controlled accurately a retarded base structure should be better for a microwave transistor than a structure exhibiting the emitter-dip effect, because the intrinsic base width can be reduced to a minimum to decrease τ_{ec} and the extrinsic base can be made wider to reduce the base resistance. This should give rise to a better figure of merit (Equation 4.2).

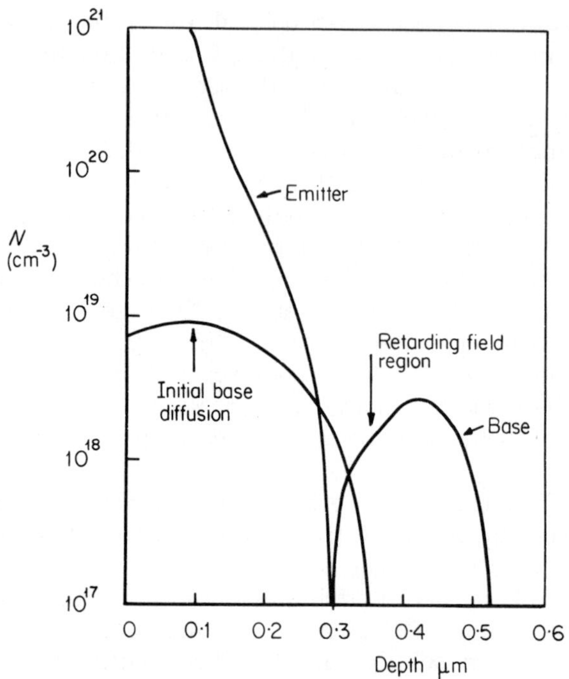

Figure 4.7 Retarding field region in the base of a phosphorus diffused emitter *n-p-n* silicon bipolar transistor

Apart from complex formation, other interactions taking place during diffusion are ion pairing, vacancy clustering or divacancy formation, and the effect of internal electric fields. Fair[57] has shown that these other interactions influence the emitter and base concentration profiles within approximately 0.1 μm of the emitter—base junction, producing a narrow dip in the base concentration profile coincident with the junction. The base doping concentration profile will have both a positive and a negative gradient, as shown by the measured profile in Figure 4.7 for a phosphorus emitter and a boron base. Part of the base will be a retarding field region for the injected electrons in transit and this will reduce the drift-field factor, n, in Equation 4.58.

Figure 4.7 shows the well known[58] kink obtained in shallow phosphorus diffusion profiles in silicon. The impurity concentration gradient through the emitter—base junction is reduced because of the faster diffusing, lower concentration, component of the phosphorus and this will increase the emitter neutral capacitance, C_{Ne} (see Equation 4.67).

In the case of shallow arsenic diffusion Fair and Weber[59] have shown

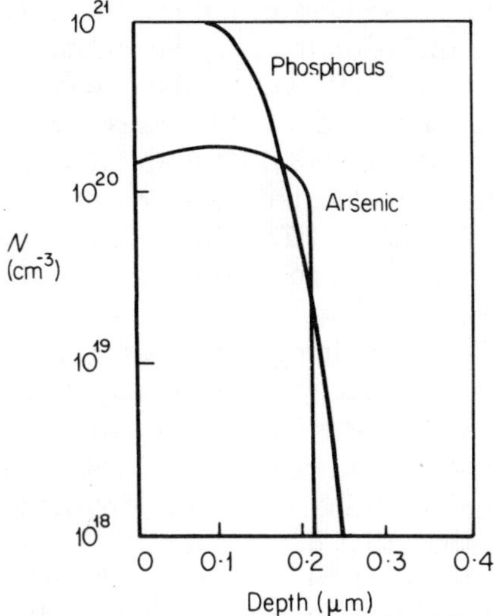

Figure 4.8 Emitter concentration profiles

that the effective diffusivity of arsenic will be concentration dependent. This result is based on the assumption that monatomic As^+ diffuses substitutionally while inactive $[V_{si}\ As_2]$ complexes are forming to reduce the flux of diffusable ions. The complexes are assumed to remain stationary in the silicon. This will make the arsenic concentration gradient very steep in the emitter junction region. Measured arsenic and phosphorus emitter concentration profiles are compared in Figure 4.8. The steep arsenic gradient will decrease C_{Ne} and should give a good emitter efficiency which will reduce τ_e (Equation 4.24), provided the doping level is not too great in the immediate vicinity of the emitter junction. For these reasons, and because of the absence of the emitter-dip effect, arsenic is preferred to phosphorus as the emitter dopant for *C*-band transistors.

Ion-implantation

Ion-implantation of impurities into semiconductors is capable of producing doped regions of accurately controlled total impurity concentration, penetration depth and profile shape.[60] In these respects it should be better theoretically than diffusion for producing the emitter and base regions of a microwave transistor. Very low noise C to X-band microwave transistors have been produced with a diffused arsenic emitter and an ion-implanted boron intrinsic base,[1] and with an ion-implanted arsenic emitter and an ion-implanted boron *base.*[1,61] The all-implanted transistors were found to have very reproducible electrical characteristics.[61]

Phosphorus is not suitable as an ion-implanted emitter dopant since it has an anomalous diffusion tail at a concentration below 10^{19} atoms cm^{-3}. The anomalous tail is thought to be due to an increased diffusion coefficient owing to vacancy formation caused by the bombarding ion beam. The enhanced phosphorus diffusion is analogous to the enhanced boron diffusion caused by an increased vacancy concentration in the emitter-dip effect (Equation 4.73). Ion-implanted arsenic also has an anomalous diffusion tail but to a lesser extent than phosphorus and only below a concentration of 5×10^{17} atoms cm^{-3}. Payne and coworkers[61] found that ion-implanted arsenic produced a good emitter junction if it was allowed to diffuse about 1,000 Å beyond the bombardment damaged region. Boron does not produce an anomalous diffusion tail because it is a light ion and its nuclear stopping power, which is a measure of its ability to produce interstitials and vacancies, is very much smaller than that of phosphorus. Thus, in losing energy to the silicon target, fewer vacancies are produced and the anomalous diffusion is greatly reduced. It is also possible that the formation of boron-vacancy complexes reduces the diffusion of vacancies into the bulk material.

Ion-implantation is a versatile method of producing the base regions of microwave transistors. For example, the impurity profile in the intrinsic base region may be shaped by using a double-implantation, that is two implants at different bombarding energy so that the peak of the profile occurs at a slightly different depth for each of them. By adjusting their overlap and the height of each peak the net base concentration and profile shape can be adjusted. In this way a more favourable value for the drift-field factor, n, in the base region may be obtained. Another illustration of ion-implantation to modify the base region is shown in Figure 4.9. In this case the intrinsic and extrinsic base regions have been implanted separately, so that the intrinsic base can be optimized to give the highest value of f_T and the extrinsic base can be optimized to give the lowest value of base resistance, r_b. A junction structure like that shown in Figure 4.9 has obvious advantages in optimizing the intrinsic and extrinsic base regions compared with the junction structure formed by the emitter-dip effect, shown in Figure 4.6. However, a junction structure similar to that shown in Figure 4.9 can also be produced by diffusion.

Contact metal

The contact metal should make a low resistance contact with the silicon and should be metallurgically stable and reliable when in contact with silicon over long periods at elevated temperature and when carrying a high current density. It should also have similar metallurgical, electrical and reliability properties with the bond wire. In addition to this it

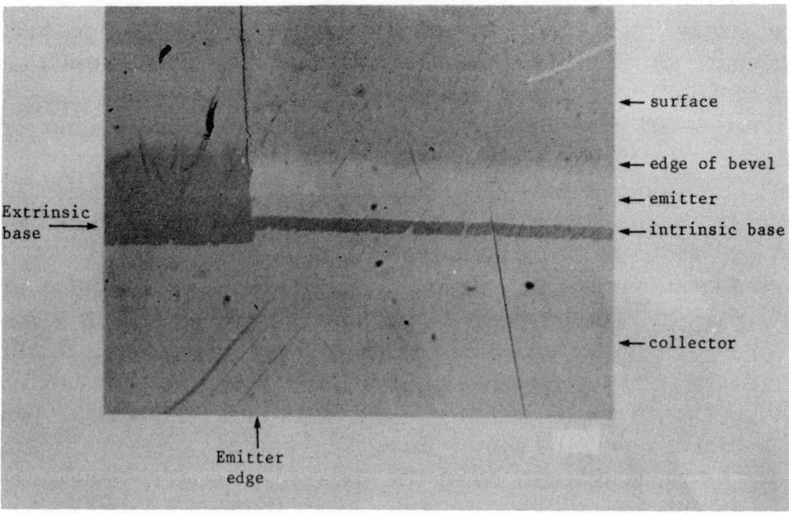

Figure 4.9 Retarded base effect in silicon *n-p-n* transistor produced by ion-implanting extrinsic and intrinsic base regions separately. Emitter depth = 0.3 μm; base width = 0.15 μm; extrinsic base depth = 0.65 μm

should be capable of definition, by chemical or sputter-etching, into very fine pattern dimensions and should not dissolve too much silicon when the contact is alloyed.

It is difficult to find one metal which possesses all of these diverse properties, but aluminium has proved to be reasonably satisfactory over a number of years. However, with reducing dimensions, recent microwave transistor designs have placed more stringent demands on the contact metal in terms of higher current density and reduced alloy penetration depth, so that it is becoming necessary to use an alloy of aluminium with about 2% of silicon, to reduce alloy penetration depth; or an Al-Si-Cu alloy which will also improve current carrying capability. An alternative to this is to use a metal system with gold as the main current carrying element. Unfortunately gold alloys readily with silicon and a barrier-metal must be interposed between them. Titanium, platinum, tungsten and molybdenum have been used as barrier metals, sometimes with platinum—silicide, nickel or chromium underneath to improve the electrical contact to the silicon.

A gold metal system can be contacted reliably with a gold lead wire and an aluminium system with an aluminium lead wire. The contact resistance of these metals to the *p*-type base region, of surface concentration about 5×10^{18} cm^{-3}, is still barely adequate and may influence the high frequency performance as discussed in Section 4.2.1.

Package design
The silicon transistor chip must be encapsulated in a package (or header) to provide it with mechanical stability, a stabilized environment and to facilitate handling and connection to the external circuit. The package is an appendage to the chip which contributes inescapable parasitic capacitance and inductance to the transistor's equivalent circuit at microwave frequencies. Microwave transistor packages are designed to minimize these parasitic elements, which degrade the gain and bandwidth of the intrinsic device, as much as possible.

At C-band and above the trend is towards microstrip circuitry and an ideal package will be designed so that it forms a low-loss transition between the microstrip circuit and the transistor chip. The best solution electrically is to mount the chip directly on the microstrip circuit, but this means that the chip has the minimum protection from its environment unless the whole circuit is hermetically sealed.

The physical dimensions of the package determine the inductance and capacitance values of its equivalent circuit, in general the smaller the package the smaller will be the values of its electrical parameters. A minimum size is imposed by the hermeticity requirement, since as the package is made smaller it becomes more difficult to make a reliable hermetic seal. Three sealing technologies are in general use: glass-to-metal, ceramic-to-metal, and plastics. The first two make reliable hermetic seals, but are expensive, plastics seals are economical but in general are non-hermetic and cause some degradation of microwave properties.

Microwave transistors are generally used in the common-emitter connection, for improved electrical stability, and in this configuration it is important to minimize the base-to-collector parasitic capacitance, since this gives rise to negative feedback in the circuit, and to reduce the emitter lead inductance to improve the stability and gain. Most circuit applications require that the collector be isolated from the ground plane, which means that the package must contain an insulator. In addition, microwave power transistors need a low thermal resistance path from the chip to the external heat sink. Beryllium oxide (BeO) is used in power transistor packages, since it is a good electrical insulator and a good thermal conductor, whereas alumina (Al_2O_3) is used in low-power transistor packages, for economic reasons, because it is a good electrical insulator but a relatively poor thermal conductor.

4.2.6 Electrical characterization

It is possible to define several characteristic frequencies, or figures of merit, for a microwave transistor. These may be specified in terms of the generalized scattering or *s*-parameters[62] or in terms of the device and package parameters expressed as lumped L, C and R components.

As the frequency approaches X-band the lumped component approach becomes less meaningful than the s-parameter method.

Scattering parameters

The s-parameters of a microwave transistor are defined in terms of the input and output currents, i_1 and i_2, the input and output impedances, Z_{in} and Z_{out}, and the positive, impedance Z_0, of a lossless transmission line system connected to the input and output ports of the transistor.[19]

s_{11} = input reflection coefficient with the output terminated in a matched load ($Z_L = Z_0$)

$$s_{11} = \frac{Z_{in} - Z_0}{Z_{in} + Z_0}$$

s_{21} = forward insertion gain with the output terminated in a matched load ($Z_L = Z_0$)

$$s_{21} = -\gamma_f \frac{i_2}{i_1}$$

s_{12} = reverse insertion gain with the input terminated in a matched load ($Z_s = Z_0$)

$$s_{12} = -\gamma_r \frac{i_1}{i_2}$$

s_{22} = output reflection coefficient with the input terminated in a matched load ($Z_s = Z_0$)

$$s_{22} = \frac{Z_{out} - Z_0}{Z_{out} + Z_0}$$

where $\gamma_f = \dfrac{2Z_0}{Z_{in} + Z_0}$ and $\gamma_r = \dfrac{2Z_0}{Z_{out} + Z_0}$

are the forward and reverse current transmission factors of the microwave transistor.

Transducer gain

For a common emitter microwave transistor the transducer gain is given by

$$G_T = |s_{21}|^2 \tag{4.75}$$

The frequency at which G_T is unity is given by

$$\omega_{trans} = \omega'_T \sqrt{\gamma_f} \tag{4.76}$$

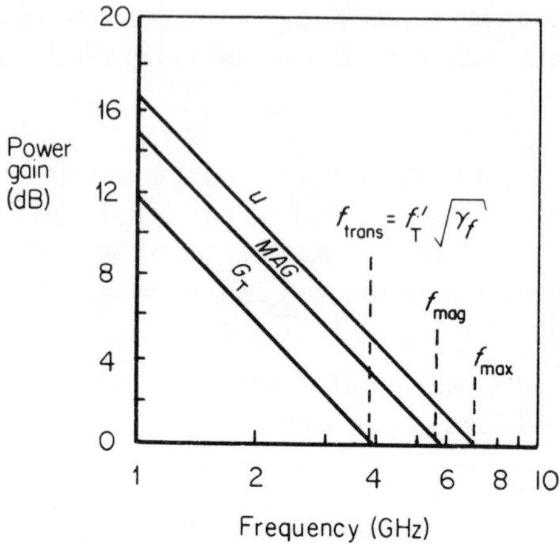

Figure 4.10 Transducer power gain, G_T, maximum available gain, MAG, and unilateral gain, U, for a silicon microwave n-p-n transistor.[19] (Reproduced by permission of Pergamon Press Ltd.)

where ω'_T, the reduced current-gain bandwidth product, is

$$\omega'_T = \frac{\omega_T}{\dfrac{1 + \omega_T C_c Z_0}{\alpha_0}} \tag{4.77}$$

where C_c is the total collector capacitance, α_0 is the low frequency common-base current gain and $\omega_T = 2\pi f_T$. In low power devices $\gamma_f \simeq 1$ and $\omega_T{}' \simeq \omega_T$, therefore $\omega_{trans} \simeq 2\pi f_T$, and the unity transducer gain frequency can be used as a measure of f_T. Figure 4.10 shows the variation of G_T with frequency for a common-emitter microwave transistor.

Maximum available power gain

There is considerable power gain at the frequency f_T, for which the grounded-emitter current gain is unity, because the ratio of the output impedance to the input impedance is greater than unity. Therefore f_T does not represent the uppermost frequency of useful operation of the transistor. The uppermost frequency of useful operation is more meaningfully described as the maximum frequency at which power amplification can be obtained. A practical way of calculating this frequency is to derive the maximum available power gain, MAG, with the idealized assumption of zero reverse insertion gain, s_{12}, while retaining the effects of internal feedback in the other s-parameters. The

MAG may be expressed in terms of the s-parameters

$$MAG = \frac{|s_{21}|^2}{(1-|s_{11}|^2)(1-|s_{22}|^2)} \tag{4.78}$$

or, in terms of the device and package parameters,

$$MAG \cong \left(\frac{f_{MAG}}{f}\right)^2 \tag{4.79}$$

where

$$f_{MAG} = \left\{\frac{\alpha_0 f_T}{8\pi(r_b + r_e + r_{se} + \omega_T L_e)C_c}\right\}^{1/2} \tag{4.80}$$

where L_e is the emitter lead wire inductance.

Equation 4.79 shows that the maximum available gain falls off at a rate of 6 dB/octave with frequency. The frequency f_{MAG}, at which $MAG = 1$, is given by Equation 4.80, which demonstrates the effect of the feedback elements r_e, r_{se} and L_e in the common-emitter branch of the equivalent circuit.

Maximum unilateral power gain

Another figure of merit for the microwave transistor is the maximum junilateral power gain, U.[63] The unilateral gain is independent of header reactances and common lead feedback elements; it is therefore a unique measure of the intrinsic device performance. The U represents the forward power gain in a feedback amplifier when the reverse power gain has been set to zero by adjustment of a lossless reciprocal feedback network around the transistor. U may be written in the form[63]

$$U \cong \left(\frac{f_{max}}{f}\right)^2 \tag{4.81}$$

where

$$f_{max} = \left(\frac{\alpha_0 f_T}{8\pi r_b C_c}\right)^{1/2} \tag{4.82}$$

f_{max} is the maximum frequency of oscillation, derived by letting the power gain fall to unity in Equation 4.1. Strictly speaking Equation 4.1 should have included α_0, but it is often assumed that $\alpha_0 = 1$.

Equation 4.81 shows that U falls with frequency at the rate of 6 dB/octave. Equations 4.81 and 4.82 may be combined to give

$$U = \frac{\alpha_0 f_T}{8\pi f^2 r_b C_c} \tag{4.83}$$

Assuming that the transistor emitter and base electrode pattern is defined in terms of electrode width and spacing, s, and length, l, and using r_0 and C_0, the base resistance and collector capacitance per unit area, then $r_b \propto r_0 s/l$ and $C_c \propto C_0, sl$. Substituting these values for r_b and C_c in Equation 4.82 gives

$$f_{max} \propto \frac{1}{s}\left(\frac{\alpha_0 f_T}{8\pi r_0 C_0}\right)^{1/2} \tag{4.84}$$

This equation is of the same form as Early's figure of merit (Equation 4.71). Both unilateral gain and maximum frequency of oscillation increase with decreasing s. This is the reason that the emitter strip width is one of the critical dimensions in a microwave transistor.

Relative values of f_{trans}, f_{MAG} and f_{max} are shown in Figure 4.10. In a transistor amplifier the feedback cannot be set to zero over a wide bandwidth and therefore the maximum available gain, MAG, is a more practical measure of power gain than the unilateral gain, U. In practice departures from the idealized 6 dB/octave slope occur, which may be attributed to second order frequency terms neglected in the calculation of MAG and U.

Stability

The transistor is not unconditionally stable over the complete frequency range up to f_{max}. Between certain frequencies it will oscillate in combination with a passive load and source admittance with no external feedback. The condition for unconditional stability (i.e. in the absence of external feedback, a passive load or source impedance will not cause oscillation) is that the Linvill[64] critical factor, C, should be less than unity. The C factor may be expressed in terms of the s-parameters

$$C = \frac{2\,|s_{12}s_{21}|}{1 + |s_{11}s_{22} - s_{12}s_{21}|^2 - |s_{11}|^2 - |s_{22}|^2} \tag{4.85}$$

In addition to $C < 1$, for complete or unconditional stability both $|s_{11}|$ and $|s_{22}|$ should be less than unity. If $C > 1$ the transistor is potentially unstable and may oscillate under certain conditions of passive load and source impedance.

The subject of stability has been discussed by Pritchard,[65] who showed that the frequency band over which the transistor is potentially unstable is narrower in the case of the common-emitter configuration than for the common-base or common-collector configuration.

4.2.7 Equivalent circuit

The equivalent circuit of a microwave transistor chip can be considered to be made up of a combination of R-C networks and a current generator. The elements of the network will be the resistors and capacitors, already discussed in the text, which represent various parts of the structure of the transistor. A new capacitance, the diffusion capacitance, C_D is introduced to represent the sum of the product of r_e with the radian frequency terms ω_e, ω_b and ω_d representing the time constants associated with the emitter base and collector depletion region. Also introduced are C_{bc} (pad), C_{ec}(pad), r_{bc}(pad) and r_{ec}(pad), which are the base and emitter bonding pad capacitances, inversely dependent upon field oxide thickness, and their associated series resistances, which are dependent upon collector epitaxial layer and substrate thicknesses and doping levels. Figure 4.11 is an attempt to give an electrical network representation of these physical parameters, it may give some insight to the device designer of their relative importance and their relationship with each other.

A simplified version of the equivalent circuit of Figure 4.11 is given in Figure 4.12. Figure 4.12 may be more useful to the circuit designer as a model for computer analysis of the transistor. In the simplified circuit the base resistance, including contact resistance, is shown in lumped form and all the capacitive elements in parallel with r_e are lumped together as C_e. C_e is equal to $1/\omega_T r_e$, where ω_T is the measured value of the bandwidth, and therefore includes C_{Te}, C_{Ne}, C_D and C_{ec}(pad). The collector capacitance is shown as an active

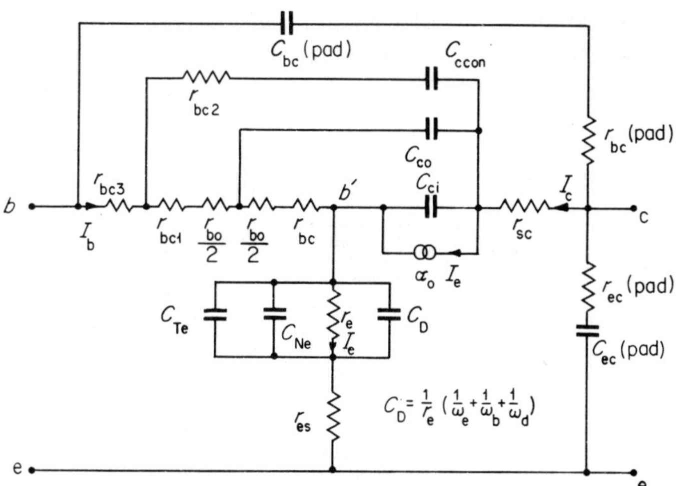

Figure 4.11 Common emitter *T*-equivalent circuit for microwave bipolar transistor chip

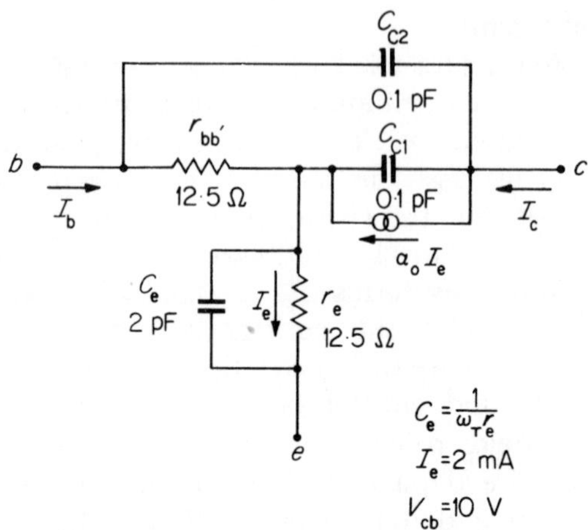

Figure 4.12 Simplified common emitter *T*-equivalent circuit for small-signal, low-noise, microwave bipolar transistor chip

capacitance C_{c1} and a parasitic capacitance C_{c2}, C_{c2} also includes the pad capacitance C_{bc}(pad). The emitter and collector series resistances have both been ignored.

Figure 4.13 shows the equivalent network of the transistor package surrounding the chip. The values shown are relevant for an alumina-insulated strip-line package, designed for small-signal use. Actual values will depend upon the particular package. The network now includes inductances of the transistor leads, as well as the distributed capacitance of the alumina insulator and various resistances which are attributed to a combination of contact resistance and the resistance of

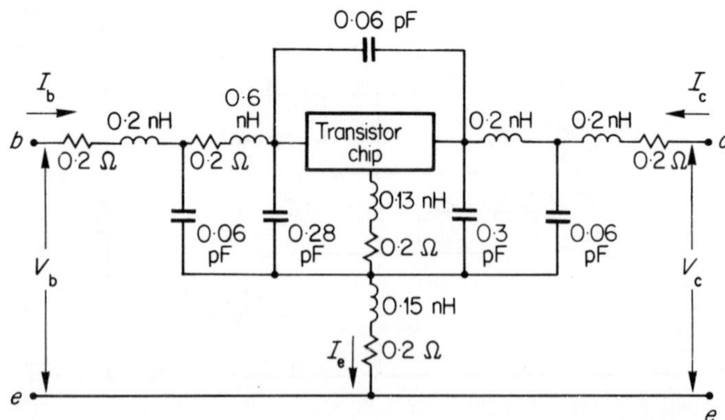

Figure 4.13 Equivalent network for transistor chip of Figure 4.12 in a small-signal, alumina insulated, strip-line package

the lead wires themselves. At microwave frequencies the effect of these resistances will be small compared with the reactance of the lead inductance and can generally be ignored.

4.3 FIELD-EFFECT TRANSISTORS

Field-effect transistors (FETs) bear a superficial similarity to bipolar transistors in their electrode arrangement, since a bipolar transistor consists basically of a long narrow emitter with a long narrow base closely spaced on either side, while a field-effect transistor consists of a long narrow gate with a long narrow source closely spaced on one side and a long narrow drain closely spaced on the other side. The analogy can be extended to their high frequency properties because a high frequency figure of merit, f_{max}, for a bipolar transistor can be shown to be inversely proportional to the emitter strip width s (Equation 4.71), whereas f_{max} for a field-effect transistor is inversely proportional to the narrow dimension, L, of the gate electrode (Equation 4.112). Both types of device are therefore critically dependent for microwave performance on the ability to define electrodes with high geometrical precision on the surface of the semiconductor material.

Three field-effect transistor structures are possible, viz. insulated gate, *p-n* junction gate and Schottky barrier gate. Insulated gate types have not yet achieved useful electrical performance in the microwave frequency range and will not be considered here. Microwave *p-n* junction gate and Schottky barrier gate FETs have been made in both silicon[66] and gallium arsenide.[67] The best microwave performance has been obtained from gallium arsenide devices with an *n*-type channel and a metal Schottky barrier gate on a semi-insulating substrate; a structure proposed by C. A. Mead.[68] A diagram of a field-effect transistor with this type of construction is shown in Figure 4.14.

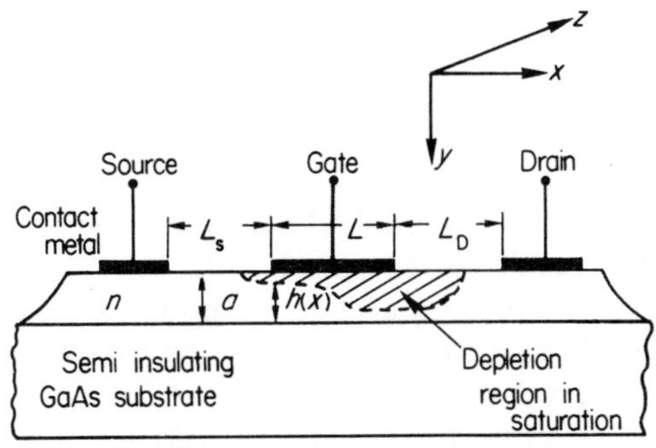

Figure 4.14 Cross-section of Schottky barrier gate GaAs field-effect transistor.[71] (Reproduced by permission of Pergamon Press Ltd.)

4.3.1 Theory of operation

In the FET of Figure 4.14 the source and drain electrodes are biased so that an electron current flows in the n-type epitaxial layer from the source, through the channel beneath the gate, to the drain electrode. The current in the channel induces a voltage drop along its length with the consequence that the Schottky barrier gate electrode becomes progressively more reverse-biased towards its drain end. This will cause a charge-depletion region to be set up in the channel to support the voltage, as required by Poisson's equation. Even with zero-voltage on the gate the charge-depletion region may extend at the drain end almost to pinch-off the channel against the semi-insulating substrate. At pinch-off the current saturates and then remains almost constant for further increase of drain voltage.

If the gate electrode is progressively reverse-biased the depletion region width will be increased in the non-pinched-off part of the channel, and towards the drain electrode. The decrease of the channel width in the non-pinched-off region will increase the channel resistance so that the drain current will be modulated by the gate voltage, giving rise to a transconductance, $g_m = dI_d/dV_g$. The family of curves of I_d versus V_d, with V_g as parameter, will resemble the family of I_c versus V_c curves, with I_e as parameter, of the bipolar transistor.

When the drain current has reached saturation the width, h, of the channel in the saturation region at the drain end will be of the order of a Debye length, L_D, since it can no longer be assumed that the transition between the channel and the depletion region is abrupt.[69] The Debye length is given by

$$L_D = \left(\frac{\epsilon\epsilon_0 kT}{e^2 N}\right)^{1/2} \tag{4.86}$$

and it can be shown that[69]

$$\frac{L_D}{a} = \frac{kT}{2eV_p}^{1/2} \tag{4.87}$$

where V_p is the voltage required to pinch-off the channel. Therefore $L_D \ll a$ for normal pinch-off voltages.

After the channel has been pinched-off a further increase in the channel-to-gate voltage results in an exponential decrease of the carrier concentration in the channel, together with an increase of carrier velocity until the remaining carriers have reached the scattering limited drift velocity, v_s. Increased drain voltage then increases the length of the velocity saturated region with almost no further change in carrier concentration in the channel. The source-to-drain current—voltage characteristic thus consists of a region where the current increases with voltage, followed by a soft knee region where the channel conductance decreases exponentially with voltage and then a current saturated plateau region.

For microwave field-effect transistors, which have very short channel lengths, velocity saturation occurs in the channel before the minimum channel width L_D is reached. The channel width at the constriction at the onset of velocity saturation is given by[70] $a(1-u)$ where

$$u^2 = \frac{V_d - V_g}{V_p} \tag{4.88}$$

V_d is the voltage between the drain end of the channel and the source. The current flowing through the constriction is therefore

$$I_m = I_0(1 - u_m) \tag{4.89}$$

where the subscript m refers to saturation and

$$I_0 = eN\, v_s\, a\, Z \tag{4.90}$$

I_0 is related to I_p, the saturation current for the non-saturated velocity case with $V_g = 0$, by the relation[71]

$$I_0 = \frac{3I_p}{z} \tag{4.91}$$

where

$$I_p = \frac{V_p eN\mu aZ}{3L} \tag{4.92}$$

and

$$z = \frac{\mu_0 V_p}{v_s L} \tag{4.93}$$

μ_0 is the low-field mobility and Z is the long dimension of the gate electrode (Figure 4.14).

In the current-saturated plateau region of the I_d—V_d characteristic the incremental drain resistance is not infinite, especially for short channel length microwave FETs. This is because[69] increased positive space charge introduced into the gate depletion layer as it extends towards the drain with increasing drain voltage will modify the edge of the depletion layer next to the channel to the left of the constricted region in Figure 4.14. This change of the channel dimensions will produce an increase of drain current, giving rise to a finite drain resistance after the knee in the I_d—V_d curve.

In Shockley's original analysis of the field-effect transistor[72] that part of the channel to the left of the constricted, or pinched-off, region, and the constricted region itself (both shown in Figure 4.14), were considered separately. In order to simplify the analysis several assumptions were made, some of which are not valid for microwave field-effect transistors with small length-to-width ratio, L/a, for the channel. The most important assumption of the Shockley theory, for devices with large L/a ratio, is the so-called 'gradual approximation'. In a field-effect transistor a voltage gradient will exist in the x-direction in the channel because of the electron current flowing from source to drain, at the same time a voltage gradient will exist in the y-direction in the gate depletion layer. The gradual approximation assumes that the x-component of the field in the depletion layer and the y-component of the field in the channel are quite small and may be neglected. This assumption enables the voltage across the gate depletion layer to be calculated from Poisson's equation, for the charge density in the depletion layer, as a function of y only, i.e.

$$\frac{d^2 V}{dy^2} = -\frac{Ne}{\epsilon\epsilon_0} \tag{4.94}$$

Since the real situation is essentially two-dimensional, Equation 4.94 implies that

$$\left|\frac{d^2 V}{dx^2}\right| \ll \frac{Ne}{\epsilon\epsilon_0} \tag{4.95}$$

The gradual approximation fails when the gate depletion layer width approaches zero, when the pinch-off condition is approached, and when $L/a \lesssim 5$.[73]

Two-dimensional solutions of Poisson's equation, with drift velocity saturation of carriers in the channel, have been obtained by computer solution for small channel length FETs[73] and the validity of the gradual channel approximation with drift velocity saturation, has been examined by Lehovec and Seeley.[74] They show that a boundary condition of the gradual channel approximation, namely that the field normal to the channel boundary is zero, can be satisfied by constructing a modified channel boundary with a reduced depletion layer width along the channel.

4.3.2 Electrical characteristics

The electrical characteristics of FETs have been described by many authors, in particular a thorough review has been given by J. R. Hauser.[69] In this chapter electrical characteristics relevant to small gate length FETs with velocity saturation in the channel will be considered, since these conditions apply to microwave Schottky barrier gallium arsenide field-effect transistors. The Schottky barrier FET is sometimes called a MESFET (metal-semiconductor field-effect transistor).

The effect on the electrical characteristics of velocity saturation in the channel was considered first by Dacey and Ross[75] for germanium FETs. The theory of small gate length GaAs FETs with velocity saturation in the channel has been developed by Turner and Wilson,[70] Lehovec and Zuleeg[71] and Hower and Bechtel.[76] Turner and Hower use a linear approximation to the velocity-field curve in GaAs, but Lehovec uses an approximation for the velocity given by

$$v = \frac{\mu_0 E}{1 + \dfrac{\mu_0 E}{v_s}} \tag{4.96}$$

which predicts that the velocity saturates at the scattering limited value, v_s. μ_0 is the low field mobility and E is the channel field.

Drain current

By combining Equations 4.88 and 4.96 Hauser[69] obtains an expression for the reduced drain current

$$\frac{I_d}{I_p} = \frac{3(u^2 - t^2) - 2(u^3 - t^3)}{1 + z(u^2 - t^2)} \tag{4.97}$$

where z is given by Equation 4.93. In Equation 4.93 the channel pinch-off voltage, V_p, is given by

$$V_p = \frac{eNa^2}{2\epsilon\epsilon_0}, \tag{4.98}$$

$$u^2 - t^2 = \frac{V_d}{V_p} \tag{4.99}$$

is the reduced drain-to-source voltage,

$$t^2 = -\frac{V_g}{V_p} \tag{4.100}$$

is the reduced gate-to-source voltage and I_p is given by Equation 4.92.

Transconductance

The transconductance[69] in the saturation regime is reduced by the factor in the denominator of Equation 4.97 from its low field value

$$g_m = g_{m0} \frac{u_m - t}{1 + z(u_m{}^2 - t^2)} \tag{4.101}$$

with

$$g_{m0} = \frac{I_0 z}{V_p} = \frac{eNa\mu Z}{L} \tag{4.102}$$

Gate capacitance

The gate capacitance is given by the change in the gate depletion layer charge, Q_D, with absolute value of gate voltage. It is given by[71]

$$\frac{C_g}{C_{g0}} = \left(\frac{z}{3}\right)\frac{u-t}{1-u}\left[\frac{4(u^3-t^3)-3(u^4-t^4)-6t}{(1-u)\left(u^2-t^2+\frac{1}{z}\right)}\right] \tag{4.103}$$

where

$$C_{g0} = \frac{\epsilon\epsilon_0 ZL}{a} = \frac{eNZLa}{2V_p} \tag{4.104}$$

is the gate capacitance of the fully depleted channel.

Cut-off frequency

The cut-off frequency of the lumped component R-C network of the channel and gate can be obtained by combining Equations 4.101 and 4.103.

$$f_{co} = \frac{g_m}{2\pi C_g} = \left(\frac{3}{\pi}\right)\frac{v_s}{Lz}$$
$$(1-u)^2\left[4(u^3-t^3)-3(u^4-t^4)-6t(1-u)\left(u^2-t^2+\frac{1}{z}\right)\right]^{-1} \tag{4.105}$$

Transit time

The transit time is given by[71] the ratio of the channel charge $(eNZLa - Q_D)$ and the channel current, I_m.

$$\tau = \frac{eNZLa - Q_D}{I_m} = \frac{L}{v_s}\frac{[1-u_m - z/6(u_m{}^4 - 4u_m t^3 + 3t^4)]}{(1-u_m)^2} \tag{4.106}$$

An identical expression for τ can be obtained from

$$\tau = \int_0^L \frac{dx}{v} = \int_0^L \frac{dx}{v_s} + \int_0^L \frac{dx}{\mu_0 E} \tag{4.107}$$

Drain conductance

The drain conductance under saturated drift velocity conditions is reduced from its low field value by the factor $1 + z(u_m^2 - t^2)$, in a similar manner to I_d (Equation 4.97) and g_m (Equation 4.101).

$$g_d \cong \frac{I_d}{E_0 L}\,[1 + z(u_m{}^2 - t^2)]^{-1} \tag{4.108}$$

where E_0 is the absolute value of the channel field at the beginning of the restricted channel region shown in Figure 4.14.

Maximum frequency of oscillation

The maximum frequency of oscillation, f_{max}, is obtained by combining Equations (4.101), (4.105) and (4.108) through the relation[77]

$$f_{max} = \frac{f_{c0}}{2}\left(\frac{g_m}{g_d}\right)^{1/2} \tag{4.109}$$

to give[71]

$$f_{max} = \frac{f_{c0}}{2}\left[\frac{\mu E_0}{v_s}\frac{(u_m - t)}{(1 - u_m)}\right]^{1/2} \tag{4.110}$$

For $t = 0$ and $z \gg 1$, so that $f_{c0} = v_s/4\pi L$ and $u_m \simeq (3/z)^{1/3} \ll 1$, Equation (4.110) reduces to

$$f_{max} = \gamma\frac{v_s}{L}\left(\frac{3}{z}\right)^{1/6} \tag{4.111}$$

for $z = 16$, $\gamma = 1/2\pi$.

It has been found empirically,[78] for gallium arsenide transistors with gate length less than 10 micrometres, that

$$f_{max} \simeq \frac{33}{L}\ (\text{GHz}) \tag{4.112}$$

where L is in micrometres. f_{max} is a figure of merit for the microwave field-effect transistor and is comparable to that for the microwave bipolar transistor (Equation 4.71).

Equation (4.111) shows that for small gate length transistors, where drift velocity saturation is important, f_{max} is proportional to L^{-1}, rather than the L^{-2}, relation of the Shockley theory. f_{max} is similar to the cut-off frequency, f_τ, determined by the limiting value of the transit time, $\tau = L/v_s$, from Equation (4.106)

$$f_\tau = \frac{1}{2\pi\tau} = \frac{v_s}{2\pi L} \tag{4.113}$$

It is apparent, from Equation (4.111), that the figure of merit of a microwave FET is directly related to the scattering limited velocity, v_s. It is for this reason that gallium arsenide is preferred to silicon for an X-band FET, since v_s (GaAs) = 2×10^7 cm s^{-1}, whereas v_s (Si) = 8×10^6 cm s^{-1}.

Extrinsic resistances

There are three extrinsic resistances in the MESFET shown in Figure (4.14). These are the source resistance, R_s, between the source contact and the source end of the channel, the drain resistance, R_d, between the drain contact and the drain end of the channel, and the gate resistance, which is the resistance of the gate metallization, R_m, along the gate in the z-direction to the gate bonding pad. Both R_s, and R_d include contact resistance and metallization resistance to their respective bonding pads.

The extrinsic resistances can be accounted for electrically by replacing the intrinsic gate, source and drain voltages in the FET equations by the following

$$\begin{aligned} V_g &= V_g' \\ V_s &= I_d' R_s \\ V_d &= V_d' - I_d' R_d \end{aligned} \qquad (4.114)$$

where the primed quantities indicate externally measured values and the unprimed quantities are the ideal FET parameters.

The effect of R_d is to make V_d' greater than V_p, while R_s decreases I_d' with respect to I_p and decreases g_m. The decrease of g_m decreases f_{c0} by adding R_s to the effective gate capacitance charging resistance of the channel.

Because the resistance of the gate metallization is distributed over the long dimension Z of the channel (Figure 4.14) the gate should be considered as a transmission line.[79] It is assumed that the metallization resistance of the source and drain contacts is so low that the transmission line character can be neglected there. The distributed gate metallization resistance behaves approximately as if a resistor $R_g = R_m/3$ were connected to the input of the transistor.

The resistance R_m is proportional to Z, but the input impedance is proportional to Z^{-1}. Therefore the ratio of R_m to the input impedance, which is a measure of the influence of R_m, varies as Z^2. This means that a decrease in the z-dimension of the gate leads to a rapid decrease of the influence of the gate metallization resistance.

The highest frequency for power gain with matched input and output, including extrinsic resistance effects, is given by[80]

$$f_{max} = \frac{f_{c0}}{2}\left(\frac{R_d}{R_s + R_g}\right)^{1/2} \qquad (4.115)$$

Power gain

The maximum stable gain is given by[81]

$$MSG \simeq \frac{g_{m0}}{\omega C_{dg}} \qquad (k \geqslant 1) \qquad (4.116)$$

where k is a stability factor similar to that for bipolar transistors (Equation 4.85).

The maximum available gain, near the frequency limit of the transistor, is given by[79]

$$MAG \doteqdot \left(\frac{\omega_{max}}{\omega}\right)^2 \tag{4.117}$$

with

$$\omega_{max} \simeq \frac{g_{m0}}{\{4C_{sg}G_d(\tau_i + \tau_s + \tau_g) + 2C_{dg}[C_{dg} + g_{m0}(\tau_t + \tau_s + 2\tau_g)]\}^{1/2}} \tag{4.118}$$

where G_d is the drain conductance, $\tau_i = R_0C_0(1-u)/3u$, $\tau_t = R_0C_0(1-u)/2u$, $\tau_s = R_sC_{sg}$, $\tau_g = R_gC_{sg}$, and R_0 and C_0 are the channel resistance and gate capacitance when $V_g = 0(u = 0)$.

The unilateral gain, near the frequency limit of the transistor, is given by[79]

$$U \simeq \left(\frac{\omega_u}{\omega}\right)^2 \tag{4.119}$$

with

$$\omega_u = \frac{g_{m0}}{[4C_{sg}G_d(\tau_i + \tau_s + \tau_g) + 4C_{dg}g_{m0}\tau_g]^{1/2}} \tag{4.120}$$

It follows from Equations (4.116), (4.117) and (4.119) that *MSG* decreases with frequency at the rate of 3 dB per octave, whereas *MAG* and *U* decrease at 6 dB per octave. *MAG* and *U* have unity gain at the maximum frequencies of oscillation ω_{max} and ω_u.

4.3.3 Equivalent circuit

An equivalent circuit of a GaAs MESFET chip, including noise sources represented by double circles, is shown in Figure 4.15. Figure 4.16 represents the location of some of the elements of Figure 4.15 within the transistor chip. The equivalent circuit and its representation within the chip are similar to those published by Wolf[79] and Baechtold.[82,83] Surrounding the chip will be an electrical network formed of inductances and capacitances associated with lead wires joining the chip to the package and of the package itself. The component values of this network will depend upon the precise details of the chip mounting in the package and the size of package used. For a small-signal, low-noise, MESFET the equivalent circuit surrounding the chip should be similar to that of Figure 4.13 for the bipolar transistor, with emitter, base and collector, replaced by source, gate and drain, respectively. However, unlike the collector, the drain of the FET will be connected to the package with a fine wire lead, therefore the inductance of 0.2 nH

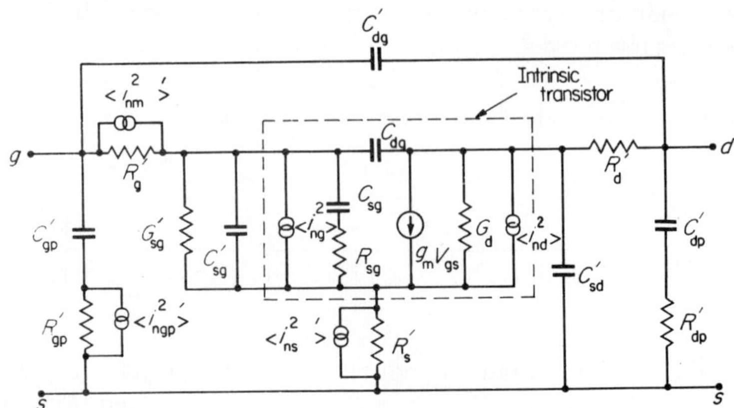

Figure 4.15 Common source equivalent circuit for GaAs MESFET chip, including noise sources.[82] (Reproduced by permission of the Institute of Electrical and Electronic Engineers, Inc.)

shown in that part of the collector lead nearest to the chip should be replaced by a 0.6 nH inductance and 0.2 Ω of contact resistance (similar to the base lead of Figure 4.13). The inductance in series with the gate and drain terminals can be absorbed in input and output matching networks.

In Figure 4.15 the intrinsic part of the transistor chip is enclosed within the dashed line. Extrinsic elements associated with the chip are indicated by primed values. Leakage conductance, G'_{sg}, associated with the reverse-biased gate, and capacitance, C'_{sg}, between the source and the gate are included as extrinsic or parasitic elements in the equivalent

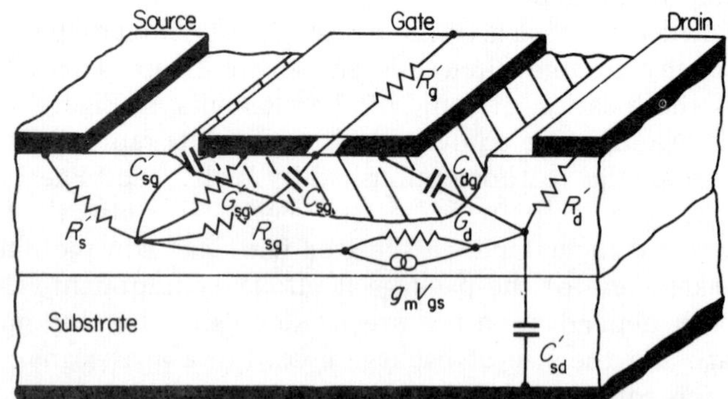

Figure 4.16 Schottky barrier gate FET showing location of some of the circuit elements in Figure 4.15 (source and substrate grounded).[82] (Reproduced by permission of the Institute of Electrical and Electronic Engineers, Inc.)

circuit. C'_{sg} would normally be the 'sidewall' capacitance of a *p-n* junction gate device and should be quite small for a Schottky gate type. The bonding pad capacitance and resistance of the gate C'_{gp} and R'_{gp}, and of the drain C'_{dp} and R'_{dp}, to the substrate (which is electrically connected to the source) are included in the equivalent circuit.

The following values for some of these equivalent circuit parameters were quoted by Liechti and coworkers[3] for a GaAs Schottky gate FET with a maximum available power gain of 11 dB at 10 GHz and an extrapolated f_{max} of 40 GHz: $C_{dg} = 0.01$ pF, $C_{sg} = 0.5$ pF, $R_{sg} = 3.5\ \Omega$, $G_d = 1.5 \times 10^{-3}\ \Omega^{-1}$, $g_{m0} = 0.033\ \Omega^{-1}$, $R'_s = 5.5\ \Omega$, $R'_d = 7\ \Omega$, $R'_g = 3\ \Omega$ and $C'_{sd} = 0.06$ pF.

The elements of the equivalent circuit are generally shown as lumped components, but the non-uniform channel resistance, R_{sg}, and the gate-to-channel non-uniform capacitance, C_{sg}, can be treated as an analogue RC transmission line.[84] Das and Schmidt[85] have considered the simplified case where the non-uniform RC transmission line can be approximated by a four-section uniform RC line with appropriately chosen capacitance and resistance for each section.

4.4 LOW-NOISE TRANSISTORS

One of the most important applications of small-signal microwave transistors is their use in low-noise amplifiers in the receivers of microwave communications and radar equipment. In this respect the noise figure of the transistor is the most important characteristic. The noise figure is a measure of the degradation in the signal-to-noise ratio that a signal undergoes when passing through the transistor. The noise figure, F, may be defined as the ratio of the signal-to-noise ratio at the input to the signal-to-noise ratio at the output. The noise of a transistor is therefore relative to the noise in the input signal generator, so that the noise figure of the transistor can never be less than unity.

Bipolar transistors have been used in low-noise amplifiers up to C-band and gallium arsenide field-effect transistors have been used up to X-band. It seems likely that the frequency range of the GaAs MESFET will be extended even further with future advances in transistor and materials technology and with improvements in microwave circuit techniques.

4.4.1 Bipolar transistors

Noise sources

The two main sources of noise in microwave bipolar transistors are thermal noise and 'shot' noise. Thermal noise is caused by thermal agitation of the current carriers in the bulk material of the transistor, giving them a random motion. Associated with the ohmic resistance, R, of the bulk material of the transistor there is a noise current, i_n, whose mean square value is

$$\overline{i_n^2} = 4kTG\,\Delta f \tag{4.121}$$

where $G = 1/R$, and Δf represents a narrow range of frequency. R is made up of the base resistance, the emitter series resistance and the collector series resistance, together with any contact resistances. Generally the other resistances are small compared with the base resistance, r_b, and may be neglected. The thermal noise due to the base resistance can also be expressed as a voltage

$$\overline{e_{nb}^2} = 4kTr_b\,\Delta f \tag{4.122}$$

Thermal, 'white', or Johnson noise, as it is called, consists of the summation of short random current pulses and is uniform over the whole frequency band.

There will be some thermal noise from the generator resistance, R_g, which also includes biasing resistors or feedback resistance in parallel with the emitter. The mean square value of this thermal noise in a narrow frequency interval is given by

$$\overline{i_{ng}^2} = 4kTG_g\,\Delta f \tag{4.123}$$

where G_g is the generator conductance.

The other main noise source in a microwave transistor, shot noise, arises from the fact that the transistor is no longer in thermal equilibrium under bias conditions and additional noise arises from the flow of electron and hole currents within the device. Shot noise is therefore current dependent, the mean square shot noise component of a current, I, being

$$\overline{i_n^2} = 2eI\,\Delta f \tag{4.124}$$

The current carriers (electrons) which surmount the potential barrier at the emitter and are injected into the base as minority carriers move towards the collector mainly by diffusion. The movement is accelerated when an internal drift field is present, but the moving carriers still undergo a large number of collisions with the crystal lattice and with the majority carriers. Hence fluctuations in the diffusion process form one of the main sources of transistor shot noise. Recently the validity of drift-diffusion transport in the analysis of very high frequency microwave transistors has been queried[86] because of their very narrow base width, which means that the minority carriers undergo few collisions in transit through the base and statistical fluctuations in the diffusion process may not be applicable to them.

Not all injected minority carriers reach the collector, since some of them recombine with majority carriers. New pairs of carriers are also generated by thermal agitation of the crystal lattice. Fluctuations in the generation—recombination process of minority charge carriers is the second fundamental source of transistor shot noise.

Random fluctuations in the minority carrier generation—recombination process will contribute to fluctuations in the saturation current of the collector junction, but the saturation current should be quite small in a well-made transistor and this source of shot noise is sometimes omitted in noise calculations.

The shot noise can be represented in an equivalent circuit of the transistor by a noise generator in the input and a noise generator in the output. The input noise generator for the common base connection will be

$$\overline{i_{ne}^2} = 2eI_e\,\Delta f \tag{4.125}$$

and for the common emitter connection will be

$$\overline{i_{nb}^2} = 2eI_b\,\Delta f \tag{4.126}$$

The output noise generator will be

$$\overline{i_{nc}^2} = 2eI_c\,\Delta f \tag{4.127}$$

In the common base connection i_{ne} and i_{nc} are correlated by the current gain, α. In the common emitter connection Fukui[87] has shown, for $f \ll f_T$, that i_{nb} and i_{nc} are not correlated, but for frequencies approaching f_T then i_{nb} and i_{nc} are correlated. Since the thermal noise of the base resistance is independent of current e_{nb} is not correlated with i_{ne}, i_{nb} or i_{nc}.

Other sources of noise are present in transistors, but are not important for the microwave transistors under consideration here. For example, fluctuations in the number of free majority carriers in the emitter, base and collector regions due to generation and recombination processes constitute possible noise sources in transistors. This so-called semiconductor current noise (or g.r. noise) is strongest in intrinsic and nearly intrinsic material The contribution of this type of noise is negligibly small under normal conditions in the extrinsic regions of the relatively heavily doped emitter, base and collector of a microwave transistor. Another type of noise, due to the variation of leakage current and surface recombination velocity with surface properties, is called flicker noise. Flicker noise is proportional to f^{-1} over a very wide frequency range and is not important in microwave transistors.

Noise figure

The noise performance of a transistor was first expressed in terms of the generally-known transistor parameters by Nielsen.[88] His equation for the noise figure is

$$F = 1 + \frac{r_b}{R_g} + \frac{r_e}{2R_g} + \frac{(r_b + r_e + R_g)^2}{2\alpha r_e R_g}\left[\left(\frac{f}{kf_T}\right)^2 + \frac{1}{h_{FE}} + \frac{I_{c0}}{I_e}\right] \quad (4.128)$$

Actually the noise figure depends upon f_α and not f_T, the factor k relates these two frequencies according to $f_\alpha = kf_T$. As already mentioned the shot noise from the collector leakage current, I_{c0}, is very small and this term is often left out of Equation 4.128. Equation 4.128 applies for both the common-emitter and common-base connections. The noise figure in the common-collector configuration is slightly different, in that a lower noise figure is predicted at very high frequencies, but this is not a common mode of operation for the transistor.

Nielsen's formula demonstrates the importance of having a low base resistance, r_b, and a high cut-off frequency, f_α, in a microwave low-noise transistor. At lower frequencies the d.c. common-emitter current gain, h_{FE}, becomes more important, since to obtain a noise figure of 1 dB or less in the lower frequency plateau region of the noise figure versus frequency curve (Figure 4.17) h_{FE} must be greater than 100 for a transistor operated with an optimum generator resistance, R_g.

Equation 4.128 may be differentiated with respect to R_g to obtain the optimum value of generator resistance, giving

$$R_g(\text{optimum}) = \left[(r_b + r_e)^2 + \frac{\alpha_0 r_e(2r_b + r_e)}{\left(\frac{f}{kf_T}\right)^2 + \frac{1}{h_{FE}}}\right]^{1/2} \quad (4.129)$$

R_g(optimum) is, most conveniently, a few tens of ohms. Equation 4.129 may then be substituted in Equation 4.128 to give the minimum value of noise figure, F_{min}. However, this results in a rather complicated expression and a simplified formula, which gives F_{min} to a good approximation, is[87,89]

$$F_{min} \simeq 1 + u + (2u + u^2)^{1/2} \quad (4.130)$$

where

$$u = \left[1 - \alpha_0 + \left(\frac{f}{f_\alpha}\right)^2\right]\frac{r_b e I_e}{\alpha_0 kT} \quad (4.131)$$

It follows that at high frequencies, where $f \gg (1-\alpha_0)^{1/2} f_\alpha$, the minimum noise figure is determined by the transistor parameters r_b and f_α, together with the emitter current, I_e. The frequency given by $(1-\alpha_0)^{1/2} f_\alpha$ corresponds approximately with the knee in the noise figure versus frequency curve (Figure 4.17). Equation 4.130 is a good

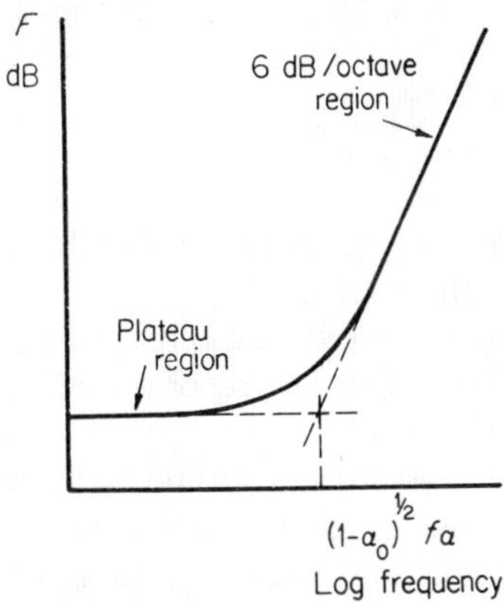

Figure 4.17 Noise figure versus frequency

approximation if the transistor has sufficiently good d.c. characteristics, i.e. if $h_{FE} > 10(f_\alpha/f)^2$.

The minimum noise figure is a function of emitter current, from Equations 4.130 and 4.131, and as the current decreases the shot noise caused by fluctuations in the collector and base currents decreases, but f_α also decreases. The result is that F_{min} becomes minimum at a certain optimum value of emitter current, I_e(optimum). Equation 4.130 can be differentiated with respect to I_e and the result equated to zero to find the optimum value of I_e for minimum noise. For constant r_b the optimum value of I_e is given by[87]

$$I_e(\text{optimum}) = \frac{2\pi\alpha_0 kTf_1 C_e}{e} \tag{4.132}$$

where f_1 is the common-emitter unity current gain frequency and C_e is the emitter transition capacitance plus the base-emitter header stray capacitance. For small-signal, low-noise, microwave transistors I_e(optimum) is typically 1—3 mA.

The conditions for minimum noise figure are not necessarily the same as the conditions for maximum power gain and it is often necessary to compromise between the two.

Effect of header parasitics

Header parasitics have little effect on F_{min} in the low microwave region up to about 2 GHz, but become important at S-band and above. The header parasitics produce a resonance in the microwave characteristics of the encapsulated transistor which effects both the gain and noise figure. Fukui[87] has given the following expression as an estimate of the resonant frequency, f_{res}, due to parasitics associated with the header

$$f_{res} \simeq \left[\frac{kT f_T}{2\pi(L_{B1} + L_{B2} + L_E)eI_c} \right]^{1/2} \tag{4.133}$$

where L_{B1} and L_{B2} are the inner and outer base lead inductances and L_E is the emitter lead inductance.

In the frequency region below one-tenth of f_{res} the effect of header parasitics can be neglected, but above one-half f_{res} the effect becomes appreciable.

Negative feedback caused by components between the collector and the base, such as the collector capacitance and the collector-base header stray capacitance, reduce the noise figure given by Equation 4.130. However, all lossy elements associated with the header have a deleterious effect on the noise figure, especially at relatively high frequencies, and tend to cancel the reduction in noise figure due to negative feedback.

4.4.2 Field-effect transistors

Intrinsic noise sources

The theory of noise in classical field-effect transistors, without charge carrier velocity saturation in the channel, was established by van der Ziel.[90,91,92] He showed that the two sources of noise in the intrinsic FET are thermal noise in the channel and induced noise at the gate. The former arises because of the usual thermal noise in the resistance of the conducting part of the channel and the latter is a consequence of the former, since every disturbance of the channel potential induces a disturbance of the gate voltage.

The channel noise is represented in the equivalent circuit of Figure 4.15 by a noise current source, i_{nd}, between the source and the drain,

$$\overline{i_{nd}^2} = 4kT\Delta f g_{m0} P \tag{4.134}$$

where g_{m0} is the magnitude of the low-frequency transconductance and P is a factor, depending upon the biasing conditions.

The induced gate noise is represented in Figure 4.15 by a noise current source, i_{ng}, between the source and the gate

$$\overline{i_{ng}{}^2} = 4kT\Delta f \frac{\omega^2 C_{sg}{}^2}{g_{m0}} R \tag{4.135}$$

where ω is the angular frequency and R is a factor depending on the biasing conditions.

The two noise currents i_{nd} and i_{ng} have the same origin, therefore they are correlated by a factor C which depends upon the biasing conditions.

$$\frac{\overline{i_{ng}{}^* i_{nd}}}{\sqrt{\overline{i_{ng}{}^2}\ \overline{i_{nd}{}^2}}} = \mathrm{j}C \tag{4.136}$$

In the van der Ziel calculation, where no velocity saturation effect is taken into account, the noise temperature of the electrons in the channel is independent of the field strength. In small gate length micro-wave field-effect transistors the high electric field in the channel will produce carrier velocity saturation near the drain end. Using a piecewise linear approximation for the velocity—field curve Baechtold has calculated the effect of the increased electron temperature on the intrinsic noise sources in silicon[82] field-effect transistors. The electron noise temperature, T_n, as a function of the electric field in silicon may be approximated by

$$\frac{T_n}{T_0} = 1 + \gamma\left(\frac{E}{E_{sat}}\right)^2 \tag{4.137}$$

where T_0 is the ambient temperature, E is the electric field, E_{sat} is the saturation field and $\gamma = 2.3$. In the case of gallium arsenide the noise temperature field curve is

$$\frac{T_n}{T_0} = 1 + \delta\left(\frac{E}{E_{sat}}\right)^3 \tag{4.138}$$

where $\delta = 6$ for the experimentally measured curve.

The increased noise temperature of the electrons in the channel will modify the intrinsic noise sources, given by Equations 4.134 and 4.135, and their correlation (Equation 4.136), so that they now become

$$\overline{i_{nd}{}^2} = 4kT\Delta f\, g_{mo}(P_1 + P_2) \tag{4.139}$$

$$\overline{i_{ng}{}^2} = 4kT\Delta f\,\frac{\omega^2 C_{sg}{}^2}{g_{mo}}(R_1 + R_2) \tag{4.140}$$

$$\overline{i_{ng}{}^* i_{nd}} = 4kT\Delta\, \omega C_{sg}\, j(Q_1 + Q_2) \tag{4.141}$$

The full expressions for the noise multiplication factors $P = P_1 + P_2$, $R = R_1 + R_2$ and the correlation coefficient

$$jC = \frac{j(Q_1 + Q_2)}{\sqrt{(P_1 + P_2)(R_1 + R_2)}}$$

are given in the papers by Baechtold.[82,83] If the noise temperature of the electrons in the channel is constant then $P_2 = R_2 = Q_2 = 0$.

The increase of the noise temperature under pronounced carrier velocity saturation has a relatively small effect on the equivalent intrinsic noise sources for silicon, but rather more effect on the channel noise for gallium arsenide.

Intervalley scattering noise in GaAs

The reason for the greater increase in the noise temperature for gallium arsenide near the saturation field, as indicated by the greater electric field dependence of Equation 4.138 than Equation 4.137, is because near the saturation field a considerable number of electrons will be scattered from the central to the satellite valleys in the conduction band, where the mobility of the electrons drops to some percent of the central valley value, and the contribution of these carriers to the total current is nearly zero. The result is similar to the generation—recombination process with its well-known frequency dependent noise, except the intervalley scattering time constants are much shorter than the time constants of generation—recombination processes so that the noise temperature does not depend on the frequency over the useful frequency range of the GaAs FET. The current fluctuations due to intervalley scattering may be represented by a noise current source

$$\overline{i_{\mathrm{iv}}^2} = 4kT_{\mathrm{iv}}g(E)\,\Delta f \tag{4.142}$$

where T_{iv} is the noise temperature due to intervalley scattering. The modified total noise temperature, T_{n}, will be the sum of T_{iv} and the temperature of the electrons in the central valley T_{e1}

$$\frac{T_{\mathrm{n}}}{T_0} = \frac{T_{\mathrm{e1}}}{T_0}(1-P) + \frac{T_{\mathrm{iv}}}{T_0} \tag{4.143}$$

where $(1-P)$ is the ratio of the number of electrons in the central valley to the total number of electrons.

To avoid this excess noise the electric field in the channel should be kept low by designing the FET to have a small pinch-off voltage and a large channel aspect ratio, L/a.

Extrinsic noise sources

Three noise sources extrinsic to the active part of the transistor are shown in Figure 4.15. These are noise sources associated with the gate metallization resistance, R_{g}, the source resistance, R_{s}, and the gate bonding pad resistance, R_{gp}. These three resistances show the normal thermal noise

$$\overline{i_{\mathrm{nT}}^{\;2}} = 4kT_{\mathrm{n}}\,\Delta f\,\frac{1}{R_{\mathrm{ext}}} \tag{4.144}$$

where R_{ext} represents the particular extrinsic resistance.

The Miller capacitance, C_{dg}, and the source bonding wire inductance have a feedback effect which reduces the noise figure slightly, but, as in the case of the bipolar transistor, lossy parasitic elements associated with the header will increase the noise figure.

In the MESFET the influence of the lossy extrinsic elements is relatively stronger than in bipolar transistors, because the intrinsic noise figure of the MESFET is so low.

Noise figure
The noise figure can be represented as a function of the source admittance, $Y_s = G_s + jB_s$, by the following equation:

$$F = F_0 + \frac{R_n}{G_s}\,[(G_s - G_{on})^2 + (B_s - B_{on})^2] \qquad (4.145)$$

where F_0 is the optimum noise figure, $Y_{on} = G_{on} + jB_{on}$, is the optimum source admittance with respect to noise, R_n is the noise resistance, and $Y_s = G_s + jB_s$ is the source admittance (G and B are the conductance and susceptance, respectively). The noise parameters F_0, $(G_{on} + jB_{on})$ and R_n of the transistor are determined by calculating the noise figure, F, for four different source admittances and applying Equation 4.145.

The optimum source admittance, Y_{og}, for maximum available power gain is not the same as Y_{on} for optimum noise figure, therefore it is not possible to match the transistor input at the same time for maximum gain and minimum noise figure. This is also true for the bipolar transistor, but to a lesser extent, since the two optimum source admittances are closer together.

The optimum noise figure does not depend very much on the drain and gate voltages.

4.5 POWER TRANSISTORS

Silicon microwave power transistors have found widespread use for broadband applications in the fields of microwave communications, microwave relay links, phased-array radar, collision avoidance systems and electronic counter-measures, as well as applications where efficiency is very important such as telemetry and satellite communications. For many applications c.w. operation is not important. Pulsed radar, aircraft transponders, navigational aids and distance measuring equipment are some of the applications which call for high pulse power at microwave frequencies. Transistors can be driven to pulse power levels which are significantly higher than the corresponding c.w. limits. The narrow pulse widths allow higher voltage operation, as the r.f. voltage breakdown limits are typically higher than the d.c. values.

Gallium arsenide microwave power transistors have not yet entered systems use, but power outputs of 1.6 W at 2 GHz[93] and 0.63 W at 6 GHz[94] have been achieved in the laboratory.

4.5.1 Bipolar transistors

Power transistors are distinguished from small-signal transistors mainly by their ability to handle higher currents. In the figure of merit expression for a transistor (Equation 4.1) the $r_b C_c$ product should be independent of area, therefore it should be possible to scale up the transistor area to handle any value of current, provided the device has a heat sink which will dissipate the heat generated within the active part of the transistor. If the current density in the transistor is not increased then f_T should ideally be independent of area, but if the current density is increased then one of the high current expressions (Equations 4.69 or 4.70) for $\tau_{ec} = 1/2\pi f_T$ could be used. The ability to increase the emitter strip length (or the strip multiplicity, which is the equivalent) at will, without deteriorating the frequency response, is the fundamental premise on which the design of high frequency, high power transistors is based. Obviously, however, the scaling up cannot go on indefinitely since technological limits exist, both in the production of fault-free expitaxial silicon and in the control of uniform processing over a relatively large area. These limitations are of a statistical nature and the relation between the density of these defects and the area of the transistor will determine the percentage yield of satisfactory transistors. As the power and hence the area of the transistor is increased the yield due to these effects must diminish.

Transistor design theory is based on relatively simple linear models which give excellent results for small-signal levels; however, at large signal levels the transistor exhibits effects which cannot be analysed by means of these simple models. These effects are generally caused by harmonic components of the voltages and currents generated by nonlinear mechanisms within the transistor. A relatively complex nonlinear model is necessary to represent this behaviour, which is then analysed by computer techniques.[95] The parameters in the small-signal equivalent circuit model of the transistor are normally assumed to have fixed values which are determined by the average level of operation, but in fact they will vary with the instantaneous signal level and, therefore, time. Nevertheless the model can be adjusted to predict small to medium signal performance quite well over wide frequency ranges. It does not, however, predict very well the characteristics of a transistor driven into saturation under large-signal conditions, where the collector—base junction is forward-biased. The reason for this is that the equivalent circuit model (Figure 4.12) does not include the conductance of a forward-biased, collector-base diode. The design theory of high power transistors is therefore on a much more empirical basis than it is for small-signal transistors.

The important performance parameters of a microwave power transistor are power output, power gain and efficiency. The power output is determined by the current and voltage handling capability of the transistor. The current handling capability is determined by the emitter periphery and epitaxial layer resistivity (because of current crowding to the edges of the emitter and the Kirk effect, respectively) and the voltage handling capability is determined by the breakdown voltage, which is limited by the resistivity of the epitaxial layer and the curvature at the edge of the base—collector junction. The power gain is determined by the ratio of f_{max} to the operating frequency, so that for high power gain a high f_{max} is required. The efficiency is the ratio of the power output to the power input and determines the heat dissipation within the device.

Geometrical considerations

The operating current and, therefore, the power output of the transistor, is dependent upon the emitter periphery because most of the current is injected near the edge of the emitter sites as a result of the transverse base current biasing effect described by Fletcher.[96] To optimize the frequency performance it is necessary to design the transistor to have the maximum emitter periphery with the minimum emitter area, contained within the minimum base area.

The requirement for minimum emitter area is essential to minimize the input capacitance, C_e of Figure 4.12, to reduce shunting of the emitter-base diode resulting in loss of injected current into the base. This is the reason for the first design requirement of a large ratio of emitter periphery to emitter area, *EP/EA*. Similarly the base area must be kept small in order to reduce the shunting effect of the output capacitance on the load resistance, which reduces the current in the load thereby decreasing the available output power. This leads to the second geometrical design requirement of a large ratio of emitter periphery to base area, *EP/BA*. A third geometrical requirement arises because of the capacitance of the base bonding pad, $C_{bc(pad)}$ in Figure 4.11, which is proportional to the base metal area over the collector. $C_{bc(pad)}$ is in parallel with the collector junction capacitance and is added to C_{Tc}, therefore the ratio of emitter periphery to pad area, *EP/PA*, should also be maximized. The effect of the base bonding pad capacitance can be reduced by increasing the oxide thickness between the metal pad and the collector region.

The current or power handling capability of the transistor is defined empirically in terms of the emitter periphery, typical values are $0.4\ \mathrm{A\ cm^{-1}}$ or $4\ \mathrm{W\ cm^{-1}}$, respectively. These values depend on the geometrical design of the transistor and the frequency.

Microwave power transistor manufacturers have used different geometrical designs in an attempt to obtain a favourable ratio of emitter periphery to emitter area, emitter periphery to base area and emitter periphery to pad area, together with ease of manufacture and reliability of the device. The three basic designs are called interdigitated, overlay and mesh. The patterns are generally orthogonal, but circular shapes have been used and a 'diamond' pattern has been proposed.[97]

Early microwave power transistors were of interdigitated design, that is, the emitter sites and the base contact sites were in the form of long narrow strips alternating with each other as shown in Figure 4.1. The strips were often called 'fingers' and because the emitter and base fingers were interlocked the structure was called interdigitated. Sometimes the emitter strips or the base strips would be connected together at one end and this was called a 'comb' structure. Later, the long emitter and base strips were placed further apart and were provided with a multiplicity of short interlocking side arms, this structure was called the 'emitter tree' or 'fishbone' pattern.

A different pattern concept was developed with the overlay, or base grid, structure. In this case the emitter sites consist of a regular pattern of small squares, rectangles or circles, each of which is surrounded by a low resistivity p^+ diffused base grid. One finger of the emitter metallization pattern can make contact with one or more columns of the emitter site pattern, so that the emitter metallization 'overlays' the base grid, being insulated from it by oxide. The base grid is connected by narrow base metal fingers alternating with the broad emitter metal fingers. A variation of this design is the metal grid structure, in which the p^+ base grid is replaced by a metal grid. The lateral diffusion of the p^+ grid is thereby eliminated and by improving the grid conductivity by an order of magnitude relative to the p^+ grid, closer emitter-to-grid spacing and longer emitter sites can be used. This technique enables a greater *EP/BA* ratio to be obtained, but requires two-level metallization.

The mesh, or emitter-grid, structure is really the inverse of the overlay structure in that the individual sites are the p^+ base contacts and the emitter is a continuous n^+ diffused grid. This results geometrically in a greater *EP/BA* ratio than an overlay of the same dimensions. However, the current handling capacity, in terms of amps per centimetre of emitter periphery, is limited by the voltage drop in parts of the emitter grid, which is greater than in structures where most of the emitter periphery is directly contacted by the emitter metallization. This results in more debiasing of those parts of the emitter periphery remote from the emitter metallization than is obtained with overlay or interdigitated structures.

Finally, if the emitter and base contact sites are arranged as an alternating pattern of small rhombus or diamond-like areas, the so-called diamond pattern, then the *EP/BA* ratio will be greater than that of a square chequer-board structure such as the emitter mesh. In common with the emitter mesh, however, parts of the emitter diamond periphery will suffer from debiasing effects owing to remoteness from the metal contact. In practice some of the geometrical advantage of the

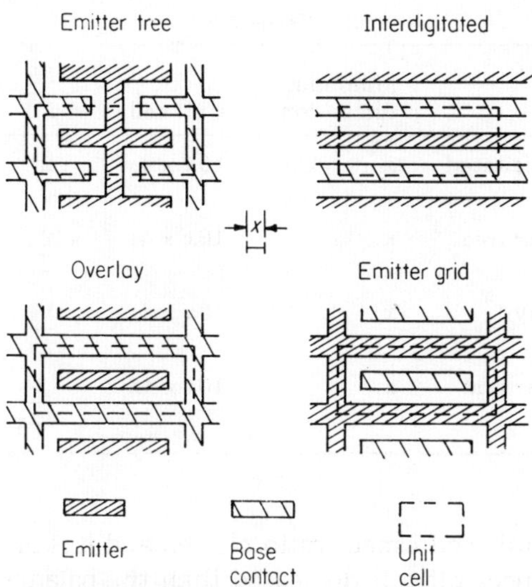

Figure 4.18 Comparison of power transistor patterns. Minimum dimension x. Unit cell ($10x \times 4x$)

diamond structure will be lost because of rounding-off of the tips of the diamonds during their definition by etching.

Four of the patterns most often used are shown in Figure 4.18 for the same unit cell area. For the sake of comparison the aspect ration of the unit cell has arbitrarily been chosen to be 2.5:1. The minimum dimension of the pattern is assumed to apply equally to emitter and base strip widths and to the emitter-to-base spacing. The minimum dimension will be decided practically by photoresist tolerances or acceptable process yields. The ratios *EP/EA* and *EP/BA* for these patterns are given in Table 4.2. The ratios are the same for the interdigitated and the emitter tree patterns and are independent of the aspect ratio of the unit cell. For a unit cell aspect ratio of 1:1 the overlay pattern has limiting values of $EP/EA = 4/x$ and $EP/BA = 1/4x$, whereas the emitter grid pattern has limiting values of $EP/EA = 12/7x$ and $EP/BA = 3/4x$. As the unit cell aspect ratio increases the *EP/EA* and *EP/BA* values for both the emitter grid and the overlay approach the fixed values applicable to the emitter tree and interdigitated patterns. For the particular geometries chosen in Figure 4.18 the values for overlay/emitter tree or interdigitated/emitter grid are in the ratio $EP/BA = 0.8{:}1{:}1.2$ and $EP/EA \simeq 1.1{:}1{:}0.9$, respectively. If a figure of merit for these patterns was chosen to be $[(EP/EA) + (EP/BA)]$ then their ratio would be approximately 0.95:1:1.05, indicating a few percent advantage for the emitter grid pattern. These values apply only for

Table 4.2

Type	Minimum dimension	Unit cell	$\frac{EP}{EA}$	$\frac{EP}{BA}$
Interdigitated	x	$10x \times 4x$	$\frac{2}{x}$	$\frac{1}{2x}$
Emitter tree	x	$10x \times 4x$	$\frac{2}{x}$	$\frac{1}{2x}$
Overlay	x	$10x \times 4x$	$\frac{16}{7x}$	$\frac{2}{5x}$
Emitter grid	x	$10x \times 4x$	$\frac{24}{13x}$	$\frac{3}{5x}$

the particular unit cell aspect ratio chosen and it is easily shown that if the unit cell aspect ratio is decreased then the balance can be swung in favour of the overlay, whereas if the unit cell aspect ratio is increased then all the designs approach the same limiting value. Evidently the intrinsic performance of microwave power transistors is almost independent of their particular geometrical pattern. The more complex patterns do have advantages over the simple interdigitated structure in terms of reliability.

In addition to maximizing the ratios EP/EA and EP/BA the structure should be designed to have the minimum emitter-to-collector MOS capacitance and the minimum base-to-collector MOS capacitance. Both of these MOS parasitic capacitances are due to bond pads over the collector oxide, hence the number of bond pads and their area should be reduced. However, the transistor must have a sufficient number of bond pads and sufficient pad area to allow the use of a number of lead wires of optimum diameter to achieve a low input Q and a low common lead inductance, otherwise the large-signal power gain will be impaired since

$$P.G = \frac{(f_T/f)^2 R_L}{4(r_b + \omega_T L_e)} \tag{4.146}$$

where L_e is the emitter lead parasitic inductance and R_L is the collector load resistance. $r_b + \omega_T L_e = R_{in}$ is the real part of the input impedance, Z_{in}.

Power—frequency product

An approximate expression can be derived relating the power output of the transistor to the frequency and the input impedance by the following considerations. Johnson has shown,[6] from the physical properties of silicon, that

$$f_T \leq \frac{4 \times 10^{11}}{V_m}\ s^{-1} \tag{4.147}$$

where V_m is the maximum allowable applied emitter-to-collector voltage. The maximum output power is obtained when the output resistance of the transistor is matched to the load resistance

$$R_L = \frac{1}{2\pi f_T C_c} \tag{4.148}$$

The output power, P, of an amplifier with a tuned load is related to R_L and the supply voltage, V_s (if the transistor just bottoms on the peaks), by the expression

$$P = \frac{V_s^2}{4R_L} \tag{4.149}$$

In practical circuit applications the maximum applied collector voltage must be kept to a value of less than one-half of V_m, since the collector voltage will swing to twice the supply voltage. Under these conditions Equation 4.149 becomes

$$P_{max} = \frac{V_m^2}{16R_L} \tag{4.150}$$

and, using the following expression for f_{max}

$$f_{max} = \left(\frac{f_T}{8\pi(r_b + \omega_T L_e)C_c} \right)^{1/2} \tag{4.151}$$

Equations 4.147, 4.148 and 4.150 give

$$P_{max} R_{in} f_{max}^2 \leq 2.5 \times 10^3\ W\ \Omega\ (\text{GHz})^2 \tag{4.152}$$

It was assumed in the derivation of Equation 4.147 that f_T was determined by collector depletion layer transit time alone, but since some of the other time constants in the f_T expression are comparable to the collector depletion layer transit time (see Section 4.2.3, for example) then f_T should be reduced by a factor of about 3. Thus Equation 4.152 is reduced to

$$P_{max} R_{in} f_{max}^2 \leq 8 \times 10^2\ W\ \Omega\ (\text{GHz})^2 \tag{4.153}$$

Practical transistors have reached about one-half of this theoretical limit. Equation 4.153 shows that at a particular frequency the power output is inversely proportional to the input resistance.

Matching networks
To achieve maximum bandwidth, efficiency and power gain, microwave power transistors are being made with an electrical network inside the package, next to the transistor chip, to transform the device input and output impedance to match a 50 Ω microwave transmission line. These internal matching networks are designed to include the inductance of the bond wires to the transistor chip and the parasitic capacitances of the package. The bond wire length must be accurately controlled and a different length for each cell of a multicell chip may be required. Aluminium wire has an advantage over gold in that its mass is sufficiently low that no supporting structure is needed to hold the wires in place in the matching network. The aluminium bond wires are sufficiently rigid to hold their preset loops, which form the inductors. On matched devices with gold lead wires rods are used to support the wires in mid-span, which is an additional complication, and if the wire is in tension severe shearing stresses occur during temperature cycling, which may lead to bond failure. An aluminium wire matching network structure is compatible with aluminium metallization on the transistor chip, similarly a gold wire network is compatible with gold metallization on the chip.

A disadvantage of the internal matching network is that it is only matched over a certain fixed frequency band, whereas the normally packaged transistor chip can be externally matched over different frequency bands.

Potential failure mechanisms
(a) Hot-spot formation The high power density and relatively large device area of power transistors leads to reliability problems which are not important in small-signal transistors. One of these problems results from thermal runaway in localized areas of the transistor structure. This type of fault is particularly troublesome in large area transistors where material or process induced variations can cause hot-spots to form. The negative temperature coefficient of the emitter—base junction voltage leads to increased current injection at these sites, resulting in thermal runaway. Recently Fair has shown[9 8] that the temperature sensitivity of current gain is less for an arsenic doped emitter than for a phosphorus doped one, so that arsenic doped emitter transistors should be less prone to thermal runaway at hot-spots. The problem of thermal runaway has been ameliorated by the use of current ballast resistors in the metallization to the multiple emitter sites. A photograph of a 2 GHz 5 W power transistor with titanium ballast resistors in each emitter finger of the interdigitated metal pattern is shown in Figure 4.19. The voltage drop developed across the ballast resistor which is passing current to a potential thermal runaway emitter site will cancel

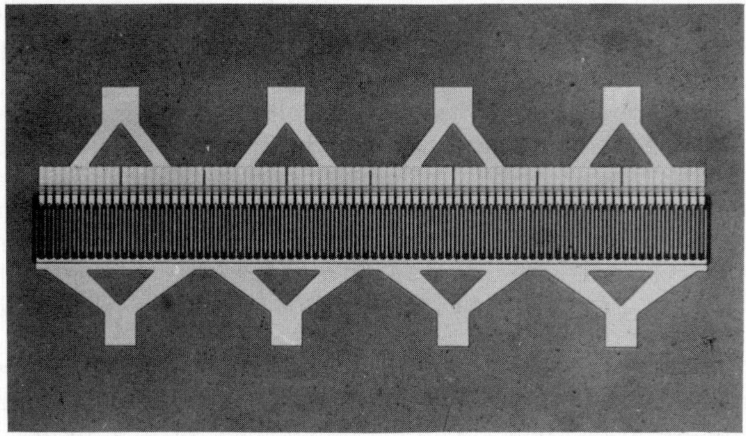

Figure 4.19 5 W 2 GHz silicon *n-p-n* transistor with titanium emitter ballast resistors

the effect of the negative temperature coefficient of the emitter–base junction voltage in that site and prevent a hot-spot from developing. Ballast resistors like those shown in Figure 4.19 are not very selective since they reduce bias along the whole of the finger, or all the emitter sites in one column for a multiple-emitter transistor. The ballast resistors can be graded with higher values in the centre of the pattern than at the edges, since this will reduce the power density in the centre relative to the edges. This will help to maintain a more uniform temperature over the entire active region of the transistor.

Individual emitter site ballasting is more efficient and this can be accomplished with the use of a high resistivity polysilicon layer, vapour-deposited between the emitter sites and the emitter metal. By controlling the doping and contacting geometry of the polysilicon to the emitter metal, ballast resistors may be placed in series with each emitter site. This technique is very effective for the overlay pattern, with its individual emitter sites, but less so for the emitter tree and emitter grid where each emitter tree, and the whole of the emitter grid, are continuous. The polysilicon layer also provides protection against alloy spike failures of the aluminium metallization through the shallow emitter and failures because of aluminium penetration through pinholes in the base oxide.

If the total active transistor area is divided into several 'cells', as is often done to even out the temperature distribution in the silicon chip, then a degree of current sharing to each cell can be promoted by an input matching network to the device. This will help to prevent local hot-spots forming owing to current hogging by individual cells.

(b) Mismatch conditions Another possible failure mode arises when the output circuit is de-tuned at full power. The resulting mismatch causes a high VSWR condition at the collector of the transistor, subjecting the transistor to instantaneous voltage peaks many times the supply voltage. Avalanching takes place within the collector depletion region, consequently there is no current steering effect by the emitter ballast resistors to reduce localized current densities. This problem can be alleviated by collector ballasting, which is provided by a thick undepleted collector layer. Transistors with both emitter and collector ballasting can usually withstand high VSWR loads at any phase angle.

(c) Second breakdown Transistor failure can also be initiated by 'second breakdown' of the collector junction. This is a transition from a high voltage low-current condition at the collector to a lower voltage high-current condition. Second breakdown may be partly due to thermal effects,[99] in which case it may be prevented by designing the transistor with ballast resistors to prevent hot-spot formation, as already described. Second breakdown is also partly attributed[100] to avalanche injection in the collector at the n-n^+ transition between the epitaxial layer and the substrate. This type of second breakdown can be avoided by a suitable choice of epitaxial layer thickness and doping level; in particular thin, lightly-doped epitaxial collector regions are susceptible to this type of second breakdown. A microwave power transistor has its allowable current and voltage regions specified to avoid operation in that part of the current-voltage regime where second breakdown is liable to occur.

(d) Electromigration Electromigration is the phenomenon of mass transport in metals at high current densities. It is a potential failure mode in microwave power transistors because the emitter contact metal film pattern may carry a high current density. The failure is an electrical open-circuit due to void formation in the conductor metal.

The mass transport of metal ions in the conductor is in the direction of current flow. The ion movement is due to diffusion via an ion-vacancy exchange mechanism with a preferred direction of flow impressed by the momentum transfer in collisions between current electrons and metal ions. Damage to the conductor occurs as a result of a change in the rate of mass transport at some point along the conductor owing to a change in the diffusion coefficient of the metal ions. The change may be initiated by a temperature gradient, an electron concentration gradient, a structural irregularity, or some combination of these factors. Under these conditions regions of metal depletion and enhancement occur which can lead to the formation of voids and hillocks. A positive gradient (in the electron flow direction) of the diffusion coefficient of the metal ions produces a region where vacancies condense to form voids. Open-circuit failure occurs if an accumulation of such voids completely fills the conductor track width.

Because the metal ions migrate by a place-exchange mechanism the concept of a median time to failure, MTF, proportional to an activation energy was proposed by Black[101,102]

$$MTF = \frac{C\,A\,\exp(\phi/kT)}{J^2} \tag{4.154}$$

where C is a constant depending on the properties of the conductor, A is the cross-sectional area of the conductor, ϕ is the activation energy of migration, k is Boltzmann's constant, T is the conductor temperature in kelvins and J is the current density.

Black established the J^{-2} relationship over a current range of 5:1, but the theory developed for contacts to bulk material, and supported by other workers[103] for thin metal films, predicts a J^{-1} relationship. However, if a temperature gradient exists in the film, a failure model with a J^{-3} dependence has been proposed.[104] Black's J^{-2} model is generally accepted in practice.

The activation energy of a given thin film conductor sample is determined by the diffusion coefficient of the predominant mechanism of mass transport. Three mechanisms have been identified: grain boundary, surface and lattice diffusion. Their relative activation energies are 0.48 eV, 0.84 eV and 1.2 eV.[101] They are characteristic of small grain crystallite films (1.2 μm), large grain crystallite films (8 μm), and large grain glass covered films, respectively. F. M. D'Heurle[104] indicates that single crystal aluminium films may have two orders of magnitude or more improvement in MTF over polycrystalline films. The predominance of a particular mechanism can, therefore, be influenced by the careful control of the metal deposition conditions or by subsequent processing of the sample. Above 275 °C lattice diffusion predominates, thus film structural effects are not important at higher temperatures. At temperatures lower than 275 °C, however, orders of magnitude improvement in conductor lifetime can be obtained through the use of large grain or large grain glassed films.

The pre-exponential factor, C, depends upon the grain size of the aluminium but can be altered by the addition of other metal ions, such as silicon[105] or copper[106] to the aluminium. For example, an alloy of Al:Cu:Si (95:4:1) is 25 times better than large grain glassed aluminium at 275 °C, but both are about the same, in terms of MTF, at 150 °C.[107] The grain size of the Al:Cu:Si alloy was not mentioned but its activation energy for metal migration is about the same as that for small grain unglassed aluminium. Fortunately the copper does not 'poison' the silicon because it is preferentially soluble in aluminium. The addition of oxygen to the aluminium[104] results in a narrower distribution of failure times, but the median failure time may be reduced.

The cross-sectional area term, A in Equation 4.154, should be such that the current density is limited to about 2×10^5 A cm^{-2}, since at higher current densities the MTF decreases rapidly.[101] Other geometrical factors influencing the failure time are the length and the width of the conductor track. Failure of the conductor occurs when there is a change in the rate of mass transport. Likely causes of such a rate change are localized temperature gradients and structural inhomogeneities (defects) in the conductor film. The magnitude or severity of the rate change determines the MTF, so the lifetime of the conductor is determined by its severest defect. If the defect density is constant then the total number of defects increases with conductor length, therefore there is a higher probability of the occurrence of a severe defect in a longer conductor track. Conductor failure occurs across the width of the track, therefore lifetime increases with width as the probability of several defects being aligned across the track becomes smaller as the width increases. In practice the track width should be several times the grain size. In this respect the emitter tree and the overlay design have an advantage over the interdigitated and emitter grid types in that the emitter metal finger can be made wider. Any localized reduction in track cross-sectional area, owing to thinning over an oxide step for example, should be avoided, since this may give rise to a thermal gradient due to Joule heating which would enhance the electromigration failure rate. The change in substrate thermal conductivity between the oxide and the silicon also introduces a temperature gradient in the conductor at the oxide step. This effect, and the reduced cross-sectional area of the conductor at the oxide step, increases the rate of electromigration in the conductor film at this point, with the resultant likelihood of void condensation into a metal break across the track.

The median failure times of conductor metal systems containing gold have been reported[108] to be better than unglassed aluminium. A lower electromigration failure rate would be expected for gold since its activation energy for self-diffusion is 1.8 eV, compared to 1.4 eV for aluminium. However, it is not possible to substitute a single layer of gold for a single layer of aluminium because gold forms a eutectic with silicon at 377 °C and also it does not adhere well to silicon dioxide. Hence composite layers of, for example Pt—Si/Ti/Pt/Au, Pt—Si/Mo/Au or Pt—Si/W/Au have been used. Basically these structures involve a low resistance contact to silicon (Pt—Si), a metal to give good adherence to the oxide (Ti), and a barrier metal (Pt, Pd, Mo, W, Ta) to prevent the gold from dissolving the silicon. These complicated structures give rise to more technological problems and, if pinholes exist in the barrier metal, to another potential failure mechanism. Whether a gold system or an aluminium system is more reliable for a microwave power transistor is, as yet, undecided.

Most electromigration tests have been made on metal films carrying a uniform current and at a uniform temperature. However, the situation in the metal conductor pattern of a microwave power transistor is more complex since both the current density and the temperature are non-uniform down the emitter strip. Thus, while the current density is highest at the end of the emitter strip nearer to the bond, the temperature will be greater near the middle of the strip. The best means of determining the median failure time of a particular transistor is by accelerated life tests on the actual device, but published electromigration failure information can be used as a guide.

(e) Microcracking When the conductor track surmounts an oxide step its cross-sectional area may be reduced. In evaporated metal films this is due to a 'shadowing' effect during deposition, which can be overcome by rotating the work table supporting the silicon wafers with respect to the metal source filament. This is the 'planetary' evaporation technique. The shadow effect can be eliminated by sputtering the metal film on to the silicon; this is most readily accomplished with gold metal systems. The cross-sectional area of the conductor track may also be reduced by the attack of chemical etching residues trapped at the oxide step, after the metal pattern definition. Either of these weaknesses in the conductor at the oxide step may result in crack formation owing to the relief of stress in the metal film.

Thermal considerations

Any power device of uniform power density will have a higher temperature towards its centre. An improved power transistor design will therefore be one in which the transistor active region is divided into smaller heat sources (cells) separated by some distance. The thermal coupling between these cells drops off logarithmically with distance. A power transistor consisting of four rectangular cells, arranged along the sides of a square, will have a fairly uniform temperature distribution across the chip because of the close coupling of the heat flux generated at the ends of each cell compared with the lighter coupling in the middle, together with the overall symmetry of the array. This cell arrangement is referred to as a 'collector mesh'.[97] Since all four transistors share a common-emitter bonding pad a large amount of parasitic capacitance is eliminated.

Because of the finite mass and length of the material between the collector *p-n* junction, where the power is dissipated in the form of heat, and the external medium, assumed to be at a constant temperature, the temperature of the transistor will increase until a steady state condition is reached. The value of the final temperature depends upon the thermal conductivity of the material between the junction and the heat sink. Under transient conditions the junction temperature also depends upon the thermal capacitance of the material.

The thermal flow is analogous to a spreading resistance, except when the chip is sufficiently thin that its thickness approaches the width of the active area of the device, when the thermal flow becomes linear. For one dimensional heat flow, in the x-direction, the equation applicable to each material between the collector junction and the heat sink is

$$K \frac{\mathrm{d}^2 \theta_\mathrm{j}}{\mathrm{d}x^2} = \rho c \frac{\mathrm{d}\theta_\mathrm{j}}{\mathrm{d}t} \tag{4.155}$$

where K is the thermal conductivity of the material, ρ is its density, c is its specific heat and t is the time. The relative junction temperature, θ_j, is the difference between the actual junction temperature and the heat sink temperature. Equation 4.155 is identical with the equation describing the voltage distribution along an RC transmission line in which a current is applied to the line at $x = 0$. The heat flow can therefore be calculated by transmission line theory.[65]

An analysis[109] of the thermal response of microwave transistors, for a four-cell transistor chip, indicates that about 90% of the temperature rise occurs within the chip, and about 60% of that occurs within the first 50 micrometres of the collector junction. The figures apply to a 0.004 inch thick silicon chip on a 0.03 inch thick beryllium oxide (BeO) chip carrier. If the bottom 0.002 inch of the chip were removed the total thermal resistance would be reduced by about 14%. If a diamond heat spreader were used, between the chip and the beryllium oxide, to spread the heat from the chip to the carrier the total thermal resistance could be reduced by about 8%. If the thickness of the chip was first reduced to 0.002 inch, and then a diamond heat spreader was used, a total reduction in thermal resistance of about 27% might be gained (compared with 14% for the thinner chip bonded directly on to the BeO carrier). In each case it is important to ensure a good thermal contact between the chip and the carrier, or the chip and the heat spreader.

4.5.2 Field-effect transistors

Gallium arsenide MESFET microwave power transistors are still in their infancy. One reason for this is the relationship between material faults and device area, which is similar to that mentioned in Section 4.5.1 for silicon bipolar transistors. Because of the greater technological difficulty of producing large areas of good quality epitaxial gallium arsenide, compared with epitaxial silicon, the yield of good gallium arsenide devices will be lower than that of silicon devices of equal area.

Nevertheless, two gallium arsenide microwave power MESFETs have been reported[93,94] with power output of 1.6 W at 2 GHz and 0.63 W at 6 GHz, respectively. Both of these devices were made with an n-type epitaxial GaAs layer grown on a semi-insulating GaAs substrate. The 2 GHz device had a mesh source geometry, rather like the emitter of an emitter-grid bipolar transistor; and the 6 GHz device had a multiple gate structure more like an interdigitated bipolar transistor. The 6 GHz type also employed epitaxial n^+ source and drain regions and a self-aligned metal gate, to reduce the input resistance.

In making a high power FET it is important to have a high breakdown voltage gate, low resistance contacts for the source and drain, and to increase the periphery of the source as much as possible without degrading the high frequency performance of the device. A high breakdown voltage gate is achieved by moderate doping (2 to 4×10^{16} cm^{-3}) of the channel layer, but it may be possible to increase the breakdown voltage by having a higher resistivity region between the gate and the drain, without degrading the transconductance in the channel. Low resistance source and drain contacts are generally achieved with alloyed Au—Ge films, but n^+ epitaxial regions may also be used. The source periphery is increased by the use of a multicell structure with a number of gate bonding pads, since as the source and gate length are increased the gate metallization resistance (R_g' in Figure 4.16) becomes a more important factor in degrading the device performance. The significance of the multicell approach was demonstrated by the 6 GHz device, since the gain remained essentially the same as more units were paralleled, but the saturation power output increased monotonically.

A multi-channel silicon FET, or 'gridistor', has been reported with a power output of 5 W at 1GHz[2] and 1 W at 2.7 GHz.[110] The buried gate grid was produced by ion-implantation.

4.6 FABRICATION TECHNOLOGY

The fabrication technology of microwave transistors is basically the same as that of lower frequency transistors except that special attention must be paid to those processes which limit the frequency performance, as discussed in the earlier parts of this chapter. Because of the diverse nature of this subject it will be possible in this section only to deal rather briefly with those techniques in general use for making silicon bipolar and gallium arsenide Schottky barrier field-effect transistors.

4.6.1 Bipolar transistors

Material

The bulk of the silicon chip consists generally of antimony doped n-type material of about 0.01 Ω cm. The single crystal boule will have been sliced into wafers, 200 μm thick, with either (100) or (111) surface planes. The wafers are polished and etched and are of low defect quality. An n-type epitaxial layer is grown on one face of the wafer by chemical vapour deposition, during which the reverse face of the wafer may be protected by silicon dioxide. The deposition conditions are chosen to minimize antimony diffusion from the substrate into the epitaxial layer, which is generally doped with phosphorus. The oxide is left on the reverse face to minimize autodoping of the epitaxial layer from the substrate during its growth. The epitaxial layer doping level and thickness depend upon the electrical specification of the transistor. The epitaxial layer requirements are discussed in Section 4.2.3 and a design chart for doping level and thickness is given in Figure 4.3. Power transistors are designed to have a higher breakdown voltage (~50 V) than low-noise transistors (~20 V). For these transistors, with shallow base diffusion, the breakdown voltage is limited by the curvature at the edge of the junction[111] and to prevent this a deeper p^+ guard-ring diffusion may be placed along the periphery of the base—collector junction. The epitaxial layer thickness must take into account the depth of the p^+ guard-ring diffusion.

Wafer cleaning
Before epitaxial growth, and all of the subsequent operations involved in the transistor's fabrication, the whole wafer, or the individual transistor chips, are subjected to a cleaning operation. This may involve wet chemical etching, chemical vapour etching, or oxide growth to remove some of the silicon surface; or merely a sequence of degreasing, removal of residual oxide, washing and drying.

Base diffusion
After the growth of approximately 0.5 μm of thermal oxide at about 1000 °C, windows are opened in it by photoresist exposure and etching to allow a p^+ diffusion for the base guard-ring (if used) and the base contact areas. The p-type source may be BN, BBr_3 or B_2H_6 and the diffusion is carried out at approximately 1100 °C to produce a resistivity of about 10 Ω/square and a depth of 0.7 to 1.0 μm. The oxide is cleared from the whole of the base area and boron is deposited in the base window at about 900 °C to give a sheet resistance in the region of 100 Ω/square. This deposited boron is diffused further, or 'driven-in', at 950—1050 °C to give a sheet resistance of 400—1000 Ω/square and a depth of 0.25—0.6 μm, depending upon the particular transistor requirement. The regrown oxide thickness in the base window will be 0.15—0.3 μm.

Emitter diffusion
Windows are opened in the base oxide and either arsenic, from a doped oxide or vapour source, or phosphorus from a $POCl_3$ or P_2O_5 source

is diffused in at 950—1050 °C to give a sheet resistance of 15—30 Ω/square and a depth of 0.15—0.4 μm. With an arsenic emitter there will be little or no emitter-dip effect in the base region underneath the emitter, but with a phosphorus emitter care must be taken to avoid an excessive emitter-dip effect.

Contact metallization

The oxide is removed from the emitter region and base contact openings are made in the base oxide to receive the contact metallization. The metallization system, which may be either aluminium based or gold based, is evaporated or sputtered on and the contact pattern is defined and etched. Ohmic contacts are obtained by sintering at about 400 °C for 5 minutes.

Final processing

Each transistor pattern on the slice is electrically tested and reject transistors are marked. The slice is thinned to about 100 μm from the back, by etching or lapping, to decrease its electrical and thermal resistance. The back of the slice may be coated with gold to facilitate chip attachment to the header. The slice is cleaved or sawn into individual chips or dice, each one containing the required transistor pattern or patterns. Good transistor units are selected and mounted on to the required header. Wires are thermocompression bonded between the transistor bonding pads and the header lead-out terminals. The transistors are baked in dry nitrogen and the header is hermetically sealed.

Alternative fabrication procedures

The bipolar transistor process described in the preceding sections is known basically as the 'diffused planar' technology. A variation of this process uses ion-implantation for the doping of the base and emitter regions. Ion-implantation has been discussed in Section 4.2.5. It is capable of producing accurately controlled emitter and base concentration profiles with a theoretically more ideal shape for transistor use than impurity profiles obtained by diffusion. It will no doubt be used more extensively in future microwave transistor production.

Another variation is the isoplanar process, which has produced the best small-signal, low-noise, bipolar transistor performance reported to date.[1] In this process both the emitter and collector junctions are located within a mesa region of the silicon, the sides of the mesa being protected by silicon dioxide. Because of the relative thickness of this oxide over the collector bulk region the parasitic bonding pad capacitances (Figure 4.11) are reduced. Owing to the method of construction it is possible to produce the part of the base region underneath the emitter (the intrinsic base) and the part of the base region between the edge of the emitter and the base contact (the extrinsic base) in separate operations. Hence the intrinsic base region may be optimized to give a high f_T and h_{FE}, while the extrinsic base may be optimized to give a low r_b. In this way the best noise performance can be achieved (see Equation 4.128). In the normal planar process the intrinsic and extrinsic base regions are not independently produced, therefore the noise performance cannot be optimized since to achieve a high f_T and h_{FE} requires a relatively lightly doped narrow base region, whereas a low r_b requires a wider, more heavily doped, base region. It is possible to produce the two base regions either by ion-implantation or diffusion; in the transistor described by Archer[1] the intrinsic base is produced by ion-implantation and the extrinsic base by diffusion. It has also been possible in this transistor to improve its high frequency performance by the lateral diffusion of the extrinsic base underneath the edges of the emitter to reduce the effective emitter width.

4.6.2 Field-effect transistors

Material

The best material for very high frequency field-effect transistors is gallium arsenide. To take advantage of the very high electron mobility in this material n-channel FETs are made. The bulk of the chip is chromium doped semi-insulating gallium arsenide, 300 μm thick, with the surface oriented 2° off a (100) crystal plane. However, some manufacturers prefer to use a (111) B surface orientation since the epitaxial growth rate on this face is slower and the thickness of the very thin epitaxial film may be controlled more accurately. The epitaxial layer is grown by the $AsCl_3$, Ga, H_2 system.[112] Before the n-type layer for the active part of the FET is grown a high resistivity, or semi-insulating, 'buffer' layer may be grown on the substrate. The buffer layer improves the crystal quality at the n-type/semi-insulating transition and prevents impurities from diffusing into the n-type layer from the substrate. Buffered n-type layers have a higher electron mobility than non-buffered ones and produce FETs with a higher gain and a lower noise figure. The buffer layer is either chromium or oxygen doped; or it may be undoped. In an analogous way an n^+ buffer layer between the n^+ substrate and the n-type active layer will provide better gallium arsenide for Gunn or LSA diodes (see Chapter 2).

The n-type epitaxial layer thickness and doping level for the active part of the FET, as a function of frequency, may be determined from Figure 4.20, which also shows the gate length and the output power. Very thin, heavily doped, n-type layers are required for X-band low-noise FETs. The growth conditions required to produce these vapour phase epitaxial gallium arsenide layers have been described by Fairman and Solomon.[113] The n-type layers are doped with either

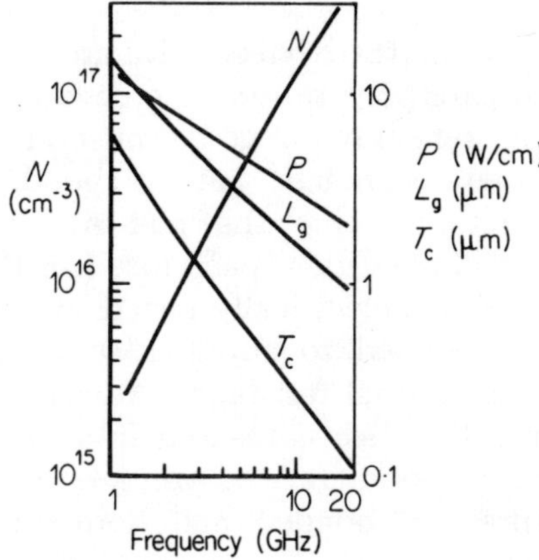

Figure 4.20 Power output P, epitaxial layer thickness T_c doping level N, and gate length L_g, for gallium arsenide MESFET.[94] (Reproduced by permission of R.C.A. Corp.)

sulphur or tin, at a doping level between 8×10^{16} cm^{-3} and 2×10^{17} cm^{-3}, and are between 0.15 and 0.35 micrometres thick. The electron mobility in the layer is in the range 3000—4500 cm^2 V^{-1} s^{-1}.

The thickness and doping level of these layers are measured by an instrument[114] which measures the rate of change of capacitance with voltage of a Schottky diode made on the surface of the layer. The C—V relationship is converted by the instrument into doping level as a function of layer thickness and the resulting impurity profile is plotted automatically.

The material requirement for field-effect transistors is much more stringent than that for bipolar transistors.

Source and drain preparation

The source and drain areas are defined in photoresist and the source and drain metal is evaporated on to the slice. The metal is generally a Au—Ge alloy, but Au—Te and Au—Te—Ge have been used. The contacts may be thickened up by a coating of Ni. The photoresist is floated off, together with the unwanted metal, and the contacts are alloyed to the GaAs with a rapid time—temperature sequence.

Gate preparation

The gate area is defined in photoresist and the gate metal is evaporated. The metal may be Cr—Ni—Au, Cr—Au, Cr—Rh or Al—Ge. The photoresist is floated off together with the unwanted metal.

Final processing

A mesa is etched to isolate the device active area either before or after the source, drain and gate preparation. If etched afterwards, a contact metallization pattern, generally of gold, must be used to bring the source, drain and gate contacts out to bonding pads over the semi-insulating substrate. Placing the pad over the semi-insulating substrate reduces the pad parasitic capacitance (see Figure 4.15).

Each transistor pattern is electrically tested and reject transistors are marked. The slice is thinned to about 150 μm from the back, by etching, to decrease its thermal resistance. Au—Ge is evaporated on the back of the slice and the slice is cleaved into individual chips, each containing a transistor. Good devices are selected, mounted on the header, thermocompression bonded and hermetically sealed in the required encapsulation.

Alternative fabrication procedures

The process described above does not use diffusion and therefore high temperature processing, which may degrade the electrical properties of the gallium arsenide, is not required. Alternative fabrication methods have been described using ion-implantation, either to produce the n-type active region by sulphur implantation,[115] or to isolate the active region by converting the surrounding part of the n-type epitaxial layer to semi-insulating gallium arsenide by proton-implantation. Ion-implantation has the disadvantage of requiring a high temperature anneal (800 °C) to electrically activate the implanted ion and to remove the implantation-produced lattice damage.

Another alternative is that the n-type epitaxial layer may be grown from a liquid solution instead of a chemical vapour.

It is an advantage to define the gate, and to align it with respect to the source, with electron beam techniques,[116] since in narrow gate length FETs the frequency performance is inversely related to the gate length and the gate to source spacing.

4.7 CONCLUDING REMARKS

In order to approach the intrinsic microwave performance theoretically available from the transistor chip it will be necessary to eliminate, if possible, the parasitic circuit elements associated with the encapsulation of the device. These package parasitics generally cause a greater degradation of performance as the frequency increases, but particular encapsulations may have resonances at certain frequencies where the performance is badly effected. Gallium arsenide field-effect transistors are able to operate at higher frequencies than silicon bipolar transistors and are therefore more liable to be degraded by package parasitics. The present trend is to integrate the unencapsulated transistor chip into a thin-film circuit. For example, the collector of the transistor may be made a part of a microstrip transmission line, thus reducing the effect of the output capacitance; or the chip may be embedded in the ground plane to reduce the common-lead inductance to the minimum associated with the bond wire length. In the case of the gallium arsenide FET 50 Ω coplanar tapered wave guides have been processed directly on the semi-insulating substrate,[3] which ensures that the device has virtually no parasitic reactances and very small coupling loss. The use of such techniques should enable bipolar transistor performance to be extended into X-band and FET performance to be extended into J- and K-bands.

Microwave power transistors will probably incorporate input and output matching networks more extensively and will be used in balanced amplifiers to give good input and output VSWR even at high signal levels. With improved device performance it will be possible to use negative feedback to stabilize the amplifier characteristics over a wider frequency band. Power transistors will need to be designed and operated to avoid electromigration in the contact metallization, the metallurgy of which will be developed to give proven reliability.

Silicon will no doubt remain the most suitable material for bipolar transistors and gallium arsenide will be favoured for X- to K-band, small-signal, field-effect transistors. It is likely that silicon will be used for medium to high power field-effect transistors in the C- to X-band region because of its better thermal conductivity than gallium arsenide. Epitaxial gallium arsenide of better quality and lower defect density will be developed which will allow larger area, high power, more reliable FETs to be made in the higher frequency range. In the longer term alloys of III—V compound semiconductors may be used for ultra high frequency FETs because some of them theoretically have a greater scattering limited electron velocity than gallium arsenide. For example[117] the electron velocity in indium arsenide phosphide ($InAs_{0.4}P_{0.6}$) approaches 6×10^7 cm s^{-1} at fields approaching 10 kV cm^{-1}, which is about three times the limiting velocity for gallium arsenide. A great practical difficulty with these more complex semiconductor alloys is likely to be the provision of reliable ohmic contacts to them.

Refinements to the transistor geometric pattern will be made with electron beam definition of resist directly on the semiconductor chip, especially for FETs which require fewer pattern alignments than bipolar transistors. In the case of the bipolar transistor, electron beam techniques may be used to write accurately defined, fault-free, intricate patterns for the mask set, which will then be exposed on the semiconductor chip by light-optical or X-ray techniques.

ACKNOWLEDGEMENT

The author would like to thank Mr E. Hughes, of the GEC Hirst Research Centre, for many helpful discussions during the preparation of this chapter.

REFERENCES

1. J. A. Archer, 'Low-noise implanted-base microwave transistors', *Solid State Electronics*, **17**, 387—393 (1974).
2. S. Kakihana, 'Current status and trends in high frequency transistors', *Microwave Systems News*, **3**, 46—50 (1973).
3. C. A. Liechti, E. Gowen and J. Cohen, 'GaAs microwave Schottky-Gate FET', IEEE International Solid-State Circuits Conference, paper THPM 14.2, *Digest of Technical Papers*, pp. 158—159, February 1972.
4. N. G. Bechtel, W. W. Hooper and D. Mock, 'X-band GaAs FET', *Microwave Journal*, **7**, 15—19 (1972).
5. W. Baechtold, K. Daetwyler, T. Forster, T. O. Mohr, W. Walter and P. Wolf, 'Si and GaAs 0.5 μ gate Schottky-barrier field-effect transistors', *Electronics Letters*, **9**, 232—234 (1973).
6. E. O. Johnson, 'Physical limitations on frequency and power parameters of transistors', *R.C.A. Rev.*, **26**, 163—177 (1965).
7. A. S. Grove, *Physics and technology of semi-conductor devices*, Wiley, 1967, p. 103.
8. S. M. Spitzer, B. Schartz and G. D. Weigle, 'Native oxide mask for zinc diffusion in gallium arsenide', *J. Electrochem Soc.*, **121**, 820—822 (1974).
9. D. W. Yarbrough, 'Status of diffusion data in binary compound semi-conductors', *Solid-State Technology*, **11**, 23—31 (1968).
10. H. Becke, D. Flatley and D. Stolnitz, 'Double-diffused gallium arsenide transistors', *Solid-State Electronics*, **8**, 255—265 (1965).
11. W. von Muench and H. Statz, 'Solid-to-solid diffusion in the gallium arsenide device technology', *Solid-State Electronics*, **9**, 939—942 (1966).
12. L. B. Valdes, *The physical theory of transistors*, McGraw-Hill, New York, 1961.
13. A. B. Phillips, *Transistor Engineering*, McGraw-Hill, New York, 1962.
14. R. L. Pritchard, 'Frequency response of grounded-base and grounded-emitter transistors', paper given at the AIEE Winter meeting, New York, N.Y., January 1954.
15. J. M. Early, 'Structure determined gain-band product of junction triode transistors', *Proc. IRE*, **46**, 1924—1927 (1958).
16. J. R. Hauser, 'The effects of distributed base potential on emitter current injection density and effective base resistance for stripe

transistor geometrics', *IEEE Transactions Electron Devices,* ED11, 238–242 (1964).

17. J. C. Irvin, 'Resistivity of bulk silicon and of diffused layers in silicon', *Bell System Tech. J.*, 41, 387–410 (1962).
18. J. A. Archer, 'Design and performance of small-signal microwave transistors', *Solid-State Electronics,* 15, 249–258 (1972).
19. M. H. White and M. O. Thurston, 'Characterisation of microwave transistors', *Solid-State Electronics,* 13, 523–542 (1970).
20. H. Lawrence and R. M. Warner Jnr., 'Diffused junction depletion layer calculations', *Bell System Tech. J.*, 39, 389–404 (1960).
21. F. W. Hewlett Jnr., F. A. Lindholm and A. J. Brodersen, 'The effect of a buried layer on the collector breakdown voltage of bipolar junction transistors', *Solid-State Electronics,* 16, 453–457 (1973).
22. T. Irie, T. Yanagawa and M. Oyama, 'Effect of external base resistance and base contact resistance on high frequency power gain of a transistor', *NEC. Res. Developm. (Japan),* 10, 56–65 (1967).
23. L. E. Terry and R. W. Wilson, 'Metallization systems for silicon integrated circuits', *Proc. IEEE,* 57, 1580–1586 (1969).
24. R. C. Hooper, J. A. Cunningham and J. G. Harper, 'Electrical contacts to silicon', *Solid-State Electronics,* 8, 831–833 (1965).
25. H. J. DeMan, 'The influence of heavy doping on the emitter efficiency of a bipolar transistor', *IEEE Transactions Electron Devices,* ED18, 833–835 (1971).
26. R. P. Mertens, H. J. DeMan and R. J. van Overstraeten, 'Calculation of the emitter efficiency of bipolar transistors', *IEEE Transactions Electron Devices,* ED20, 722–778 (1973).
27. M. S. Mock, 'On heavy doping effects and the injection efficiency of silicon transistors', *Solid-State Electronics,* 17, 819–824 (1974).
28. R. J. van Overstraeten, H. J. DeMan and R. Mertens, 'Transport equations in heavy doped silicon', *IEEE Transactions Electron Devices,* ED20, 290–298 (1973).
29. H. J. DeMan, R. Mertens and R. J. van Overstraeten, 'Influence of heavy doping effects on the f_T prediction of transistors', *Electronics Letters,* 9, 174–176 (1973).
30. J. C. Henderson and R. J. D. Scarbrough, 'The influence of emitter–base junction depth and emitter doping level on the calculation of f_T and h_{FE}', paper presented at ESSDERC, Munich, September 1973.
31. A. B. Bhattacharya and T. N. Basavaraj, 'Transition capacitance calculations for double-diffused *p-n* junctions', *Solid-State Electronics,* 16, 467–476 (1973).

32. B. R. Chawla and H. K. Gummel, 'Transistion region capacitance of diffused *p-n* junctions', *IEEE Transactions Electron Devices*, ED18, 178—195 (1971).
33. H. Kroemer, 'The drift transistor', *Transistors I*, R.C.A. Laboratories, Princeton, N.J., 1956, pp. 202—220.
34. J. L. Moll and I. M. Ross, 'The dependence of transistor parameters on the distribution of base layer resistivity', *Proc. IRE*, 44, 72—78 (1956).
35. A. H. Marshak, 'Optimum doping distribution for minimum base transit time', *IEEE Transactions Electron Devices*, ED14, 190—194 (1967).
36. J. te Winkel, 'Drift transistor: simplified electrical characterisation', *Electronic Radio Eng.*, 36, 280—288 (1959).
37. M. B. Das and J. M. Humenick, 'Transient limitations of bipolar transistors under a step-function emitter-base voltage excitation', *Proc. IEEE*, 61, 1164—1165 (1973).
38. J. Lindmayer and C. Wrigley, 'The high injection level operation of drift transistors', *Solid-State Electronics*, 2, 79—84 (1961).
39. H. F. Cooke, 'Microwave transistors: Theory and design, *Proc. IEEE*, 59, 1163—1181 (1971).
40. T. C. Lo, 'Base transport factor calculations for transistors with complementary error function and Gaussian base doping profiles', *IEEE Transactions Electron Devices*, ED18, 243—248 (1971).
41. A. Gover, J. Grinberg and A. Seidman, 'Computation of bipolar transistor base parameters for general distribution of impurities in base', *IEEE Transactions Electron Devices*, ED19, 967—975 (1972).
42. A. W. Matz, 'A modification of the theory of the variation of junction transistor current gain with operating point and frequency', *J. Electr. and Control*, 7, 133—152 (1959).
43. C. T. Kirk Jnr., 'A theory of transistor cut-off frequency (f_T) fall-off at high current density', *IRE Transactions Electron Devices*, ED9, 164—174 (1962).
44. J. M. Early, '*pnip* and *npin* junction transistor triodes', *Bell System Tech. J.*, 33, 517—534 (1954).
45. J. L. Moll, J. L. Su and A. C. M. Wang, 'Multiplication in collector junctions of silicon *n-p-n* and *p-n-p* transistors', *IEEE Transactions Electron Devices*, ED17, 420—423 (1970).
46. R. J. Whittier and D. A. Tremere, 'Current gain and cut-off frequency fall-off at high currents', *IEEE Transactions Electron Devices*, ED16, 39—57 (1969).
47. G. Messenger, 'An analysis of switching effects in high power diffused base silicon transistors', presented at the IEEE International Electron Devices Meeting, Washington, D.C. October 30th, 1959.
48. A van der Ziel and D. Agouridis, 'The cut-off frequency fall-off in

UHF transistors at high currents', *Proc. IEEE (Letters)*, 54, 411–412 (1966).

49. D. L. Bowler and F. A. Lindholm, 'High current regimes in transistor collector regions', *IEEE Transactions Electron Devices*, ED20, 257–263 (1973).
50. J. R. A. Beale and J. A. G. Slatter, 'The equivalent circuit of a transistor with a highly doped collector operating in saturation', *Solid-State Electronics*, 11, 241–252 (1968).
51. W. H. Eckton Jnr., 'An analysis of the capabilities of germanium and silicon in the design of microwave transistors', *NEREM Record*, 8, 130–131, 1966.
52. B. E. Deal, 'The current understanding of charges in the thermally oxidised silicon structure', *J. Electrochem. Soc.*, 121, 198c–205c (1974).
53. J. W. Lathrop, 'Semiconductor-network technology — 1964', *Proc. IEEE*, 52, 1430–1444 (1964).
54. B. I. Boltaks, *Diffusion in semiconductors*, Infosearch, London, 1963.
55. R. M. Burger and R. P. Donovan, *Fundamentals of silicon integrated technology*, Vol. 1, Prentice-Hall, New Jersey, 1967.
56. A. F. W. Willoughby, 'Anomalous diffusion effects in silicon (A review)', *Journal of Materials Science*, 3, 89–98 (1968).
57. R. B. Fair, 'Quantitative theory of retarded base diffusion in silicon *n-p-n* structures with aresenic emitters', *J. Appl. Phys.*, 44, 283–291 (1963).
58. J. C. C. Tsai, 'Shallow phosphorus diffusion profiles in silicon', *Proc. IEEE*, 57, 1499–1506 (1969).
59. R. B. Fair and G. R. Weber, 'Relationship between resistivity and total arsenic concentration in heavily doped *n*- and *p*-type silicon', *J. Appl. Phys.*, 44, 280–282 (1973).
60. L. N. Large, 'Ion implantation — A new method of doping semiconductors. Parts 1 and 2', *Contemporary Physics*, 10, 277–298 and 505–532 (1969).
61. R. S. Payne, R. J. Scavuzzo, K. H. Olson, J. M. Nacci and R. A. Moline, 'Fully ion-implanted bipolar transistors', *IEEE Transactions Electron Devices*, ED21, 273–278 (1974).
62. K. Kurokawa, 'Power waves and the scattering matrix', *IEEE Trans. Microwave Theory and Techniques*, MTT13 194–202 (1965).
63. S. J. Mason, 'Power gain in feedback amplifiers', *IRE Trans. Circuit Theory*, CT1, 20–25 (1954).
64. J. G. Linvill and L. G. Schimpf, 'The design of tetrode transistor amplifiers', *Bell System Tech. Journ.*, 35, 813–840 (1956).
65. R. L. Pritchard, *Electrical characteristics of transistors*, McGraw-Hill, 1967.

66. K. E. Drangeid, R. Jaggi, S. Middelhoek, T. Mohr, A. Moser, G. Sasso, R. Sommerhalder and P. Wolf, 'Microwave silicon Schottky barrier field-effect transistor', *Electron Lett.*, 4, 362—363 (1968).
67. W. W. Hooper and W. I. Lehrer, 'An epitaxial GaAs field effect transistor', *Proc. IEEE*, 55, 1237—1238 (1967).
68. C. A. Mead, 'Schottky-barrier gate field-effect transistor', *Proc. IEEE*, 54, 307—308 (1966).
69. J. R. Hauser, *Fundamentals of silicon integrated device technology* (Eds. R. M. Burger and R. P. Donovan), Vol. 2, Prentice-Hall, 1968.
70. J. A. Turner and B. L. H. Wilson, 'Implications of carrier velocity saturation in a gallium arsenide field-effect transistor', *Proc. Symp. GaAs, Institute of Physics and Physical Society Conf. Series No. 7*, 195—204 (1968).
71. K. Lehovec and R. Zuleeg, 'Voltage—current characteristics of GaAs J-FETS in the hot electron range', *Solid-State Electronics*, 13, 1415—1426 (1970).
72. W. Shockley, 'A unipolar field-effect transistor', *Proc. IRE* 40, 1365—1376 (1952).
73. D. P. Kennedy and R. R. O'Brien, 'Computer-aided two-dimensional analysis of the junction field-effect transistor', *IBM. J. Res. Develop.*, 14, 95—116 (1970).
74. K. Lehovec and W. G. Seeley, 'On the validity of the gradual channel approximation for junction field-effect transistors with drift velocity saturation', *Solid-State Electronics*, 16, 1047—1054 (1973).
75. G. C. Dacey and I. M. Ross, 'The field-effect transistor', *Bell System Tech. J.*, 34, 1149—1189 (1955).
76. P. L. Hower and N. G. Bechtel, 'Current saturation and small-signal characteristics of GaAs Field-effect transistors', *IEEE Transactions Electron Devices*, ED20, 213—220 (1973).
77. P. L. Hower, W. W. Hooper, D. A. Tremere, W. Lehrer and C. A. Bittmann, 'The Schottky-barrier gallium arsenide field-effect transistor', *Proc. Symp. GaAs, Institute of Physics and Physical Society Conf. Series No. 7*, 187—194, 1968.
78. R. Zuleeg and K. Lehovec, 'High frequency and temperature characteristics of GaAs junction field-effect transistors in the hot electron range', *Proc. Symp. GaAs, Institute of Physics Conf. Series No. 9*, 240—250, 1970.
79. P. Wolf, 'Microwave properties of Schottky-barrier field-effect transistors', *IBM. J. Res. Develop.*,14, 125—141 (1970).
80. K. E. Drangeid and R. Sommerhalder, 'Dynamic performance of Schottky-barrier field-effect transistors', *IBM. J. Res. Develop.*, 14, 82—94 (1970).

81. J. M. Rollett, 'Stability and power-gain invariants of linear two-ports', *IRE Transactions*, CT9, 29–32 (1962).
82. W. Baechtold, 'Noise behaviour of Schottky-barrier gate field-effect transistors at microwave frequencies', *IEEE Transactions Electron Devices*, ED18, 97–104 (1971).
83. W. Baechtold, 'Noise behaviour of GaAs field-effect transistors with short gate lengths', *IEEE Transactions Electron Devices*, ED19, 674–680 (1972).
84. A. van der Ziel and J. W. Ero, 'Small-signal, high-frequency theory of field-effect transistors', *IEEE Transactions Electron Devices*, ED11, 128–135 (1964).
85. M. B. Das and P. Schmidt, 'High-frequency limitations of abrupt-junction FETs', *IEEE Transactions Electron Devices*, ED20 779–792 (1973).
86. P. Rohr, F. A. Lindholm and K. R. Allen, 'Questionability of drift-diffusion transport in the analysis of small semiconductor devices', *Solid-State Electronics*, 17, 729–734 (1974).
87. H. Fukui, 'The noise performance of microwave transistors', *IEEE Transactions Electron Devices*, ED13, 329–341 (1966).
88. E. G. Nielsen, 'Behaviour of noise figure in junction transistors', *Proc. IRE*, 45, 957–963 (1957).
89. S. D. Malaviya and A. van der Ziel, 'A simplified approach to noise in microwave transistors', *Solid-State Electronics*, 13, 1511–1518 (1970).
90. A. van der Ziel, 'Thermal noise in field-effect transistors', *Proc. IRE*, 50, 1808–1812 (1962).
91. A. van der Ziel, 'Gate noise in field-effect transistors at moderately high frequencies', *Proc. IEEE*, 51, 461–467 (1963).
92. A. van der Ziel, 'Noise in solid-state devices and lasers', *Proc. IEEE*, 58, 1178–1206 (1970).
93. M. Fukuta, T. Mimura, I. Tujimura and A. Furumoto, 'IEEE International solid-state circuits conference, Paper THAM 7.6', *Digest of Technical Papers*, pp. 84–85, February 1973.
94. L. S. Napoli, J. J. Hughes, W. F. Reichert and S. Jolly, 'GaAs FET for high power amplifiers at microwave frequencies', *R.C.A. Review*, 34, 608–615 (1973).
95. R. L. Bailey, 'Large-signal nonlinear analysis of a high-power high-frequency junction transistor', *IEEE Transactions Electron Devices*, ED17, 108–119 (1970).
96. N. H. Fletcher, 'The high current limit for semiconductor junction devices', *Proc. IRE*, 45, 862–872 (1957).
97. J. A. Benjamin, 'New design concepts for microwave power transistors', *Microwave Journal*, 15, 39–64 (1972).
98. R. B. Fair, 'Optimum low-level injection efficiency of silicon

transistors with shallow arsenic emitters', *IEEE Transactions Electron Devices*, **ED20**, 642–647 (1973).

99. H. A. Schafft, 'Second breakdown — A comprehensive review', *Proc. IEEE*, 55, 1272–1288 (1967).
100. P. L. Hower and V. G. K. Reddi, 'Avalanche injection and second breakdown in transistors', *IEEE Transactions Electron Devices*, **ED17**, 320–335 (1970).
101. J. R. Black, 'Electromigration — A brief survey and some recent results', *IEEE Transactions Electron Devices*, **ED16**, 338 – 347 (1969).
102. J. R. Black, 'Electromigration failure modes in aluminium metallisation for semiconductor devices', *Proc. IEEE*, **57**, 1587–1594 (1969).
103. G. Hoffman and H. M. Breitling, 'On the current density dependence of electromigration in thin films', *Proc. IEEE*, **58**, 833 (1970).
104. F. M. D'Heurle, 'Electromigration and failure in electronics: An introduction', *Proc. IEEE*, 59, 1409–1418 (1971).
105. G. J. van Gurp, 'Electromigration in aluminium films containing silicon', *Applied Physics Letters*, **19**, 476–478 (1971).
106. A. K. Kahar, 'Electromigration studies on aluminium–copper stripes', *Solid-State Technology*, **16**, 47–62 (1973).
107. T. Moutoux, 'Let's take another look at transistor reliability', *Microwaves*, 12, 52–58 (1973).
108. L. E. Terry and R. W. Wilson, 'Metallization systems for silicon integrated circuits', *Proc. IEEE*, 57, 1580–1586 (1969).
109. G. K. Baxter, 'Thermal response of microwave transistors under pulsed power operation', *IEEE Transactions on Parts, Hybrids and Packaging*, **PHP9**, 185–193 (1973).
110. D. P. Lecrosnier and G. P. Pelous, 'Ion-implanted FET for power applications', *IEEE Transactions Electron Devices*, **ED21**, 113–118 (1974).
111. S. M. Sze and G. Gibbons, 'Effect of junction curvature on breakdown voltage in semiconductors', *Solid-State Electronics*, **9**, 831–845 (1966). See also O. Leistiko and A. S. Grove, 'Breakdown voltage of planar silicon junctions', *Solid-State Electronics*, 9, 847–857 (1966).
112. J. R. Knight, D. Effer and P. R. Evans, 'The preparation of high purity gallium arsenide by vapour phase epitaxial growth', *Solid-State Electronics*, 8, 178–180 (1965).
113. R. D. Fairman and R. Solomon, 'Submicron epitaxial films for GaAs field-effect transistors', *J. Electrochem Soc.*, **120**, 541–544 (1973).
114. P. J. Baxandall, D. J. Colliver and A. F. Fray, 'An instrument for

the rapid determination of semiconductor impurity profiles', *Journal of Physics E: Scientific Instruments*, 4, 213—221 (1971).

115. R. G. Hunsperger and N. Hirsch, 'GaAs field-effect transistors with ion-implanted channels', *Electronics Letters*, 9, 577—578 (1973).
116. J. A. Turner, A. J. Waller, R. Bennett and D. Parker, 'An electron beam fabricated GaAs microwave field-effect transistor', *Proc. Symp. GaAs, Institute of Physics Conf. Series No. 9*, 234—239, 1970.
117. W. Fawcett, C. Hilsum and H. D. Rees, 'Optimum semiconductor for microwave devices', *Electronics Letters*, 5, 313—314 (1969).

CHAPTER 5

Microwave Solid State Oscillator Circuits

K. KUROKAWA

Microwave solid state oscillators convert d.c. power into r.f. power by means of an interaction between the circuit and solid state devices such as transistors, transferred electron devices, impatt or tunnel diodes. These devices are called active devices, for they generate r.f. power. The actual circuit–device interaction in oscillators is inherently nonlinear. If it were linear, no steady state would be reached with a finite r.f. amplitude. The general behaviour of linear passive networks is well understood. On the other hand, the general behaviour of networks containing active nonlinear elements is scarcely known. In the field of nonlinear active networks, microwave oscillators constitute a small part in which a relatively fruitful study has been carried out. The reason for the success lies in the fact that either the current through, or the voltage across, the active device is kept almost sinusoidal by the high Q resonator in the oscillator circuit.

This chapter will be divided into three parts. In the first part, the circuit–device interaction in microwave solid state oscillators will be discussed in detail under the assumption that the device current is almost sinusoidal. In view of the dual property between voltage and current, the discussion can be easily extended to the case in which the device voltage is almost sinusoidal by changing all the impedances in the equations to the corresponding admittances and interchanging voltage and current.

In the second part, the results of the first part will be used to understand the behaviour of practical oscillator circuits which have appeared in the literature. The discussion includes the behaviour of multiple-device oscillators. When the output power of a single device is not sufficient, we attempt to combine the outputs from two or more devices. When the number of active devices is two, it is fairly easy to design and operate such a power combiner. However, if the number of devices to be combined exceeds two, we usually run into difficulty. Small changes in the load condition or in the bias supply, or even in the environmental temperature, bring about sudden changes in the mode of operation. The output power as well as oscillation frequency literally jumps from one value to another. This is the well-known mode problem of multiple-device oscillators. Section 5.13 will present recent solutions to this problem.

In the final part, the measurements of important characteristics of microwave oscillators will be discussed. The theory of microwave oscillators is most conveniently developed by referring to the device terminals. However, measurements are usually done at the output port. For this reason, whenever appropriate, a brief discussion is given of how to relate the measured result to the theory.

The references given at the end of this chapter are by no means exhaustive. Only those directly related to our discussions are listed. Many classical papers are left out. Some of the important papers in the field of microwave oscillator circuits may be found in Reference 8.

5.1 THEORY

5.1.1 Device impedance

A microwave solid state oscillator is illustrated in Figure 5.1. In this schematic diagram, the oscillator is represented by a resonant cavity containing an active device. Let us first consider the steady state. The current flowing through the active device is a periodic function of time with the fundamental frequency $f = \omega/2\pi$. In ordinary microwave oscillators, the harmonic components in the current are small because of the filtering action of the resonant cavity. So, we will assume that this is the case. Then, the current is given by $A \cos(\omega t + \varphi)$ plus small harmonic components, i.e.

$$i(t) = \mathrm{Re}[Ae^{j(\omega t+\varphi)}] + \text{(small harmonic components)} \tag{5.1}$$

where A is the amplitude and φ is the phase of the fundamental component. The corresponding device voltage $v_\mathrm{d}(t)$ may have in-phase and out-of-phase components. They will be expressed by $-\bar{R}(A)A \cos(\omega t + \varphi)$ and $\bar{X}(A)A \sin(\omega t + \varphi)$, respectively. In addition to these fundamental components, the device voltage may contain harmonic components. Although the harmonic components in $i(t)$ are assumed to be small, those in $v_\mathrm{d}(t)$ may not be small. Thus, we have

$$v_\mathrm{d}(t) = \mathrm{Re}[-\bar{Z}(A)Ae^{j(\omega t+\varphi)}] + \text{(harmonic components)} \tag{5.2}$$

where

$$\bar{Z}(A) = \bar{R}(A) - j\bar{X}(A) \tag{5.3}$$

$-\bar{Z}(A)$ is called the device impedance, since it is the device voltage divided by the device current, both being at the fundamental frequency. Strictly speaking, $\bar{R}(A)$ and $\bar{X}(A)$ will depend on the amplitudes and relative phases of all the harmonic components in $i(t)$ in addition to A. However, if the harmonic components in $i(t)$ are all kept small, as we have assumed, their influence on $v_\mathrm{d}(t)$ will be small com-

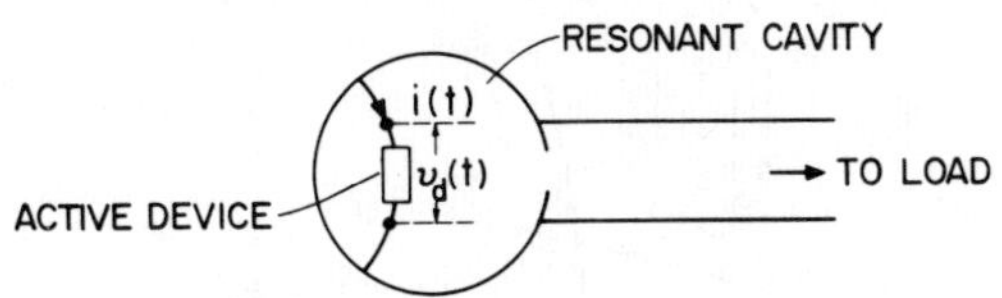

Figure 5.1 Schematic diagram of a microwave oscillator

pared to that of A. So, $\bar{R}(A)$ and $\bar{X}(A)$ and hence $\bar{Z}(A)$ are expressed as functions of A only. A word of caution is in order. If the condition of small harmonic components in $i(t)$ is violated as in the case of second harmonic tuning, the following discussion will not be applicable. A different analysis becomes necessary for each particular case. However, we will not discuss such cases with significantly large harmonics in $i(t)$.

Let $Z(\omega) = R(\omega) + jX(\omega)$ be the circuit impedance seen by the device. The voltage developed across this impedance due to $i(t)$ is given by

$$v_c(t) = \text{Re}[Z(\omega)\, Ae^{j(\omega t+\varphi)}] + (\text{harmonic components}) \tag{5.4}$$

Although the harmonic components in $i(t)$ are assumed to be small, depending on the magnitude of $R(n\omega)$ and $X(n\omega)$, the nth harmonic component in $v_c(t)$ may not be small. The harmonic components in $v_c(t)$ are, therefore, not indicated as being small. For free-running oscillators, the sum of $v_c(t)$ and $v_d(t)$ must be equal to zero since no external voltage is applied:

$$v_c(t) + v_d(t) = 0 \tag{5.5}$$

Substituting (5.2) and (5.4) into (5.5), multiplying by $\cos(\omega t+\varphi)$ and $\sin(\omega t+\varphi)$, respectively, and integrating over one r.f. cycle, because of the orthogonality between trigonometric functions, we have

$$[R(\omega) - \bar{R}(A)]A = 0 \tag{5.6}$$

$$[X(\omega) + \bar{X}(A)]A = 0 \tag{5.7}$$

or, equivalently,

$$[Z(\omega) - \bar{Z}(A)]I = 0 \tag{5.8}$$

where

$$I = Ae^{j(\omega t+\varphi)} \tag{5.9}$$

Instead of $\cos(\omega t+\varphi)$ and $\sin(\omega t+\varphi)$, if we multiply (5.5) by $\cos(n\omega t+\varphi_n)$ and $\sin(n\omega t+\varphi_n)$, respectively, and integrate over one r.f. cycle, we obtain a corresponding relation for the nth harmonic component, which we shall not be interested in. Since I is finite, Equation 5.8 indicates that $Z(\omega)$ must be equal to $\bar{Z}(A)$ for a steady state free

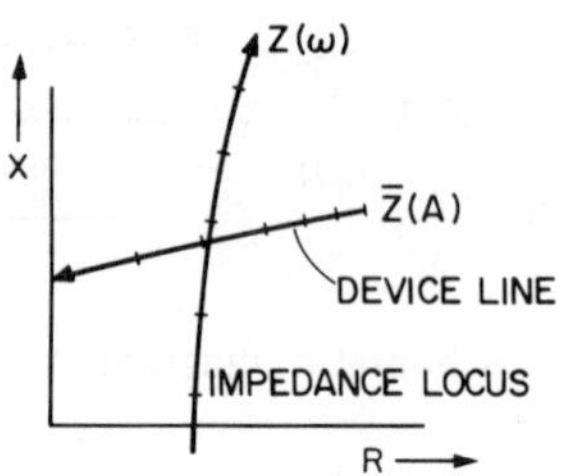

Figure 5.2 Device line and impedance locus

running oscillation:

$$Z(\omega) = \bar{Z}(A) \tag{5.10}$$

The oscillation frequency and amplitude can be determined by (5.10) as follows. Let us first draw the locus of the circuit impedance and that of the negative of the device impedance on the complex plane by varying ω and A, respectively, as shown in Figure 5.2. The locus of $Z(\omega)$ will be called the impedance locus. The arrowhead attached to the locus indicates the direction of increasing ω and the scale along the locus indicates equally spaced frequencies. The locus of $\bar{Z}(A)$ will be called the device line. When the r.f. current amplitude flowing through the device at the fundamental frequency is A, the point corresponding to $\bar{Z}(A)$ will be called the operating point. The arrowhead attached to the device line indicates the direction of increasing A. The scale along the device line indicates equal increments of the r.f. current amplitude A. The intersection of these two loci satisfies (5.10). Consequently, the operating point in the steady state is located at the intersection between the impedance locus and the device line and the oscillation frequency and amplitude can be determined by the scales on the loci at the intersection.[1]

In the above discussion, we neglected the frequency dependence of $\bar{Z}(A)$. This is usually justifiable since the device impedance is a slowly varying function of ω compared to the circuit impedance $Z(\omega)$. However, when we consider a wide frequency range, $\bar{Z}(A)$ cannot be considered to be independent of ω. In such a case, we may divide the frequency range of interest into several narrow frequency ranges and in each narrow range we may assume that $\bar{Z}(A)$ is independent of ω and determine the steady state operating point as before.

5.1.2 Differential equations[2]

Let us now turn our attention to the case in which both A and φ are allowed to be slowly varying functions of time. We assume that (5.1) still holds, namely, the harmonic components in $i(t)$ are kept small.

Then (5.2) will be also valid. Note that $\bar{R}(A)$ and $\bar{X}(A)$ change with A. The harmonic components of $v_d(t)$ depend primarily on A and φ. Since A and φ are assumed to be slowly varying functions of time, those harmonic components which depend primarily on A and φ are also slowly varying functions of time. When multiplied by $\cos(\omega t+\varphi)$ and $\sin(\omega t+\varphi)$, respectively, and integrated over one r.f. cycle, their contributions will vanish to the first order approximation. Equation 5.4 requires a deeper discussion. $Z(\omega)$ is originally obtained by first replacing d/dt by $j\omega$ in the differential equation expressing the relation between the voltage and current and then calculating the ratio between them. This is justifiable when A and φ are constant, since

$$\frac{d^n}{dt^n}\,\mathrm{Re}[Ae^{j(\omega t+\varphi)}] = \mathrm{Re}[(j\omega)^n Ae^{j(\omega t+\varphi)}]$$

However, when A and φ vary with time, the time derivative is given by

$$\frac{d}{dt}\,\mathrm{Re}[Ae^{j(\omega t+\varphi)}] = \mathrm{Re}\left\{\left[j\left(\omega+\frac{d\varphi}{dt}\right)+\frac{1}{A}\frac{dA}{dt}\right]Ae^{j(\omega t+\varphi)}\right\}$$

Similarly, when A and φ are slowly varying functions of time, the nth derivative is given by

$$\frac{d^n}{dt^n}\,\mathrm{Re}[Ae^{j(\omega t+\varphi)}] = \mathrm{Re}\left\{\left[j\left(\omega+\frac{d\varphi}{dt}\right)+\frac{1}{A}\frac{dA}{dt}\right]^n Ae^{j(\omega t+\varphi)}\right\} \tag{5.11}$$

to the first order approximation. Equation 5.11 indicates that if ω in $Z(\omega)$ is replaced by

$$\left(\omega+\frac{d\varphi}{dt}\right)-j\frac{1}{A}\frac{dA}{dt},$$

then (5.4) will give an approximate expression for $v_c(t)$.

To make the following discussion somewhat general, let us assume that a voltage source $e(t)$ is added in the oscillator circuit. Equation (5.5) is replaced by

$$v_c(t)+v_d(t)=e(t) \tag{5.12}$$

For free-running oscillations, $e(t)$ is zero. However, $e(t)$ will represent the noise voltage source at the device terminals when the noise behaviour of oscillators is studied and $e(t)$ will represent the injection signal voltage seen from the device terminals when injection locking is considered. Substituting (5.2) and the modified (5.4) into (5.12), multiplying by $\cos(\omega t+\varphi)$ and $\sin(\omega t+\varphi)$, respectively, and integrating over one r.f. cycle, we obtain

$$\mathrm{Re}\left\{\left[Z\left(\omega+\frac{\mathrm{d}\varphi}{\mathrm{d}t}-j\frac{1}{A}\frac{\mathrm{d}A}{\mathrm{d}t}\right)-\bar{Z}(A)\right]A\right\}=e_{\mathrm{c}}(t) \tag{5.13}$$

$$-\mathrm{Im}\left\{\left[Z\left(\omega+\frac{\mathrm{d}\varphi}{\mathrm{d}t}-j\frac{1}{A}\frac{\mathrm{d}A}{\mathrm{d}t}\right)-\bar{Z}(A)\right]A\right\}=e_{\mathrm{s}}(t) \tag{5.14}$$

where

$$e_{\mathrm{c}}(t)=\frac{2}{\tau}\int_{t-\tau}^{t}e(t)\cos(\omega t+\varphi)\,\mathrm{d}t \tag{5.15}$$

$$e_{\mathrm{s}}(t)=\frac{2}{\tau}\int_{t-\tau}^{t}e(t)\sin(\omega t+\varphi)\,\mathrm{d}t \tag{5.16}$$

and τ is the period of one r.f. cycle.

To obtain (5.13) and (5.14), we have used the fact that all the harmonic components are slowly varying functions of time and hence the orthogonality between trigonometric functions can be used to eliminate the effects of the harmonic components as we mentioned before. Equations 5.13 and 5.14 are the differential equations which govern the behaviour of amplitude A and phase φ. They usually contain the higher order terms of $d\varphi/\mathrm{d}t$ and $(1/A)(\mathrm{d}A/\mathrm{d}t)$. Consequently, it is difficult to solve them. To simplify the problem, let us assume that $d\varphi/\mathrm{d}t$ and $(1/A)(\mathrm{d}A/\mathrm{d}t)$ are sufficiently small compared to ω and the first two terms of the Taylor expansion around ω can approximate

$$Z\left(\omega+\frac{\mathrm{d}\varphi}{\mathrm{d}t}-j\frac{1}{A}\frac{\mathrm{d}A}{\mathrm{d}t}\right),$$

namely,

$$Z\left(\omega+\frac{\mathrm{d}\varphi}{\mathrm{d}t}-j\frac{1}{A}\frac{\mathrm{d}A}{\mathrm{d}t}\right)\simeq Z(\omega)+Z'(\omega)\left(\frac{\mathrm{d}\varphi}{\mathrm{d}t}-j\frac{1}{A}\frac{\mathrm{d}A}{\mathrm{d}t}\right) \tag{5.17}$$

where the prime indicates the derivative with respect to ω. Substituting (5.17) into (5.13) and (5.14) gives

$$R(\omega)-\bar{R}(A)+R'(\omega)\frac{\mathrm{d}\varphi}{\mathrm{d}t}+X'(\omega)\frac{1}{A}\frac{\mathrm{d}A}{\mathrm{d}t}=\frac{1}{A}e_{\mathrm{c}}(t) \tag{5.18}$$

$$-X(\omega)-\bar{X}(A)-X'(\omega)\frac{\mathrm{d}\varphi}{\mathrm{d}t}+R'(\omega)\frac{1}{A}\frac{\mathrm{d}A}{\mathrm{d}t}=\frac{1}{A}e_{\mathrm{s}}(t) \tag{5.19}$$

These are the equations which we will use in the study of oscillator stability, noise and injection locking.

5.1.3 Stability of operating points

Let us consider free-running oscillations again. In this case,

$$e_c(t) = e_s(t) = 0 \tag{5.20}$$

Furthermore, in the steady state

$$\frac{d\varphi}{dt} = \frac{dA}{dt} = 0 \tag{5.21}$$

Substituting (5.20) and (5.21) into (5.18) and (5.19), we have

$$R(\omega) - \bar{R}(A) = 0, \quad X(\omega) + \bar{X}(A) = 0 \tag{5.22}$$

which are equivalent to (5.10) as is expected. Equation 5.22 determines the steady state amplitude A_0 and the oscillation frequency ω_0. Now, suppose that A varies by a small amount δA from the steady state value A_0. If δA decays with time, the solution of (5.22), namely, the corresponding intersection between the impedance locus and the device line represents a stable operating point. If δA increases with time, the corresponding intersection is not realizable as a steady state operating point. So, let us investigate the behaviour of δA using (5.18) and (5.19). The first two terms of (5.18) are given by

$$R(\omega) - \bar{R}(A) = R(\omega_0) - \bar{R}(A_0 + \delta A) = -\frac{\partial \bar{R}(A_0)}{\partial A}\,\delta A \tag{5.23}$$

If we define the saturation factor of the negative resistance by

$$s = -\frac{A_0}{\bar{R}(A_0)}\,\frac{\partial \bar{R}(A_0)}{\partial A} \tag{5.24}$$

the first two terms of (5.18) becomes

$$R(\omega) - \bar{R}(A) = \frac{\delta A}{A_0}\, s\bar{R}(A_0) \tag{5.25}$$

Similarly, the first two terms of (5.19) can be expressed as

$$X(\omega) + \bar{X}(A) = \frac{\delta A}{A_0}\, r\bar{R}(A_0) \tag{5.26}$$

where r is the saturation factor of the device reactance defined by

$$r = \frac{A_0}{\bar{R}(A_0)}\,\frac{\partial \bar{X}(A_0)}{\partial A} \tag{5.27}$$

Substituting (5.25) and (5.26) into (5.18) and (5.19), we have

$$\frac{\delta A}{A_0}\, s\bar{R}(A_0) + R'(\omega_0)\frac{d\varphi}{dt} + X'(\omega_0)\frac{1}{A_0}\frac{d\delta A}{dt} = 0$$

$$-\frac{\delta A}{A_0}\, r\bar{R}(A_0) - X'(\omega_0)\frac{d\varphi}{dt} + R'(\omega_0)\frac{1}{A_0}\frac{d\delta A}{dt} = 0$$

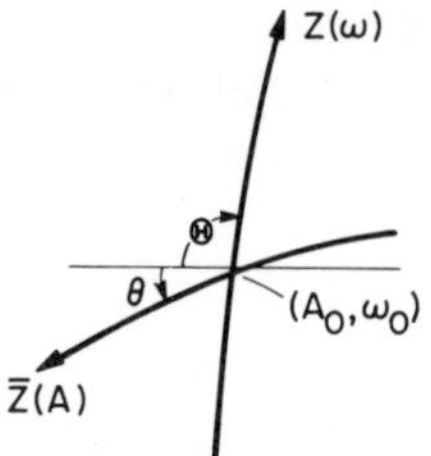

Figure 5.3 Defining θ and Θ

where (5.20) is used. Eliminating $\mathrm{d}\varphi/\mathrm{d}t$ from these two equations gives

$$[sX'(\omega_0) - rR'(\omega_0)]\bar{R}(A_0)\delta A + |Z'(\omega_0)|^2 \frac{\mathrm{d}\delta A}{\mathrm{d}t} = 0 \tag{5.28}$$

If δA is to decrease with time, the condition

$$[sX'(\omega_0) - rR'(\omega_0)]\bar{R}(A_0) > 0 \tag{5.29}$$

must be satisfied.

Referring to Figure 5.3, the saturation factors s and r can be expressed as follows:

$$s = -\frac{A_0}{\bar{R}(A_0)} \frac{\partial}{\partial A} \operatorname{Re} \bar{Z}(A_0) = \frac{A_0}{\bar{R}(A_0)} \left| \frac{\partial \bar{Z}(A_0)}{\partial A} \right| \cos\theta \tag{5.30}$$

$$r = -\frac{A_0}{\bar{R}(A_0)} \frac{\partial}{\partial A} \operatorname{Im} \bar{Z}(A_0) = \frac{A_0}{\bar{R}(A_0)} \left| \frac{\partial \bar{Z}(A_0)}{\partial A} \right| \sin\theta \tag{5.31}$$

Similarly, $R'(\omega_0)$ and $X'(\omega_0)$ can be expressed as

$$R'(\omega_0) = -|Z'(\omega_0)| \cos\Theta \tag{5.32}$$

$$X'(\omega_0) = |Z'(\omega_0)| \sin\Theta \tag{5.33}$$

Substituting (5.30) to (5.33) into (5.29), we obtain

$$A_0 \left| \frac{\partial \bar{Z}(A_0)}{\partial A} \right| |Z'(\omega_0)| \sin(\theta + \Theta) > 0 \tag{5.34}$$

The inequality (5.34) shows that $\sin(\theta+\Theta)$ must be positive for the intersection between the device line and the impedance locus to represent a stable operating point. In other words, the intersecting angle measured clockwise from the device line arrow direction to the impedance locus arrow direction must be less that 180° for a stable operation. As an example, suppose the impedance locus draws a loop as shown in Figure 5.4 and the device line intersects the impedance locus at three locations. The intersections P_1 and P_2 represent possible stable free-running operating points. On the other hand, P_3 is not realizable as

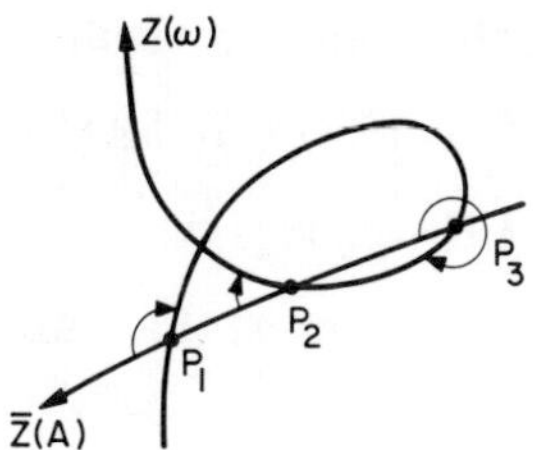

Figure 5.4 Explaining the stabilities of operating points

a stable operating point since the intersecting angle measured as stated above exceeds 180°. Which of the two possible points P_1 and P_2 represents the actual operating point is determined by the history of the oscillation as we shall discuss later.

5.1.4 Noise[(2–4)]

Because of the noise produced in the device and circuit, the amplitude and phase of the r.f. current in the oscillator fluctuate with time. Whatever the origin of the noise is, its effect can be represented by a noise voltage source in series with the device impedance. So, let $e(t)$ in (5.12) represent the noise voltage. The corresponding $e_c(t)$ and $e_s(t)$ will be written as

$$e_c(t) = n_1(t), \quad e_s(t) = n_2(t) \tag{5.35}$$

If $e(t)$ is random, there is no correlation between $n_1(t)$ and $n_2(t)$ and

$$\overline{n_1(t)^2} = \overline{2e(t)^2}, \quad \overline{n_2(t)^2} = \overline{2e(t)^2} \tag{5.36}$$

The factor 2 in (5.36) comes from the fact that the factor 2 in front of the integrals in (5.15) and (5.16) gives a factor of 4 when it is squared and the cosine and sine in the integrands give a factor of 1/2 when square-averaged. Following the procedure presented in the previous section, (5.18) and (5.19) give

$$\delta A\ s\bar{R}(A_0) + R'(\omega_0)\,A_0\,\frac{\mathrm{d}\varphi}{\mathrm{d}t} + X'(\omega_0)\,\frac{\mathrm{d}\delta A}{\mathrm{d}t} = n_1(t) \tag{5.37}$$

$$-\,\delta A\ r\bar{R}(A_0) - X'(\omega_0)\,A_0\,\frac{\mathrm{d}\varphi}{\mathrm{d}t} + R'(\omega_0)\,\frac{\mathrm{d}\delta A}{\mathrm{d}t} = n_2(t) \tag{5.38}$$

Eliminating $\mathrm{d}\varphi/\mathrm{d}t$ from these two equations, we have

$$[s\bar{R}(A_0)X'(\omega_0) - r\bar{R}(A_0)R'(\omega_0)]\ \delta A + |\,Z'(\omega_0)\,|^2\ \frac{\mathrm{d}\delta A}{\mathrm{d}t}$$

$$= X'(\omega_0)n_1(t) + R'(\omega_0)n_2(t)$$

In the frequency domain, the above equation reduces to

$$[s\bar{R}(A_0)X'(\omega_0) - r\bar{R}(A_0)R'(\omega_0)]\ \delta A(f) + |Z'(\omega_0)|^2\ j\omega\delta A(f)$$
$$= X'(\omega_0)n_1(f) + R'(\omega_0)n_2(f)$$

Consequently, the power spectrum of the amplitude fluctuation is given by

$$|\delta A(f)|^2 = \frac{2\,|Z'(\omega_0)|^2\ |e(f)|^2}{\omega^2\ |Z'(\omega_0)|^4 + [s\bar{R}(A_0)X'(\omega_0) - r\bar{R}(A_0)R'(\omega_0)]^2} \tag{5.39}$$

where use is made of (5.36) and no correlation between $n_1(t)$ and $n_2(t)$ is assumed. Note that ω in the denominator represents the base band frequency (angular) where the fluctuation is measured and not the r.f. frequency.

The power spectrum of the time derivation of the phase can be obtained in a manner similar to the above from (5.37) and (5.38). The result is given by

$$\left|\frac{d}{dt}\varphi(f)\right|^2 = \frac{2\ |e(f)|^2}{A_0{}^2}\ \frac{\omega^2\ |Z'(\omega_0)|^2 + (s^2 + r^2)\bar{R}(A_0)^2}{\omega^2\ |Z'(\omega_0)|^4 + [s\bar{R}(A_0)X'(\omega_0) - r\bar{R}(A_0)R'(\omega_0)]^2} \tag{5.40}$$

Since the time derivative of the phase corresponds to the angular frequency, (5.40) gives the power spectrum of the frequency fluctuation of the oscillator when divided by $(2\pi)^2$.

When the intersecting angle between the device line and the circuit locus becomes sharp, $s\bar{R}(A_0)X'(\omega_0)-r\bar{R}(A_0)R'(\omega_0)$ in the denominators of (5.39) and (5.40) becomes small since it is proportional to the sine of the intersecting angle. Then, the low frequency components of both the amplitude and frequency fluctuations become large. In other words, the oscillator gets noisy as the intersecting angle becomes sharp (close to 0° or 180°).

In the above discussion, the correlation between $n_1(t)$ and $n_2(t)$ is neglected. In practice, a finite correlation may exist and additional terms appear on the right-hand sides of (5.39) and (5.40). However, it remains true that the oscillator gets noisy as the intersecting angle between the device line and the impedance locus becomes sharp.

5.1.5 Hysteresis in tuning

The impedance locus of a typical multiple-tuned oscillator contains a loop as shown in Figure 5.4. Now, suppose a continuous adjustment of a tuning element brings the loop downwards, as illustrated in Figure

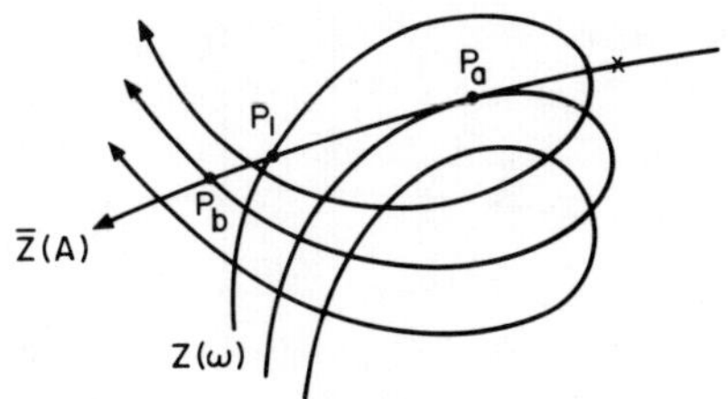

Figure 5.5 Explaining a jump of the operating point

5.5, and the maximum power is generated at the operating point marked by x along the device line. If the initial operating point is located at P_1, as the loop moves downwards, the operating point shifts to the right on the device line. The frequency of oscillation increases since the intersection moves along the impedance locus in the arrow direction. At the same time, the output power will increase since the operating P point approaches the maximum power point indicated by x on the device line. As the operating point approaches the upper edge of the loop, the oscillation gets noisy since the intersecting angle between the device line and the impedance locus becomes sharp. When the upper edge of the loop separates from the device line, the operating point jumps from P_a to P_b in the figure, since with the disappearance of the intersection P_a, no steady state free-running oscillation is possible near the upper edge of the loop. The oscillation frequency suddenly increases and the power drops as shown in Figure 5.6. If we reverse the adjustment of the tuning element, the frequency gradually decreases

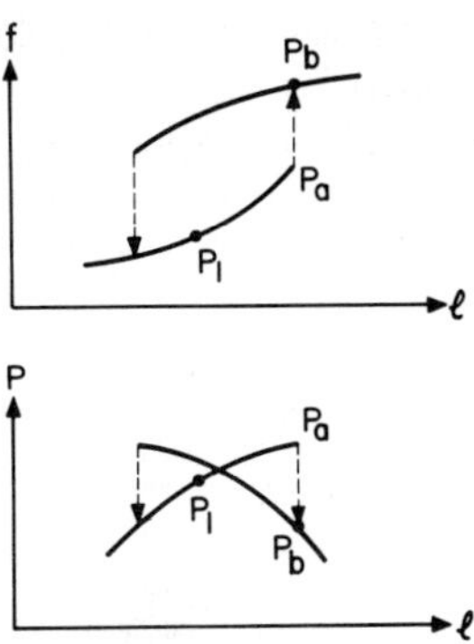

Figure 5.6 Typical frequency and power tuning hysteresis of a multiple-tuned oscillator

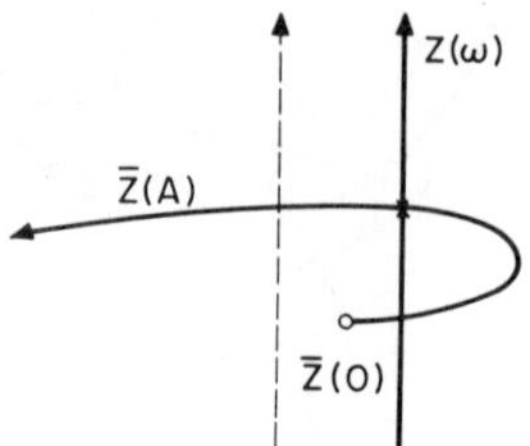

Figure 5.7 Typical switch-on hysteresis

and the power increases. When the lower edge of the loop separates from the device line, another jump of the operating point takes place. The oscillation frequency as well as the power suddenly decreases. If the original position of the tuning element is restored after this jump, the operating point comes back to the original position P_1. We have clearly seen hysteresis in the tuning.

In practice, the impedance locus may have many loops and the loops change their sizes and positions during tuning. So, very complicated and frustrating tuning behaviour can be experienced with multiple-tuned oscillators.

Another interesting type of hysteresis will be observed if the device line has a shape like the one illustrated in Figure 5.7. Suppose that the impedance locus is initially located at the position shown by the dotted vertical line and the maximum power is generated at the point marked by x on the device line. The operating point is located at the intersection of the device line and the dotted impedance locus. Now, by changing the coupling to the load, the impedance locus can be brought to the position shown by the solid vertical line thus achieving the maximum power. Then, suppose the d.c. power supply is turned off. When the power supply is turned on the next time, no oscillation will start since the small signal device impedance $\bar{Z}(0)$ is not on the right-hand side of the impedance locus and hence the total resistance in the

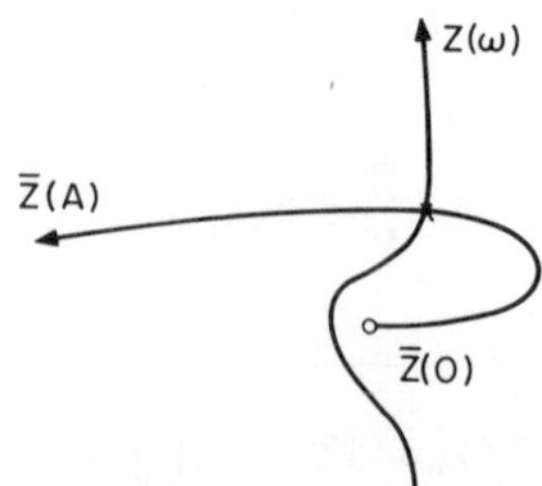

Figure 5.8 A method of correcting the situation in Figure 5.7

circuit is positive for a small signal. In order to start the oscillation, the impedance locus must be moved sufficiently towards the left so that $\overline{Z}(0)$ appears on the right-hand side of the impedance locus. In other words, a retuning is necessary every time, after the power supply is turned off, if the maximum power is to be achieved. This kind of tuning behaviour is not at all unusual with Gunn diodes over a certain frequency range slightly above the transit time frequency. The only way to avoid retuning with this kind of device is to devise a special circuit with an impedance locus like the one shown in Figure 5.8.

5.1.6 Efficiency[5]

So far, we have studied the behaviour of oscillators at the device terminals. The generated power P is given by $\frac{1}{2}\overline{R}(A_0)A_0^2$. However, the output power P_0 is generally smaller than the generated power because of the circuit loss which we have not considered so far. The ratio of the output power to the generated power is called the circuit efficiency, η. We shall calculate the circuit efficiency and derive a few related formulas which we will use in the later discussions. Let the load resistance be R_L and the oscillator circuit be represented by a two-port pair network connecting the device impedance and R_L as shown in Figure 5.9. The relations between the terminal voltages and currents are

$$V_1 = Z_{11}I_1 + Z_{12}I_2 \tag{5.41}$$

$$V_2 = Z_{12}I_1 + Z_{22}I_2 \tag{5.42}$$

The network is assumed to be reciprocal. Since $V_1 = (\overline{R} - j\overline{X})I_1$ and $V_2 = -R_L I_2$, we obtain

$$I_2 = \frac{-Z_{12}}{Z_{11} - \overline{R} + j\overline{X}} \tag{5.43}$$

$$I_2 = \frac{-Z_{12}}{Z_{22} + R_L} I_1 \tag{5.44}$$

from (5.41) and (5.42), respectively. Combining these two equations,

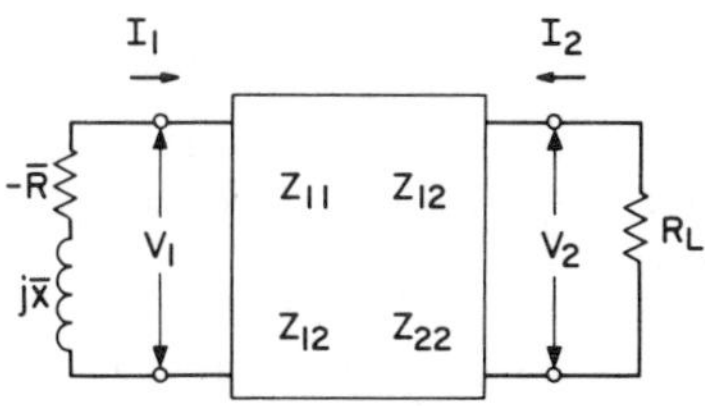

Figure 5.9 Two-terminal network representing the oscillator circuit

we have

$$Z_{12}{}^2 = (Z_{11} - \bar{R} + j\bar{X})(Z_{22} + R_L) \tag{5.45}$$

The efficiency is defined as

$$\eta = \frac{P_0}{P} = \frac{\frac{1}{2}R_L \; |I_2|^2}{\frac{1}{2}\bar{R}(A_0)\,|I_1|^2}$$

Substituting (5.44) for I_2 and then using (5.45), we have

$$\eta = \frac{R_L}{\bar{R}(A_0)} \left| \frac{Z_{12}}{Z_{22} + R_L} \right|^2 = \frac{R_L}{\bar{R}(A_0)} \left| \frac{Z_{11} - \bar{R} + j\bar{X}}{Z_{22} + R_L} \right| \tag{5.46}$$

Let us next consider the injection of a signal from the load into the oscillator. The result will be used in the discussion of injection locking. The equivalent circuit is given by Figure 5.10. The injection signal is represented by a voltage source E_L. Since

$$V_2 = Z_{22}I_2 + Z_{12}I_1 = E_L - R_LI_2,$$

we have

$$I_2 = \frac{E_L - Z_{12}I_1}{Z_{22} + R_L} \tag{5.47}$$

Substituting (5.47) for I_2 in (5.41), we obtain

$$V_1 = \left(Z_{11} - \frac{Z_{12}{}^2}{Z_{22} + R_L} \right) I_1 + \frac{E_L Z_{12}}{Z_{22} + R_L} \tag{5.48}$$

Equation (5.48) indicates that E_L is transformed by the circuit to

$$E = \frac{E_L Z_{12}}{Z_{22} + R_L} \tag{5.49}$$

and presented to the device through a series impedance $Z_{11} - Z_{12}^2/(Z_{22} + R_L)$ as illustrated in Figure 5.11. The injection signal power is given by

$$P_i = |E_L|^2/8R_L$$

which is equal to the available power from the load. The corresponding power at the device terminals is given by $|E|^2$ divided by $8\text{Re}[Z_{11} - Z_{12}^2/(Z_{22} + R_L)]$. Since this is equal to $8\bar{R}(A_0)$, the injection signal power at the device terminals is given by

$$\frac{|E|^2}{8\bar{R}(A_0)} = \frac{|E_L|^2}{8R_L} \frac{R_L}{\bar{R}(A_0)} \left| \frac{Z_{12}}{Z_{22} + R_L} \cdot \right|^2 = \eta P_i \tag{5.50}$$

where use is made of (5.49) and the first equality in (5.46). Equation

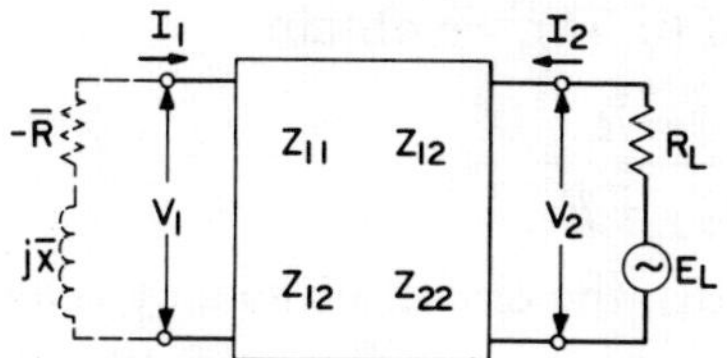

Figure 5.10 Calculation of the injection signal power

(5.50) indicates that the injection signal power undergoes the same circuit loss as the generated power when it goes through the circuit in the reverse direction.

Using (5.4) and (5.44) the circuit impedance the device sees is given by

$$Z(\omega) = Z_{11} - \frac{Z_{12}^{2}}{Z_{22} + R_{\text{L}}} \tag{5.51}$$

Differentiating with respect to ω at ω_0, we have

$$\begin{aligned} Z'(\omega_0) &= Z'_{11} + \frac{Z'_{22}Z_{12}^{2}}{(Z_{22} + R_{\text{L}})^2} - \frac{2Z'_{12}Z_{12}}{Z_{22} + R_{\text{L}}} \\ &= Z'_{11} + \frac{Z'_{22}(Z_{11} - \bar{R} + j\bar{X}) - 2Z_{12}Z'_{12}}{Z_{22} + R_{\text{L}}} \end{aligned} \tag{5.52}$$

where use is made of (5.45). Similarly, from (5.42) and (5.43) the output impedance of the oscillator is given by

$$Z_{\text{out}}(\omega) = Z_{22} - \frac{Z_{12}^{2}}{Z_{11} - \bar{R} + j\bar{X}} \tag{5.53}$$

Differentiating with respect to ω at ω_0, we have

$$Z'_{\text{out}}(\omega_0) = Z'_{22} + \frac{Z'_{11}(Z_{22} + R_{\text{L}}) - 2Z_{12}Z'_{12}}{Z_{11} - \bar{R} + j\bar{X}} \tag{5.54}$$

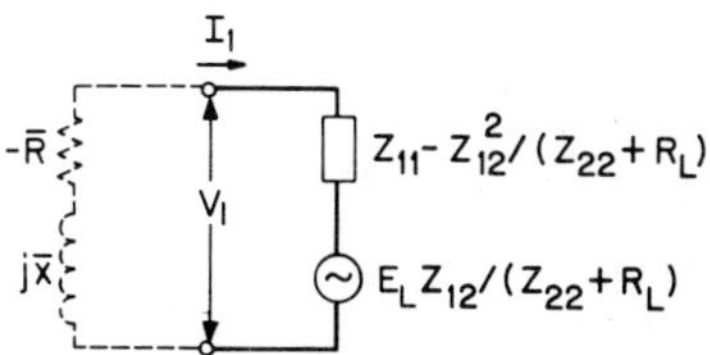

Figure 5.11 The circuit impedance with the injection signal seen by the device

Combining (5.52) and (5.54), we obtain

$$\frac{Z'(\omega_0)}{Z'_{out}(\omega_0)} = \frac{Z_{11} - \bar{R} + j\bar{X}}{Z_{22} + R_L} \tag{5.55}$$

Substituting into (5.46), the circuit efficiency is also expressed as

$$\eta = \frac{R_L}{\bar{R}(A_0)} \left| \frac{Z'(\omega_0)}{Z'_{out}(\omega_0)} \right| \tag{5.56}$$

Note that $|Z'(\omega_0)|/\bar{R}(A_0)$ and $|Z'_{out}(\omega_0)|/R_L$ indicate how quickly the normalized circuit and output impedance vary with ω, respectively. Their ratio gives the circuit efficiency.

Finally, let us consider the partial derivative of Z_{out} with respect to A at A_0.

$$\frac{\partial Z_{out}}{\partial A} = \frac{Z_{12}{}^2 \left(-\frac{\partial \bar{R}}{\partial A} + j \frac{\partial \bar{X}}{\partial A} \right)}{(Z_{11} - \bar{R} + j\bar{X})^2} = \frac{Z_{22} + R_L}{Z_{11} - \bar{R} + j\bar{X}} \left(-\frac{\partial \bar{R}}{\partial A} + j \frac{\partial \bar{X}}{\partial A} \right)$$

$$= \frac{Z'_{out}(\omega_0)}{Z'(\omega_0)} \left(-\frac{\partial \bar{R}}{\partial A} + j \frac{\partial \bar{X}}{\partial A} \right) \tag{5.57}$$

Substituting this into (5.56), the circuit efficiency can also be expressed as

$$\eta = \frac{R_L}{\bar{R}(A_0)} \frac{\left| -\frac{\partial \bar{R}}{\partial A} + j \frac{\partial \bar{X}}{\partial A} \right|}{\left| \frac{\partial Z_{out}}{\partial A} \right|} \tag{5.58}$$

Note that

$$\left| -\frac{\partial \bar{R}}{\partial A} + j \frac{\partial \bar{X}}{\partial A} \right| \Big/ \bar{R}(A_0) \text{ and } \left| \frac{\partial Z_{out}}{\partial A} \right| \Big/ R_L$$

indicate how quickly the normalized device impedance and the normalized output impedance vary with A and their ratio gives the circuit efficiency. Equation 5.57 can be rewritten as follows

$$\frac{\dfrac{\partial Z_{out}}{\partial A}}{\dfrac{\partial Z_{out}}{\partial \omega}} = -\frac{\left(\dfrac{\partial \bar{R}}{\partial A} - j \dfrac{\partial \bar{X}}{\partial A} \right)}{\dfrac{\partial Z}{\partial \omega}} \tag{5.59}$$

Equation 5.59 indicates that the angle between the device line and the impedance locus is equal to the angle between the constant frequency and constant A loci of the output impedance of the oscillator.

5.1.7 Injection locking[1,2,6–8]

When a small signal is injected into an oscillator, the oscillation frequency is synchronized to the frequency of the injected signal provided that it is sufficiently close to the free-running frequency of the oscillator. This phenomenon is called injection locking. The small locking signal is usually applied to the oscillator through a circulator as shown in Figure 5.12. The circuit impedance is the same as that of the free-running oscillator. The injection signal can be represented by a series voltage source

$$e(t) = |E| \cos \omega_i t \tag{5.60}$$

Equation 5.18 and 5.19 become

$$R(\omega_i) - \bar{R}(A) + R'(\omega_i) \frac{d\varphi}{dt} + X'(\omega_i) \frac{1}{A} \frac{dA}{dt} = \frac{1}{A} |E| \cos \varphi \tag{5.61}$$

$$-X(\omega_i) - \bar{X}(A) - X'(\omega_i) \frac{d\varphi}{dt} + R'(\omega_i) \frac{1}{A} \frac{dA}{dt} = \frac{1}{A} |E| \sin \varphi \tag{5.62}$$

Although ω is now replaced by ω_i, the above equations do not necessarily assume the oscillation to be locked since the oscillation frequency is given by $\omega_i + (d\varphi/dt)$ and $d\varphi/dt$ is yet to be determined.

Now assume that the oscillator is locked. In the steady state, $dA/dt = d\varphi/dt = 0$. Therefore, (5.61) and (5.62) reduce to

$$R(\omega_i) - \bar{R}(A) = \frac{1}{A} |E| \cos \varphi \tag{5.63}$$

$$-X(\omega_i) - \bar{X}(A) = \frac{1}{A} |E| \sin \varphi \tag{5.64}$$

If $|E|$ is small, A is expected to stay close to the free-running value A_0. Then, A on the right-hand side can be replaced by A_0 to the first order approximation. The above equations are now equivalent to

$$Z(\omega_i) - \bar{Z}(A) = \frac{|E|}{A_0} e^{-j\varphi} \tag{5.65}$$

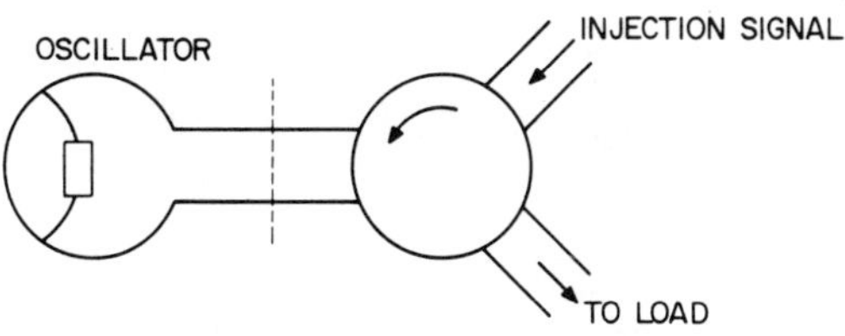

Figure 5.12 Injection locking experiment of a microwave oscillator

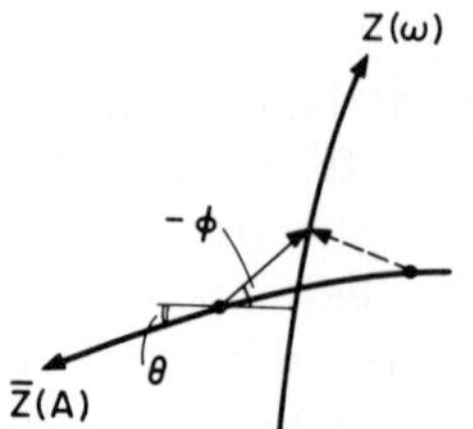

Figure 5.13 Definition of the injection vector

where φ is the phase of the device current referred to the injection signal. Equation (5.65) indicates that the vector drawn from the operating point $\bar{Z}(A)$ on the device line to the ω_i point on the impedance locus should have the amplitude given by $|E|/A_0$. There are in general two possible vectors which satisfy this condition as shown in Figure 5.13 by the solid and dotted arrows. However, only the solid arrow represents the stable locking as we shall see next.

Let us assume that A and φ are slightly perturbed and become $A_0 + \delta A$ *and* $\varphi + \mathrm{d}\varphi$ respectively. Then, (5.61) and (5.62) become

$$\frac{\delta A}{A_0} s\bar{R}(A_0) + R'(\omega_i)\frac{\mathrm{d}\delta\varphi}{\mathrm{d}t} + X'(\omega_i)\frac{1}{A_0}\frac{\mathrm{d}\delta A}{\mathrm{d}t} = -\frac{|E|}{A_0}\sin\varphi\delta\varphi$$

$$-\frac{\delta A}{A_0} r\bar{R}(A_0) - X'(\omega_i)\frac{\mathrm{d}\delta\varphi}{\mathrm{d}t} + R'(\omega_i)\frac{1}{A_0}\frac{\mathrm{d}\delta A}{\mathrm{d}t} = \frac{|E|}{A_0}\cos\varphi\delta\varphi$$

where s and r are defined by (5.24) and (5.27), respectively. Eliminating δA from the above equations gives

$$|Z'|^2\frac{\mathrm{d}^2\delta\varphi}{\mathrm{d}t^2} + \left[\bar{R}(A_0)(sX' - rR') + \frac{|E|}{A_0}(R'\sin\varphi + X'\cos\varphi)\right]$$
$$\frac{\mathrm{d}\delta\varphi}{\mathrm{d}t} + \bar{R}(A_0)\frac{|E|}{A_0}\sqrt{s^2 + r^2}\cos(\varphi + \theta)\,\delta\varphi = 0$$

For stable operation, $\delta\varphi$ must decay with time. This requires that the coefficients in the above differential equation be positive:

$$\bar{R}(A_0)(sX' - rR') + \frac{|E|}{A_0}(R'\sin\varphi + X'\cos\varphi) > 0 \tag{5.66}$$

$$\cos(\varphi + \theta) > 0 \tag{5.67}$$

When $|E|$ is small as we have assumed, the first condition is usually satisfied because of the stable free-running condition (5.29) and $\omega_i \simeq \omega_0$. The second condition (5.67) determines whether the locking is stable. The solid arrow satisfies (5.67) but the dotted arrow does not. So, we will consider only the solid arrow which represents the stable locking. This arrow is called the injection vector. Its magnitude is equal to $|E|/A_0$ and its angle represents the phase of the injection signal relative to that of the device current. From (5.50), we have

$$|E|^2 = 8\bar{R}(A_0)\eta P_i \tag{5.68}$$

The output power P_0 is η times the generated power $\frac{1}{2}\bar{R}(A_0)A_0^2$;

$$P_0 = \tfrac{1}{2}\eta\bar{R}(A_0)A_0^2 \tag{5.69}$$

From (5.68) and (5.69), the magnitude of the injection vector is calculated to be

$$\frac{|E|}{A_0} = 2\eta\bar{R}(A_0)\sqrt{\frac{P_i}{P_0}} \tag{5.70}$$

If the injection signal frequency ω_i is varied while keeping the power P_i constant, the injection vector moves as shown in Figure 5.14. Below ω_1 and above ω_2, the locking does not take place since the vector relation (5.65) is not satisfied. The distance from the device line to the corresponding point $Z(\omega)$ for $\omega < \omega_1$ and $\omega > \omega_2$ becomes larger than $|E|/A_0 = 2\eta\bar{R}(A_0)\sqrt{P_i/P_0}$. Between ω_1 and ω_2, (5.65) is satisfied and the locking takes place. From ω_1 to ω_2, the phase $(-\varphi)$ changes 180° from $-90^\circ+\theta$ to $90^\circ+\theta$. Near ω_1 in the locking range, φ is positive and the phase of the device current leads the injection signal phase while near ω_2 the phase of the device current lags behind the injection signal. This is understandable since the free-running frequency of the oscillator is ω_0 which is higher than ω_1 but lower than ω_2. Near ω_1, the oscillator tries to oscillate at ω_0 and the phase tends to become faster but the injection signal is holding back the phase for synchronization. Near ω_2, the injection signal is pulling the phase forward for synchronization. Near ω_0, this simple picture of pulling forward and

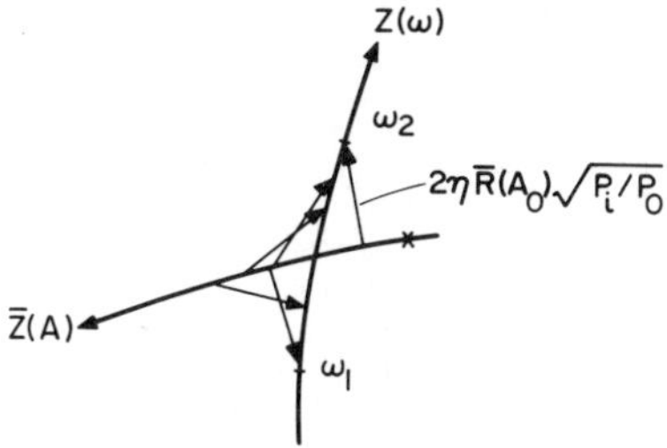

Figure 5.14 Motion of the injection vector with ω_i

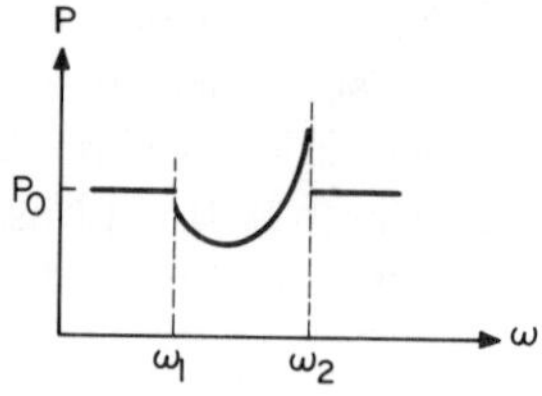

Figure 5.15 Output power versus injection signal frequency

holding back the phase fails since the effect of the device reactance becomes predominant there.

Suppose the maximum power point is located at the point marked by x on the device line. As ω varies in the locking range, the power changes as shown in Figure 5.15. As soon as the locking takes place at ω_1, the power decreases because the corresponding operating point is now located farther away from the maximum power point than the free-running point. As ω increases from ω_1, the power first decreases and then starts increasing because the operating point now moves towards the maximum power point. As the unlocking takes place at ω_2, the power returns to the free-running value.

If the maximum power point is located on the other side of the impedance locus, the power curve will become convex rather than concave. If the maximum power point is located close to the free-running operating point, the first order variation of the output power becomes zero as ω_i varies near ω_0 and independent of $|E|$. This indicates that an excellent amplitude limiter can be built using a locked oscillator.

The locking range of an oscillator can be calculated as follows. At the extremes of the locking range, the injection vector will be located as shown in Figure 5.16. From this figure, the frequencies $\omega_0 \pm \Delta\omega$ at the

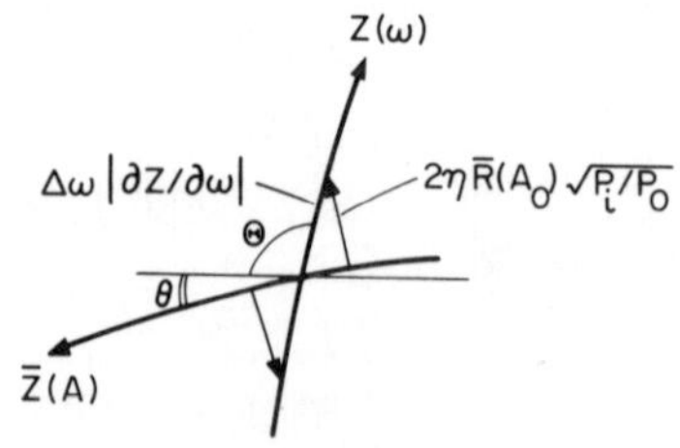

Figure 5.16 Calculation of the locking range

extremes can be obtained. The frequency increment $\Delta\omega$ satisfies

$$\Delta\omega \left|\frac{\partial Z}{\partial \omega}\right| \sin(\Theta + \theta) = 2\eta\bar{R}(A_0)\sqrt{\frac{P_i}{P_0}}$$

or equivalently

$$\frac{\Delta\omega}{\omega_0} = \frac{2\eta\bar{R}(A_0)}{\omega_0 \left|\frac{\partial Z}{\partial \omega}\right| \sin(\Theta + \theta)}\sqrt{\frac{P_i}{P_0}} \tag{5.71}$$

For single-tuned oscillator circuits,

$$Z = j\left(\omega L - \frac{1}{\omega C}\right) + R_a + R_L \simeq 2j(\omega - \omega_a)L + R_a + R_L \tag{5.72}$$

where LC represents the resonator, R_a the resonator loss, R_L the load and

$$\omega_a = \frac{1}{\sqrt{LC}}$$

Since, in this case

$$\frac{\partial Z}{\partial \omega} \simeq 2L, \quad \Theta = 90^\circ, \quad \eta = \frac{R_L}{R_L + R_a} = \frac{R_L}{\bar{R}(A_0)}$$

(5.71) reduces to

$$\frac{\Delta\omega}{\omega_0} = \frac{1}{Q_{ext}}\sqrt{\frac{P_i}{P_0}}\frac{1}{\cos\theta} \tag{5.73}$$

where

$$Q_{ext} = \frac{\omega_0 L}{R_L} \tag{5.74}$$

The total locking range is given by $2\Delta\omega$. Note that R_a does not appear in the formula. The Q in the locking formula is not the loaded Q but the external Q of the resonator. The locking range is proportional to the square root of the injection signal power and inversely proportional to the external Q of the cavity. When the device line is not perpendicular to the impedance locus, the locking range increases by a factor of $1/\cos\theta$.

The above discussion assumes that the injection signal and hence the perturbation given by the injection signal to the oscillator is small, allowing the first order approximation. If the injection signal becomes large, however, the approximation used in the above discussion becomes no longer valid and a different approach must be employed. The interested reader is referred to Reference 8.

5.2 PRACTICAL OSCILLATOR CIRCUITS

5.2.1 Coaxial mounts

One of the most popular circuits used for device evaluation is the coaxial mount with slug tuners as illustrated in Figure 5.17. The positions of the slug tuners can be adjusted by trial and error until the impedance locus intersects the device line in the proper direction for a stable oscillator. Once, oscillation takes place, the output power and frequency can be monitored to optimize the slug tuner positions for a desired frequency or a maximum power. The coaxial mount is popular since it is versatile and its design requires essentially no prior knowledge of the device impedance. A wide range of device impedance is acceptable to this mount. However, it is difficult to obtain the maximum power, desired frequency and acceptable noise property all simultaneously. Discontinuous changes in the frequency and amplitude discussed in Section 5.5 will be observed with this mount and hysteresis in tuning makes it virtually impossible to have confidence that the best performance for a given device has been achieved. Often, the adjustment is trapped near a local optimum. In the extreme case, one may get an optimum with second harmonic tuning without realizing it.

A slightly more controllable behaviour is achieved with the circuit in Figure 5.18. A quarter-wavelength transformer is inserted near the device. The spacing l between the coaxial short and the transformer primarily determines the oscillation frequency and the transformer ratio determines the loading or, equivalently, the operating point along the device line. Several transformers with different diameters may be needed. They will give different transformer ratios and hence different operating points, one of which is hopefully near-optimum. A slide screw tuner can be inserted between the oscillator and the load for fine tuning. However, the penetration of the screw into the coaxial line must be kept minimum if hysteresis in tuning is to be avoided.

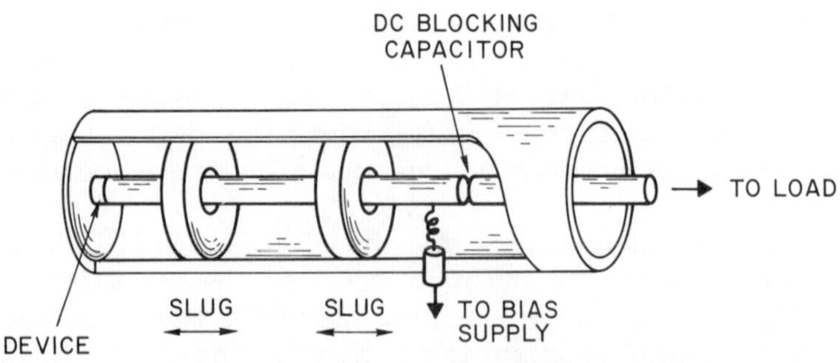

Figure 5.17 Oscillator circuit with slug tuners

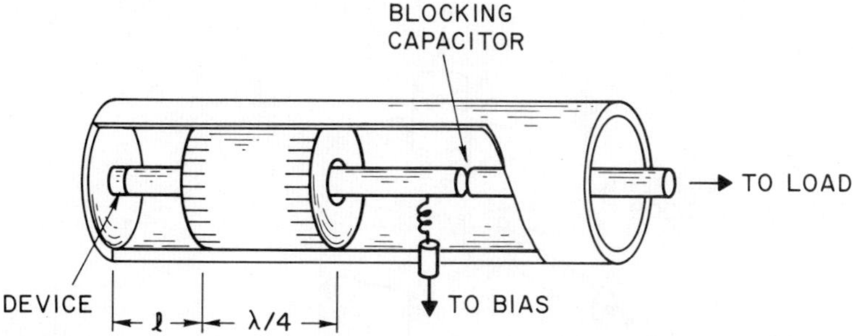

Figure 5.18 Oscillator circuit with a quarter-wavelength transformer

5.2.2 Cavity oscillators

Coaxial cavities like the one shown in Figure 5.19 are also widely used at low microwave frequencies. The oscillation frequency can be adjusted by inserting a screw through the side wall and the loading can be adjusted by rotating the coupling loop to the load. The coupling to the load can also be accomplished by a probe, of which the penetration will determine the loading. Figure 5.20 shows a semi-coaxial cavity. The oscillation frequency is adjusted by changing the capacitance between the centre-conductor and the device. When a waveguide output is desired, a waveguide cavity may be advantageous. Figure 5.21 illustrates such an oscillator. The window opening determines the loading. The oscillation frequency can be adjusted by inserting a dielectric rod or a metallic rod into the cavity. A dielectric rod is often preferred because the cut-off waveguide effect of the thin dielectric rod itself can be utilized to prevent r.f. leakage. Metallic contacts often show erratic behaviour at microwave frequencies and a choke arrangement may become necessary when a metallic rod is used for the frequency

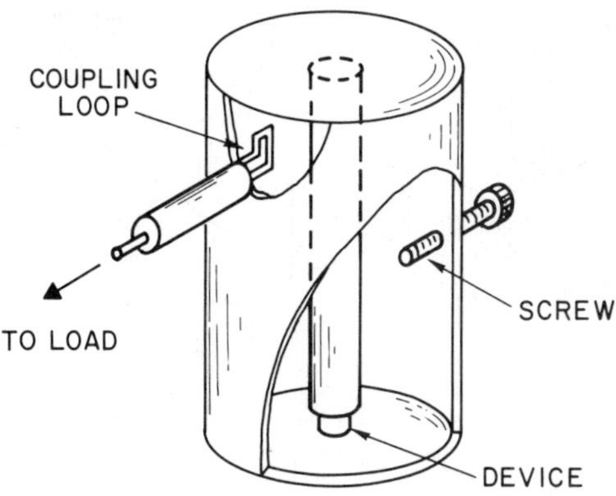

Figure 5.19 Coaxial cavity oscillator

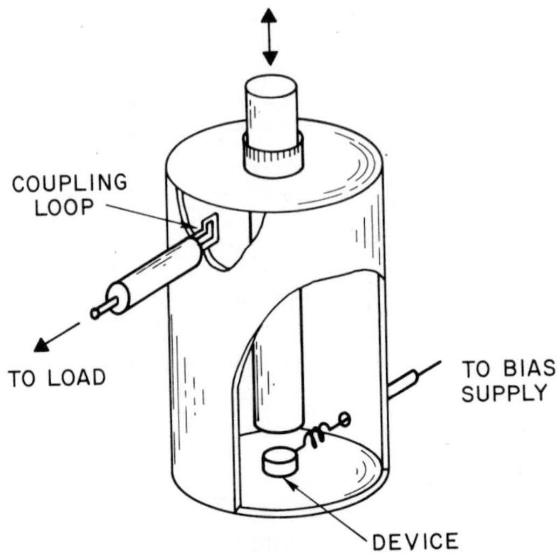

Figure 5.20 Semi-coaxial cavity oscillator

adjustment. It may be worth noting that the distance from the window to the device rather than the cavity length is approximately half a guided wavelength.

5.2.3 Resonant cap circuits[9,10]

At millimetre wave frequencies, the cavity dimensions become so small that it becomes difficult to keep the circuit loss low. Since the circuit

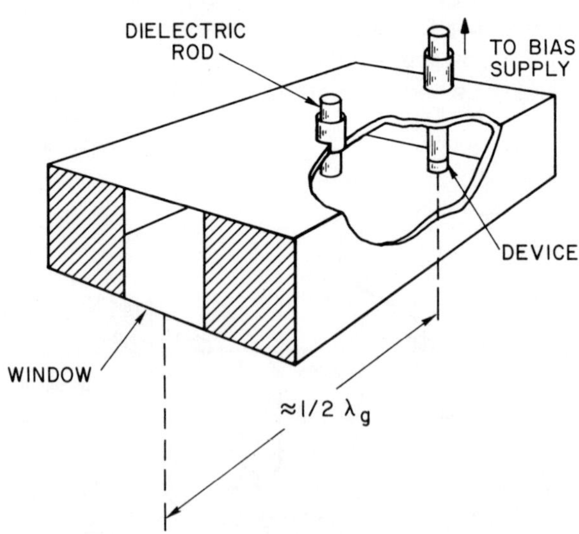

Figure 5.21 Waveguide cavity oscillator

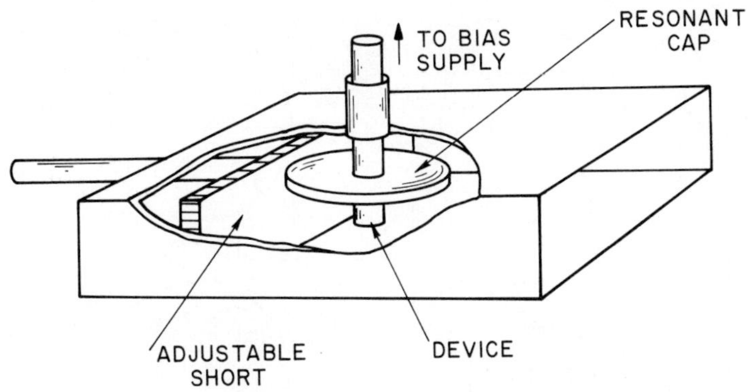

Figure 5.22 Resonant cap oscillator

loss brings the impedance locus away from the periphery of the Smith chart, the impedance locus will not intersect the device line when the latter is short. This makes oscillation often unattainable in the available range of circuit adjustment. This problem of circuit loss is crucial for low impedance impatt diodes. To overcome the loss problem, resonant cap circuits such as the one illustrated in Figure 5.22 are often employed.[11] An equivalent circuit of the resonant cap circuit is shown in Figure 5.23. The lower side of the cap works as a low impedance radial transmission line approximately a quarter wavelength long. It is represented by a uniform transmission line and a series inductance X_C in the equivalent circuit. The device is connected in series with X_C. The other end of the transmission line is connected to an inductance X_L approximately representing the post above the cap in the waveguide. The inductance is then connected to a short-circuited transmission line representing the waveguide. The distance l from the centre of the cap to the waveguide short is adjustable.

To simplify our discussion, let us first neglect the frequency dependence of all the components except that of the quarter-wavelength transmission line representing the cap. The impedance of

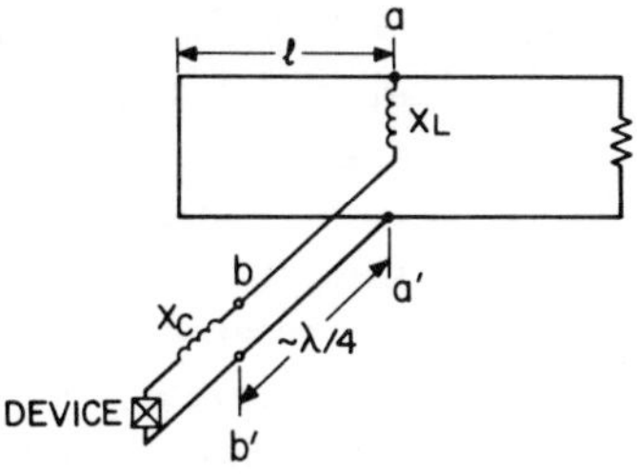

Figure 5.23 An equivalent circuit of the resonant cap oscillator

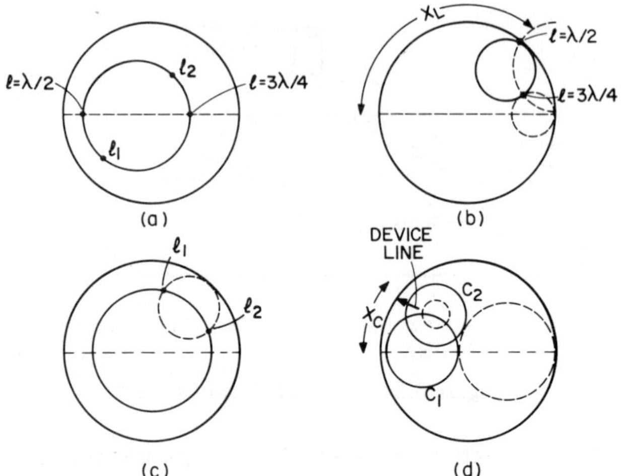

Figure 5.24 Explaining the tuning characteristic of the resonant cap oscillator

the waveguide seen from a—a′ varies as shown in Figure 5.24(a) as l varies. The reason why the locus does not touch the periphery of the Smith chart is that the waveguide has finite circuit loss. The Smith chart is normalized to the characteristic impedance of the transmission line representing the cap. Adding jX_L, the impedance locus becomes the solid line in Figure 5.24(b). The impedance locus looking at b—b′ becomes a circle of which the diameter is determined by the waveguide short position l. The loci for l_1 and l_2 have the same diameters as shown in Figure 5.24(c). If the Smith chart is normalized to 50 Ω, the impedance locus becomes a circle such as C_1 in Figure 5.24(d). Adding X_C, the impedance locus seen by the device is obtained. This is indicated by C_2. The intersection with the device line determines the oscillation frequency and the power. Since the operating points corresponding to two different short positions l_1 and l_2 are identical, the corresponding generated powers are the same although the oscillation frequencies determined by the scales on the impedance locus are different. The contributuion of the circuit loss to the total resistance seen by the device is, however, greater for l_1 than for l_2, as one can easily see from Figure 5.24(a). Consequently, the circuit efficiency and hence the output power is higher for l_2 than for l_1. When l is equal to $\lambda_g/2$, the diameter of the impedance locus C_2 becomes large and almost all the generated power will be consumed in the circuit loss and little output power is obtained to the waveguide load. Beyond a certain range of l, the impedance locus will be so far away from the periphery of the Smith chart, as exemplified by the dotted circle in Figure 5.24(d), that the intersection with the device

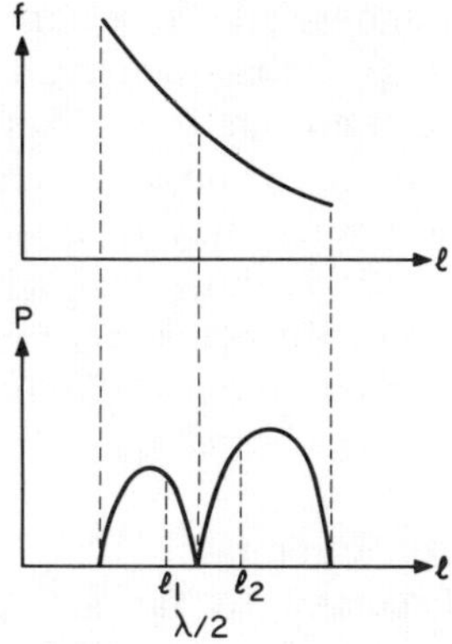

Figure 5.25 Frequency and power versus the distance l from the centre of the cap to the waveguide short

line may cease to exist. As a result, the oscillation frequency and output power may vary as shown in Figure 5.25. Note that a fairly high waveguide impedance is transformed to a fairly low impedance by the low impedance transmission line representing the resonant cap. This is the primary reason why the resonant cap circuit can be advantageously used with low impedance devices such as impatt diodes at millimetre wave frequencies. So far, we have neglected the frequency dependencies of X_L, X_C and the short-circuited section l of the waveguide. When they are taken into account, the impedance locus with a fixed l may look like the one illustrated in Figure 5.26. This is because the electrical length of the short-circuited section as well as X_L and X_C increases with increasing f. Furthermore additional loops may appear in the

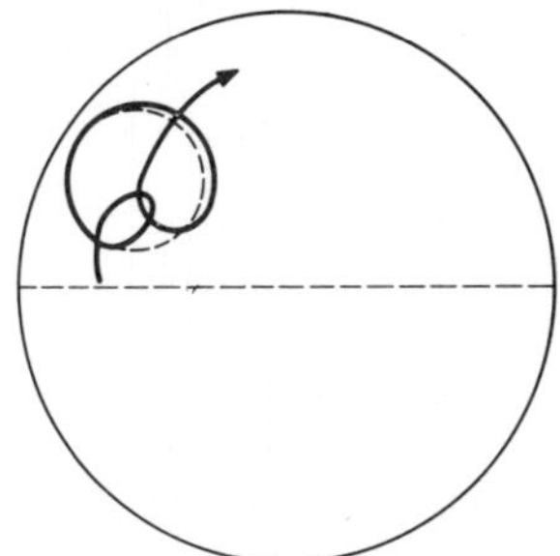

Figure 5.26 Impedance locus of the resonant cap circuit

impedance locus due to higher-order resonant modes within the space underneath the resonant cap. Consequently, a typical impedance locus will have many loops, which change their sizes as well as their positions as l varies. This means a good possibility of hysteresis in tuning. Nevertheless, the oscillation frequency of a well designed resonant cap circuit can be varied over 10%—20% frequency range by adjusting l without hysteresis. Since the contribution of circuit losses can be easily minimized, the resonant cap circuits are widely used with impatt diodes at millimetre wave frequencies.

5.2.4 Single-tuned oscillator circuits[12–14]

The oscillator circuits discussed so far are all designed to have the smallest possible circuit loss in order to get the maximum output power from a given device. This is done without paying due attention to complications in tuning. As a result, hysteresis in tuning accompanied by frequency and amplitude jumps is frequently observed. If a small circuit loss is accepted from the beginning, however, oscillators with well-behaved tuning characteristics can be designed. One such circuit is illustrated in Figure 5.27. In this circuit, the device is mounted at one end of a coaxial line coupled to a waveguide cavity which is in turn coupled to the load through a rotary joint. The other end of the coaxial line is r.f. terminated but d.c. insulated and the d.c. power is supplied through the centre conductor.[15,16] The equivalent circuit of the

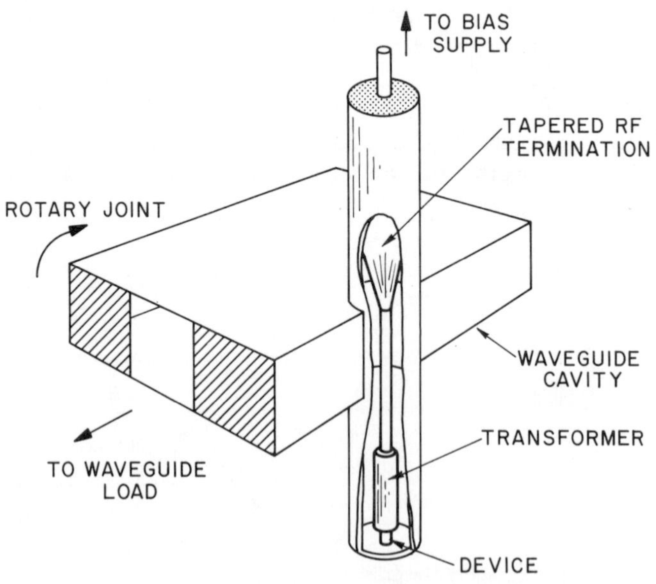

Figure 5.27 A single-tuned oscillator

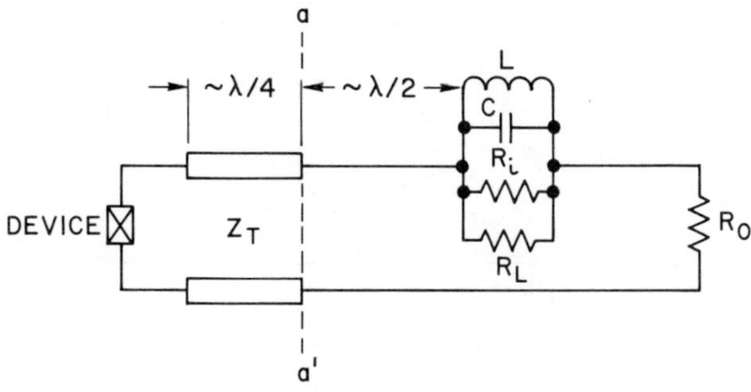

Figure 5.28 An equivalent circuit of the single-tuned oscillator shown in Figure 5.27

oscillator is shown in Figure 5.28. The *LC* resonator represents the waveguide cavity, R_L the load resistance transformed by the window and rotary joint, R_i the internal loss of the cavity, R_0 the coaxial termination, Z_T the characteristic impedance of the transformer section in the coaxial line. The impedance seen looking toward the cavity from a—a′ draws a circle on the Smith chart as shown in Figure 5.29. At frequencies far away from the resonant frequency of the cavity, the impedance is essentially equal to R_0, which corresponds to the centre of the Smith chart. At the resonant frequency, the impedance passes through the point $R_0 + R_i R_L/(R_L + R_i)$. The magnitude of R_L and hence the diameter of the circle can be adjusted by the rotary joint which changes the coupling between the waveguide load and the cavity. When the same impedance is seen through the transformer, the

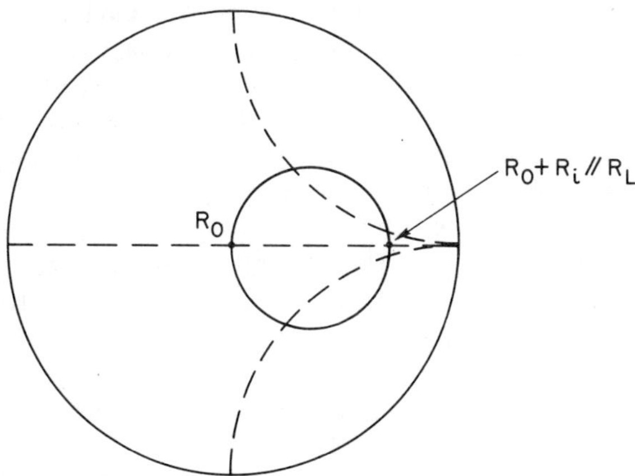

Figure 5.29 Cavity impedance seen from a-a′ in Figure 5.28

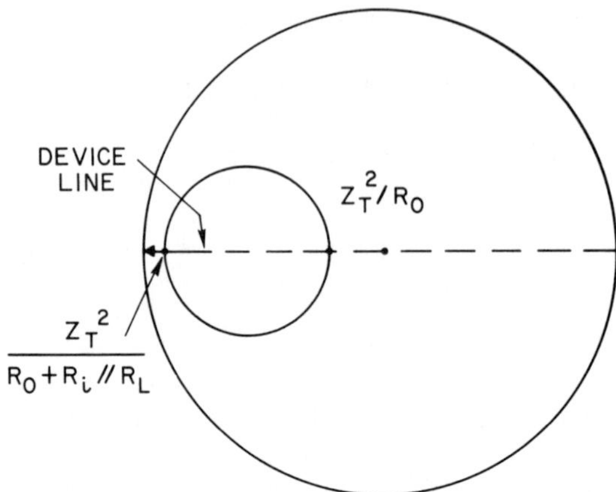

Figure 5.30 Impedance locus and device line of the single-tuned oscillator

impedance locus will become as shown in Figure 5.30. If the impedance locus intersects the device line as illustrated in the figure, oscillation will take place. In practice, the length of the transformer section has to be chosen so as to ensure the intersection is near the resonant frequency of the cavity. Once the length is properly chosen, the operating point on the device line can be adjusted by the rotary joint and the oscillation frequency can be adjusted by changing the resonant frequency of the cavity. In this way, the power and frequency can be adjusted independently of each other with this circuit.

The cavity will resonate in many different modes at different frequencies. The impedance circle corresponding to each of these resonances appears at a different position on the Smith chart determined by the electrical length from the cavity to the device. Consequently, it is unlikely that additional intersections will take place when the device line is short on the Smith chart. If the device line is long, such additional intersections may take place, but it is relatively easy to eliminate them by changing the cavity dimensions or by incorporating appropriate mode suppressors in the cavity. Thus, the single-tuned characteristic without hysteresis can be secured.

The above advantages of the circuit, namely, the independent adjustability of the frequency and power and the single-tuned characteristic, do not come free of charge. They are obtained at the expense of the circuit efficiency. The generated power is divided into R_0 and the parallel connection of R_i and R_L. So, the circuit efficiency is reduced by a factor of $R_iR_L/[R_iR_L + R_0(R_i + R_L)]$ owing to the existence of the r.f. termination in series with the resonant circuit. Of the power consumed in the parallel connection of R_i and R_L, only that part consumed in R_L becomes the output power. Thus, the circuit efficiency is given by

$$\eta = \frac{R_i}{R_i + R_L} \frac{R_i R_L}{R_i R_L + R_0(R_i + R_L)} \tag{5.75}$$

To make our discussion simple, let us assume that $\overline{Z}(A)$ is given by $\overline{R}$ at the operating point and the oscillation is at the resonant frequency of the cavity. Then, we have

$$\frac{Z_T{}^2}{\overline{R}} = R_0 + \frac{R_i R_L}{R_i + R_L} \tag{5.76}$$

Substituting (5.76) into (5.75), a little manipulation shows that

$$\eta = \left[1 - \frac{R_0}{R_i}\left(\frac{Z_T{}^2}{\overline{R}R_0} - 1\right)\right]\left(1 - \frac{\overline{R}R_0}{Z_T{}^2}\right) \tag{5.77}$$

For a given set of $\overline{R}$, R_i and R_0, the maximum efficiency is calculated to be

$$\eta_{max} = 1 - \frac{2R_0}{R_i}\left(\sqrt{\frac{R_i}{R_0} + 1} - 1\right) \tag{5.78}$$

The maximum efficiency occurs when

$$Z_T{}^2 = \overline{R}R_0\sqrt{\frac{R_i}{R_0} + 1} \tag{5.79}$$

The ratio of the external Q of the oscillator to the unloaded Q of the cavity is given by

$$\frac{Q_{ext}}{Q_0} = \frac{R_L}{R_i} = \frac{\sqrt{\frac{R_i}{R_0} + 1} - 1}{\frac{R_i}{R_0} - \sqrt{\frac{R_i}{R_0} + 1} + 1} \tag{5.80}$$

where an assumption is made that the energy stored in the coaxial line is negligible compared to that stored in the cavity. Let $Q_{ext,c}$ be the external Q of the cavity formed by the coupling into the coaxial line terminated at both ends. Then, we have

$$\frac{Q_{ext,c}}{Q_0} = \frac{2R_0}{R_i} \tag{5.81}$$

Suppose that $Q_0 = 50\, Q_{ext,c}$, then η_{max} is calculated to be 82% with $Z_T^2 \simeq 10\overline{R}R_0$ and $Q_{ext} \simeq 0.1\, Q_0$. Of the 18% circuit loss, approximately one half goes into R_0 and the balance is consumed in the cavity. For low noise applications, a higher Q_{ext} may be desired. In such a

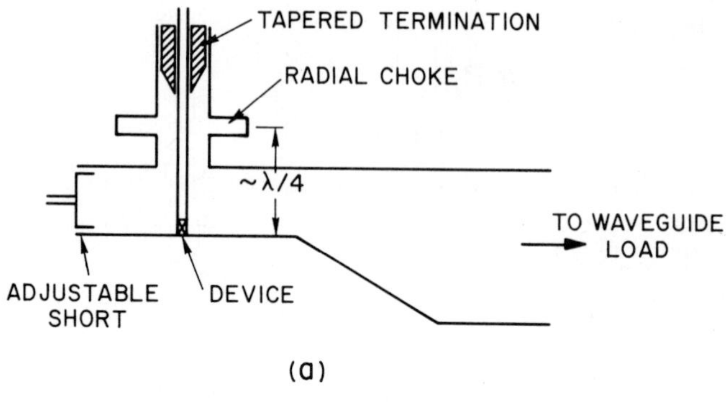

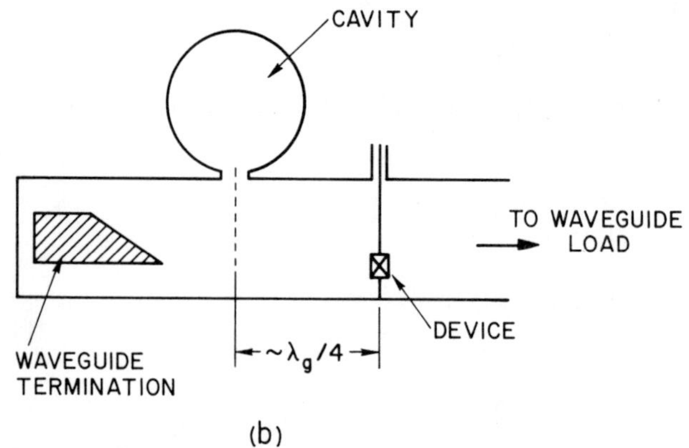

Figure 5.31 Two examples of single-tuned oscillator circuits

case, the contribution of R_0 to the circuit loss becomes smaller than the above numerical example indicates. In any case, the decrease of the circuit efficiency is small, considering the advantages obtainable with this single-tuned circuit.

Once it is known that a deliberate introduction of small circuit loss may drastically improve the tuning characteristic of an oscillator, a variety of circuits can be devised. Figure 5.31 illustrates just two of them. Figure 5.31(a) uses a radial choke to determine the oscillation frequency.[17] The adjustable short primarily change the loading. Figure 5.31(b) uses a high Q cavity.[18] At the resonant frequency of the cavity, the waveguide is essentially open-circuited at the cavity position and the termination behind the cavity is decoupled. This enables the device to oscillate with the waveguide load on the right-hand side (not shown). At the detuned frequencies of the cavity, the cavity is essentially short-circuited and the termination prevents oscillation.

5.2.5 Sweep generators[19–21]

Most sweep generators use YIG spheres for tuning. When a two-terminal device such as a Gunn diode is used, the circuit illustrated in Figure 5.32 is often employed. Two orthogonal coils couple to a YIG sphere. One coil is connected to the active device and the other to the load resistance R_0. When no d.c. magnetic field is applied to the YIG sphere, the device and load are decoupled to each other. When the d.c. magnetic field is applied, the coupling between the device and load takes place through the YIG resonance. The impedance seen by the device is given by

$$Z = j\omega L_a + \frac{\left(\dfrac{R_0}{Q_{ext}}\right)}{j\left(\dfrac{\omega}{\omega_a'} - \dfrac{\omega_a'}{\omega}\right) + \dfrac{1}{Q_L}} \tag{5.82}$$

where L_a is the inductance of the coil, Q_{ext} is the external Q, Q_L is the loaded Q and ω_a' is the resonant frequency of the loaded YIG sphere. Q_L^{-1} and ω_a' are given by

$$\frac{1}{Q_L} = \frac{1}{Q_a} + \frac{1}{Q_{ext}} \frac{1}{1 + \left(\dfrac{\omega_a L_a}{R_0}\right)^2} \tag{5.83}$$

$$\omega_a' = \omega_a \left[1 + \frac{1}{2Q_{ext}} \frac{\left(\dfrac{\omega_a L_a}{R_0}\right)}{1 + \left(\dfrac{\omega_a L_a}{R_0}\right)^2} \right] \tag{5.84}$$

where Q_a is the unloaded Q and ω_a is the natural resonant frequency of the YIG sphere. The resonant frequency ω_a is proportional to H_{DC}. ω_a'

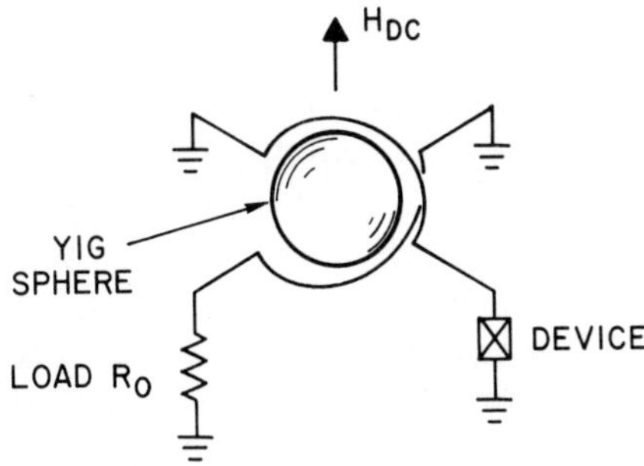

Figure 5.32 A sweep generator using an YIG sphere

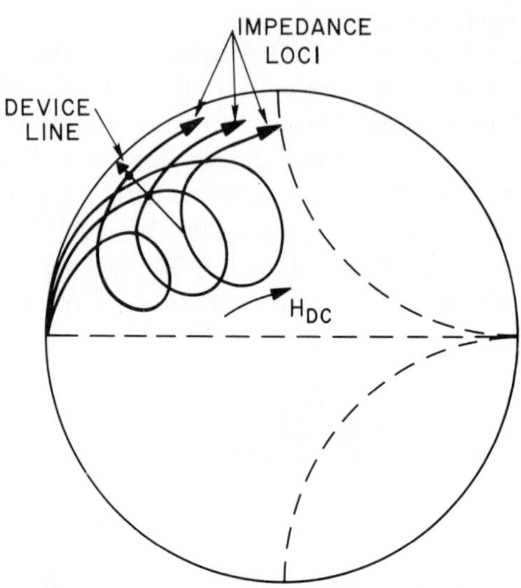

Figure 5.33 Impedance locus and device line of the sweep generator shown in Figure 5.32

is different from ω_a because of the loading effect of the coil connected to the load resistance.

The impedance locus on the Smith chart moves with H_{DC} as illustrated in Figure 5.33. As H_{DC} increases, the operating point moves from the outer periphery toward the centre of the Smith chart, passing through the maximum power point. Consequently, the power initially increases with H_{DC} but beyond a certain H_{DC} it starts to decrease as shown in Figure 5.34(a). On the other hand, the oscillation frequency monotonically increases with H_{DC} as shown in Figure 5.34(b). Toward the low frequency end, the linearity of the oscillation frequency as a function of H_{DC} degrades because of the coil inductance L_a. The oscillation frequency asymptotically approaches that of the free oscillation frequency without the YIG sphere as H_{DC} decreases. If H_{DC} exceeds a certain limit, the impedance loop separates from the device line. Then, the intended intersection disappears and the operating point jumps to the other intersection. This creates hysteresis in tuning as shown in Figure 5.35. Whether the jumps take place or not is determined by the excursion of H_{DC} as well as the diameter of the impedance loop. If the circuit parameters are properly chosen, the sweep range may be extended over one octave or more without a frequency jump.

For the detailed design of a sweeper, the motion of the device line with frequency, which has been neglected so far, has to be taken into

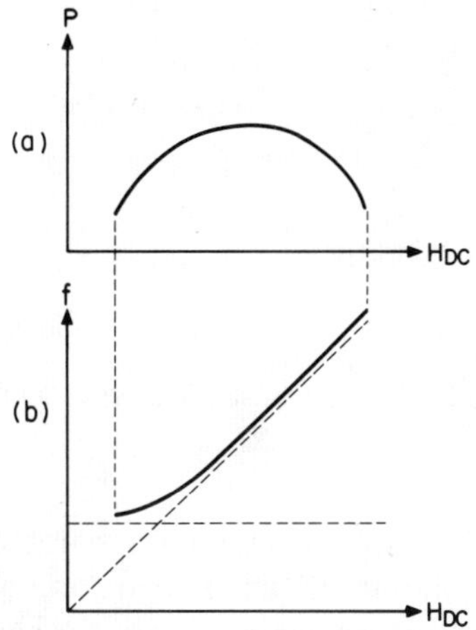

Figure 5.34 Tuning characteristics of the sweep generator

account, because the frequency range involved is large. In general, the device line moves counter-clockwise with increasing frequency, while the impedance locus moves clockwise with increasing H_x. This means that the intersection disappears at a lower value of H_{DC} than the stationary device line may indicate. In practice, the YIG sphere may show a number of spurious resonances in addition to the desired one. The spurious resonances tend to create hysteresis in tuning. To minimize such tuning complications, care must be taken to keep the YIG sphere as spherically symmetrical as possible and the d.c. magnetic field as uniform as possible. For example, the mechanical support of the YIG sphere needs careful design since the support tends to destroy the desired symmetry of the r.f. field.

When a transistor is used as the active device for a sweep generator,

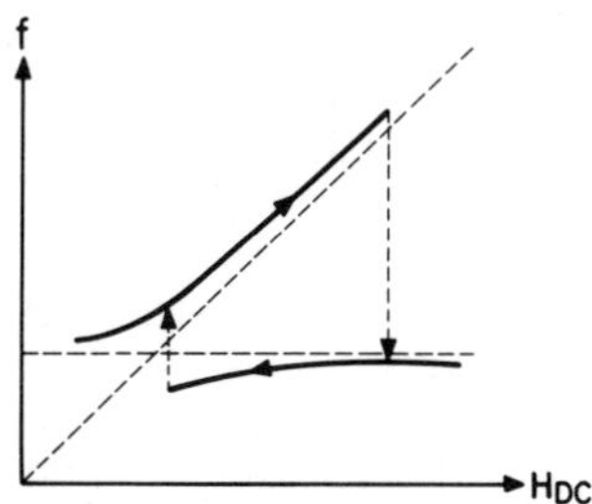

Figure 5.35 Possible tuning hysteresis of the sweep generator

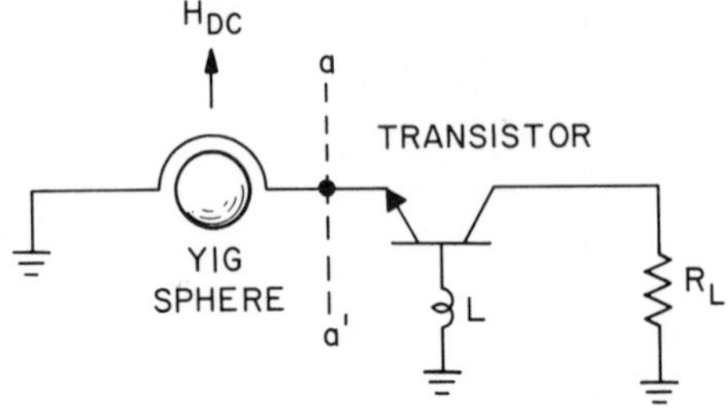

Figure 5.36 A transistor sweep generator

the output and the YIG sphere can be separated as shown in Figure 5.36. Let us consider the admittance seen looking into the emitter of the transistor. The simplest equivalent circuit is shown in Figure 5.37. C represents the base-collector capacitance, L the inductance from the base to the ground and R_L the load resistance. When α is approximately equal to 1, a simple calculation shows that

$$Y_e \simeq \left(1 - \frac{\omega_c^2}{\omega^2}\right) \frac{1}{R_L} + \frac{1}{j\omega L} \qquad (5.85)$$

where

$$\omega_c = \frac{1}{\sqrt{LC}} \qquad (5.86)$$

When $\omega < \omega_c$, the real part of Y_e is negative. As the r.f. amplitude increases, the magnitude of α decreases and $-Y_e$ varies on the Smith chart as shown by the dotted arrow in Figure 5.38.

The reciprocal of the dotted arrow is the device line, which is shown by the solid arrow in the same figure. The circuit impedance of the YIG

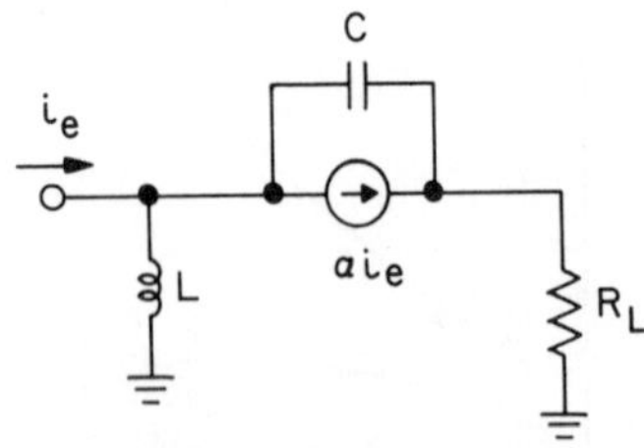

Figure 5.37 A simplified equivalent circuit of the transistor seen from a-a′ in Figure 5.36

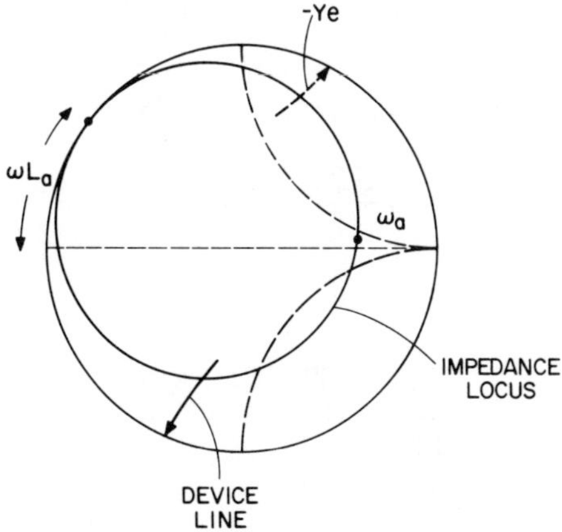

Figure 5.38 Impedance locus and device line of the transistor sweep generator

resonator seen from a—a′ is given by

$$Z = j\omega L_a + \frac{\left(\dfrac{R_0}{Q_{ext}}\right)}{j\left(\dfrac{\omega}{\omega_a} - \dfrac{\omega_a}{\omega}\right) + \dfrac{1}{Q_a}} \tag{5.87}$$

Note that ω_a is the resonant frequency of the YIG sphere and Q_a is the unloaded Q. The impedance locus is a large circle on the Smith chart and the intersection with the device line determines the oscillation frequency and amplitude. Since Q_a is high, the oscillation frequency is almost equal to ω_a which is proportional to H_{DC}. Consequently, the frequency linearity of a transistor sweep generator is generally far superior to that of a Gunn diode sweep generator.

5.2.6 Multiple device oscillators[22–25]

For simplicity, let us first consider two similar oscillators connected through a 3 dB coupler as shown in Figure 5.39. If the 3 dB coupler and the terminations are perfect, each oscillator sees the reference resistance R_0 regardless of the frequency. The two oscillators are completely independent of each other. Now, suppose that a small reflection is introduced in one output port as illustrated in Figure 5.40. Then, a part of the generated power from oscillator 1 will be reflected from the movable short and will return to oscillator 2 and vice versa. Consequently, the two oscillators lock together just like the injection locking discussed in Section 5.1.7. By changing the position of the movable short, we can vary the relative phase of the two oscillations

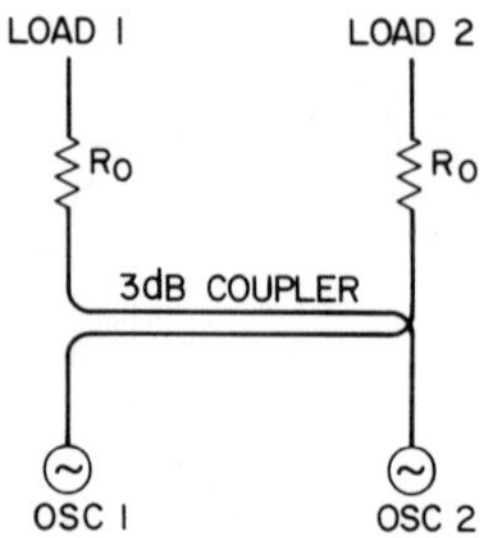

Figure 5.39 Connecting two oscillators by a 3 dB coupler

and hence the ratio of the power going into load 1 and load 2. Suppose that we have adjusted the short position in such a way that most of the generated power goes into load 1 and almost none to load 2. Under this condition, the power into load 1 will be almost equal to twice the power each oscillator delivers to R_0. We can easily accomplish this adjustment under laboratory conditions. However, the combined oscillator constructed in this way is sensitive to small variations in the load condition. If a small reflection comparable to the one in load 2 is introduced in load 1, the original phase relation between the two oscillators will be completely upset and only a small fraction of the original power may reach load 1 and the balance goes to load 2. In addition to the load condition, small variations in the oscillators due to variations in the environmental temperature or the power supply voltage may cause large variations in the output power into load 1. The desired phase relation could be established only by a delicate balance between the two oscillators interacting with each other. The balance is easily disturbed by small changes in the load condition, environmental temperature and power supply. If we extend the above concept and combine many oscillators through 3 dB couplers, the stability of the

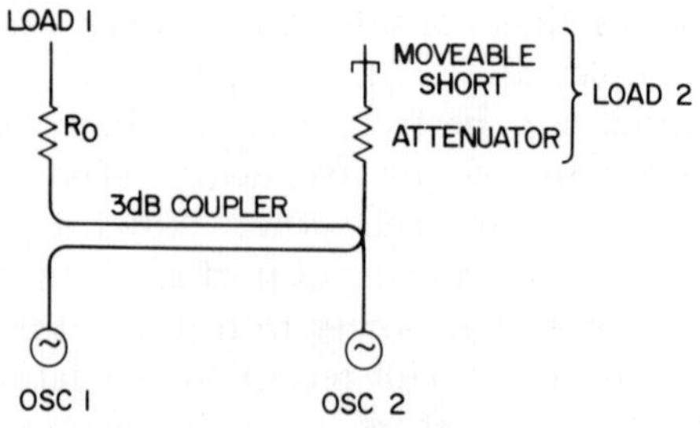

Figure 5.40 A power combiner of two oscillators using a 3 dB coupler

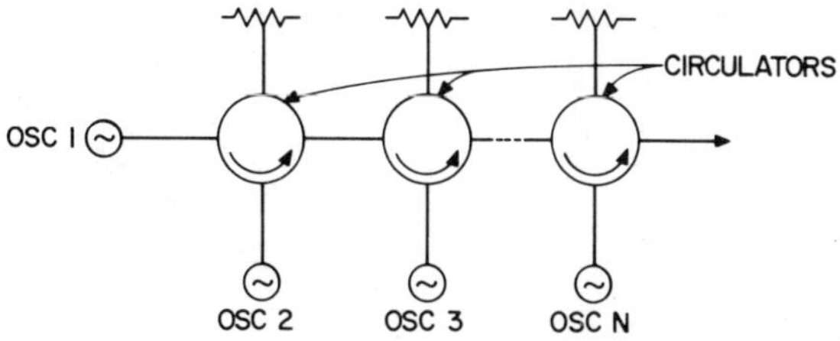

Figure 5.41 A power combiner using circulators

combined oscillator may well become difficult to achieve. Although power combiners of this kind have been constructed and successfully operated under laboratory conditions, it is worth noting that the adjustment of such circuits has been critical and the stability marginal.

To alleviate the difficulty arising from the interactions among component oscillators, an application of non-reciprocal elements is suggested as shown in Figure 5.41. In this scheme, a number of oscillators (or negative resistance amplifiers) are connected through circulators and injection locked by a master oscillator. The scheme works very well if wide-band circulators are used and the interactions between active devices in the reverse direction are well suppressed over the frequency range in which the devices show negative resistance. As the number N of the oscillators to be combined increases, however, the loss in the circulator will eventually cancel almost all the added power in the stage. Then, a further addition of oscillators will become of no use. The relation between the output and the number of stages is illustrated in Figure 5.42. It is easily seen that long before the maximum power is reached, the overall efficiency becomes so poor that such a power combiner with large N becomes uninteresting for practical applications.

Rucker's symmetrical oscillator may be the first multiple-device oscillator of practical interest having well-understandable performance at microwave frequencies. It consists of five coaxial transmission lines each about a quarter wave-length long terminated by a device and arranged radially about a common bias network and a common output circuit as shown in Figure 5.43. The outer conductors of the coaxial lines are not explicitly shown in the figure, but they form the common grounding circuit and give an excellent shielding effect between the adjacent coaxial lines. A resistor $R_{,\mathrm{STAB}}$ is incorporated into each coaxial centre conductor. The capacitance C between the output coupling disc and each coaxial centre conductor provides the necessary coupling to the common load R_{L}. The bypass capacitor in the bias network is located about a quarter-wavelength away from the hub of the oscillator. To see why Rucker's oscillator gives stable oscillation, let us study the oscillation condition in detail.

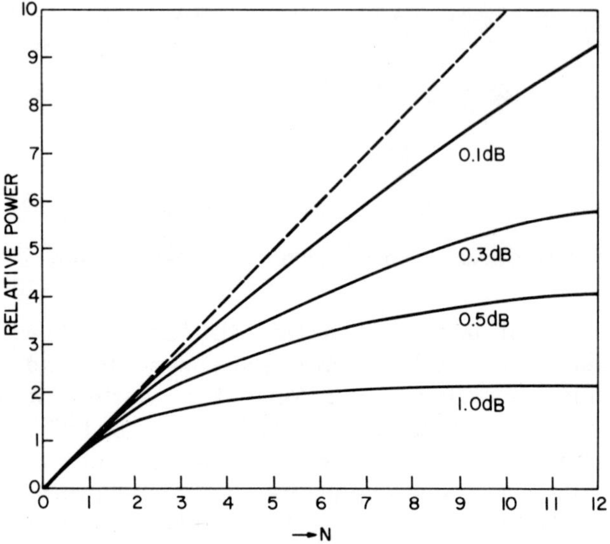

Figure 5.42 Output power versus the number of oscillators to be combined. The parameter is the insertion loss of the four-port circulator per pass

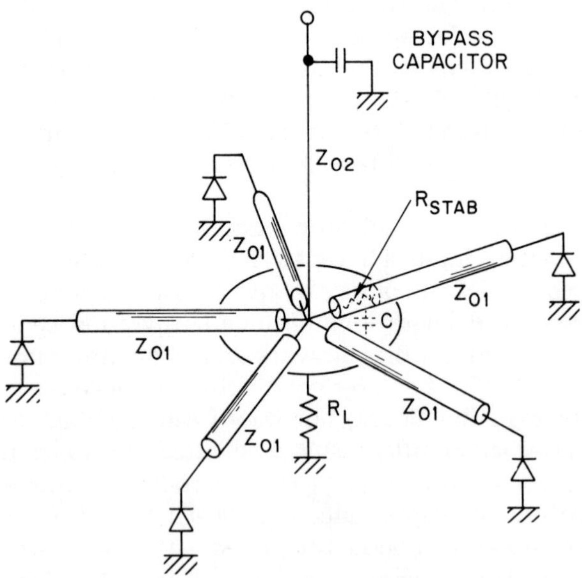

Figure 5.43 Rucker's multiple-device symmetrical oscillator

Let Z be the circuit impedance matrix seen from the device terminals. Z is a 5 x 5 matrix with its components being $Z_{ij}(i, j = 1$ to $5)$. From the reciprocity relation, we have $Z_{ij} = Z_{ji}$. For Rucker's oscillator, the relation between port 1 and port 2 is identical to the relation between port 1 and any other port since the coupling between adjacent coaxial lines is well shielded. From this consideration, we have $Z_{12} = Z_{13} = Z_{14} = Z_{15}$. Furthermore from the symmetry of the configuration, $Z_{11} = Z_{22} = Z_{33} = Z_{44} = Z_{55}$. Consequently, the impedance matrix takes the following form:

$$\mathbf{Z} = \begin{bmatrix} Z_{11} & Z_{12} & Z_{12} & Z_{12} & Z_{12} \\ Z_{12} & Z_{11} & Z_{12} & Z_{12} & Z_{12} \\ Z_{12} & Z_{12} & Z_{11} & Z_{12} & Z_{12} \\ Z_{12} & Z_{12} & Z_{12} & Z_{11} & Z_{12} \\ Z_{12} & Z_{12} & Z_{12} & Z_{12} & Z_{11} \end{bmatrix} \tag{5.88}$$

The nth eigenvector of $\mathbf{Z}$ is given by

$$\mathbf{x}_n = \frac{1}{\sqrt{5}} \begin{bmatrix} 1 \\ \exp(jna) \\ \exp(j2na) \\ \exp(j3na) \\ \exp(j4na) \end{bmatrix} \tag{5.89}$$

where n runs from 0 to 4 and

$$a = 2\pi/5 \tag{5.90}$$

The corresponding eigenvalues are given by

$$\lambda_0 = Z_{11} + 4Z_{12} \tag{5.91}$$

$$\lambda_n = Z_{11} - Z_{12} \qquad (n \neq 0) \tag{5.92}$$

Corresponding to $\overline{Z}(A)$, we have a diagonal matrix defined by

$$\overline{\mathbf{Z}} = \text{diag}[\overline{Z}(A_1), \overline{Z}(A_2), \overline{Z}(A_3), \overline{Z}(A_4), \overline{Z}(A_5)] \tag{5.93}$$

where A_i is the current amplitude of the ith device. The condition for free-running oscillation is given by

$$(\mathbf{Z} - \overline{\mathbf{Z}})\mathbf{i} = 0 \tag{5.94}$$

where $\mathbf{i}$ is the vector representing the device currents. By introducing another set of row vectors

$$\tilde{\mathbf{x}}_m = \frac{1}{\sqrt{5}} [1, \exp(-jma), \exp(-j2ma), \exp(-j3ma), \exp(-j4ma)] \tag{5.95}$$

where m runs from 0 to 4, we can express $\mathbf{i}$ and $\overline{\mathbf{Z}}\mathbf{i}$ as follows:

$$\mathbf{i} = \sum_n (\tilde{\mathbf{x}}_n \cdot \mathbf{i})\, \mathbf{x}_n \tag{5.96}$$

$$\overline{\mathbf{Z}}\mathbf{i} = \sum_n (\tilde{\mathbf{x}}_n \cdot \overline{\mathbf{Z}}\mathbf{i})\, \mathbf{x}_n \tag{5.97}$$

where the orthonormal relations between $\tilde{\mathbf{x}}_m$ s and $\mathbf{x}_n$ s are used. Substituting these expressions into (5.94) and equating terms with the same eigenvector, we obtain

$$\tilde{\mathbf{x}}_n \cdot (\lambda_n \mathbf{I} - \overline{\mathbf{Z}})\, \mathbf{i} = 0 \tag{5.98}$$

where n runs from 0 to 4. By inspection, we find that the possible solutions are given by

$$\mathbf{i} = \sqrt{5}\, A \mathbf{x}_m \tag{5.99}$$

provided that

$$\lambda_m = \overline{Z}(A) \tag{5.100}$$

In order to understand the physical meaning of (5.100), let us consider λ_m. It is the impedance seen from one of the device terminals when the mth mode is excited. The equivalent circuit of λ_0 is given in Figure 5.44. The reason why R_L and Z_{02} are multiplied by 5 and the bypass capacitor is divided by 5 is as follows. The equivalent circuit considers only one device terminal while the voltages developed across these elements are due to the combined effect of the currents flowing through five device terminals.

The equivalent circuit of $\lambda_n (n \neq 0)$ is given in Figure 5.45. The bypass capacitor, R_L and Z_{02} disappear from the equivalent circuit. This is because the voltages developed across these elements are the sums of the contributions from the currents flowing through five device terminals and they cancel out each other because of the phase relation between the currents.

Referring to Figure 5.45, we can easily see that $|\lambda_n|$ $(n \neq 0)$ can be made considerably larger than $|\overline{Z}(A)|$ by selecting a proper R_{STAB}.

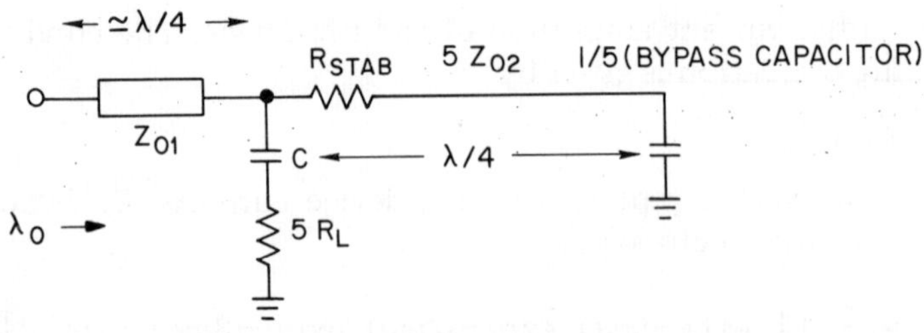

Figure 5.44 The equivalent circuit of λ_0

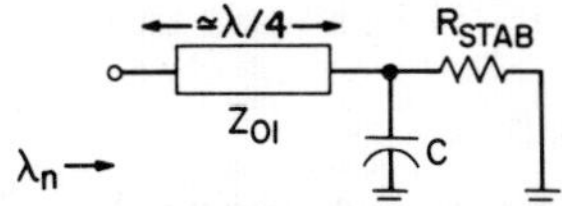

Figure 5.45 The equivalent circuit of λ_n $(n \neq 0)$

Suppose a proper R_{STAB} is selected in this way, then (5.100) is not satisfied. This means that the oscillation in the nth mode $(n \neq 0)$ is prohibited. On the other hand, λ_0 will vary with ω as shown in Figure 5.46. If the circuit parameters are properly chosen, the impedance locus will intersect this device line and (5.100) is satisfied and the oscillation in the 0th mode will take place. The intersection determines the oscillation amplitude and frequency just as in the case of single device oscillators. Note that the currents through the devices are in-phase for the 0th mode. Also note that R_{STAB} will not waste the generated power because the quarter-wavelength transmission short-circuited by the bypass capacitance gives an open circuit to R_{STAB}.

In the above discussion, the solutions are found by inspection. For these solutions, the amplitudes of the device currents are all equal to each other. A real question is whether other solutions with the amplitudes different from each other can exist or not. In the case of Rucker's oscillator, the following discussion shows that the 0th mode is the only possible mode of oscillation for all practical purposes. From (5.98), we have

$$\lambda_1 \mathbf{i} - \bar{\mathbf{Z}}\mathbf{i} = K\mathbf{x}_0 \tag{5.101}$$

where K is a constant. Note that if (5.101) is not satisfied the four equations corresponding to $n = 1$ to 4 in (5.98) cannot be satisfied

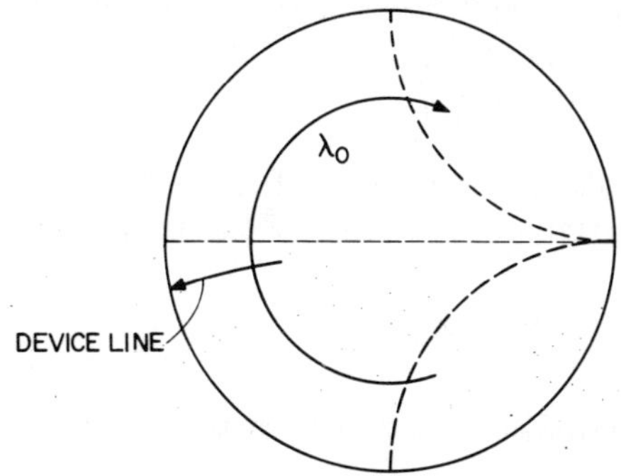

Figure 5.46 Relation between the device line and impedance locus for the 0th mode of Rucker's oscillator

simultaneously. Equation (5.101) shows that

$$\mathbf{i} = (K/\lambda_1)\mathbf{x}_0 + (\overline{\mathbf{Z}}/\lambda_1)\mathbf{i} \tag{5.102}$$

Since λ_1 is assumed to be large compared to $|\,\overline{Z}(A)\,|$, the magnitude of each component of the last term is small compared to the magnitude of the corresponding component of i or, in short, the last term is small compared to i itself. As a result, when i is expressed in the form

$$\mathbf{i} = \sqrt{5}\,Ax_0 + \boldsymbol{\delta} \tag{5.103}$$

where $\tilde{\mathbf{x}}_0 \cdot \boldsymbol{\delta} = 0$, $\boldsymbol{\delta}$ is small compared to **i**. Let us write $\overline{\mathbf{Z}}$ in the form

$$\overline{\mathbf{Z}} = \overline{\mathbf{Z}}_0\mathbf{I} + \Delta\overline{\mathbf{Z}} \tag{5.104}$$

where $\Delta\overline{\mathbf{Z}}$ satisfies

$$\text{Trace}(\Delta\overline{\mathbf{Z}}) = 0 \tag{5.105}$$

Note that $\Delta\overline{\mathbf{Z}}$ is small since $\boldsymbol{\delta}$ is small. Equation 5.98 corresponding to $n = 0$ becomes

$$(\lambda_0 - \overline{Z}_0)\sqrt{5}A_0 - \tilde{\mathbf{x}}_0 \cdot \Delta\overline{\mathbf{Z}} \cdot \boldsymbol{\delta} = 0 \tag{5.106}$$

From (5.106), we see that

$$\lambda_0 = \overline{Z}_0 \tag{5.107}$$

determines the current amplitude A and the oscillation frequency f_0 to the first-order approximation. Equation 5.101 becomes

$$\lambda_1\delta - \overline{Z}_0\delta - \sqrt{5}\,A\,\Delta\overline{\mathbf{Z}} \cdot \mathbf{x}_0 - \Delta\overline{\mathbf{Z}} \cdot \delta = 0$$

which gives

$$\delta = \frac{\sqrt{5}\,A\,\Delta\overline{\mathbf{Z}} \cdot \mathbf{x}_0}{\lambda_1 - \overline{Z}_0} = \frac{\sqrt{5}\,A\,\Delta\overline{\mathbf{Z}} \cdot \mathbf{x}_0}{\lambda_1 - \lambda_0}$$

to the first-order approximation. Substituting into (5.103) gives

$$\mathbf{i} = \sqrt{5}\,A\left(\mathbf{x}_0 + \frac{\Delta\overline{\mathbf{Z}} \cdot \mathbf{x}_0}{\lambda_1 - \lambda_0}\right) \tag{5.108}$$

To be consistent with the variation of $\overline{Z}(A)$ due to small change in the current amplitude, each component $\Delta\overline{Z}$ has to satisfy

$$\frac{\partial\overline{Z}(A)}{\partial A}(\,|\,i_k\,| - A) = \Delta\overline{Z}_k$$

From (5.108), this equation is equivalent to

$$\frac{\partial\overline{Z}(A)}{\partial A}\,\frac{A\Delta\overline{Z}_k}{\lambda_1 - \lambda_0} = \Delta\overline{Z}_k$$

which requires

$$A \frac{\partial \bar{Z}(A)}{\partial A} = \lambda_1 - \lambda_0 \tag{5.109}$$

The left-hand side is a device parameter and the right-hand side depends on the circuit. Since f_0 and A are determined by (5.107), the left-hand side and the right-hand side are independently fixed for the given device and circuit. Consequently, (5.109) cannot be satisfied except by a most unusual coincidence. From this, we conclude that if $|\lambda_1| \gg |\bar{Z}(A)|$ in the whole range of interest, the 0th mode is the only possible mode of oscillation for all practical purposes.

Summarizing the above discussion, there are at least as many possible modes of oscillation as the number of devices in the symmetrical oscillator. However, by properly choosing the circuit parameters all but the 0th mode can be prohibited. The operating point of the 0th mode can be determined from the intersection between λ_0 and the device line, where λ_0 is the 0th eigenvalue of the impedance matrix representing the circuit.

Once the above principle is understood, it is not difficult to devise other multiple-device oscillators free from the mode problem. One such circuit uses three packaged devices in the circuit configuration illustrated in Figure 5.18. Instead of just one diode, three diodes are symmetrically arranged at the coaxial end. With three diodes, three different modes are possible corresponding to three eigenvectors $(1, 1, 1)$, $(1, e^{ja}, e^{-ja})$, $(1, e^{-ja}, e^{ja})$ where $a = 2\pi/3$. To suppress the last two modes, slits are cut in the centre conductor and three resistors are inserted at the open ends as illustrated in Figure 5.47.[26] The three resistors each R Ω have no effect on the 0th mode. On the other hand, for the other two modes, each device sees (R/3) Ω in parallel with a short-circuited transmission line representing the slit. So, by properly choosing R as well as the geometry of the slit, these two modes can be suppressed ensuring stable oscillation in the 0th mode. It may be worth mentioning that in the 0th mode each device sees three times larger circuit impedance than the case shown in Figure 5.18. This means that the transformer ratio must be reselected so as to attain a desired operating point for this new oscillator.

The single-tuned oscillator shown in Figure 5.27 can also be extended to multiple-device oscillators as shown in Figure 5.48. The coaxial lines are coupled to the waveguide cavity at the locations where the magnetic field becomes maximum. There are two different causes which bring about mode instabilities. One is involved with undesired resonant modes of the cavity and the other is inherent to multiple-device oscillators, as discussed above. The oscillation in the undesired resonant modes are all suppressed in this circuit configuration, because

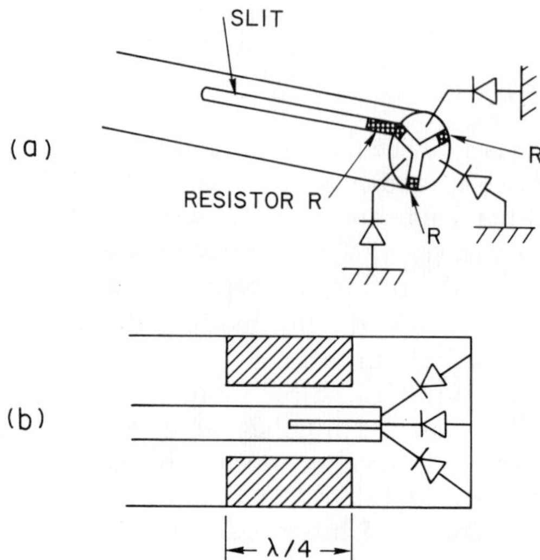

Figure 5.47 A power combiner of three devices using the coaxial configuration shown in Figure 5.18

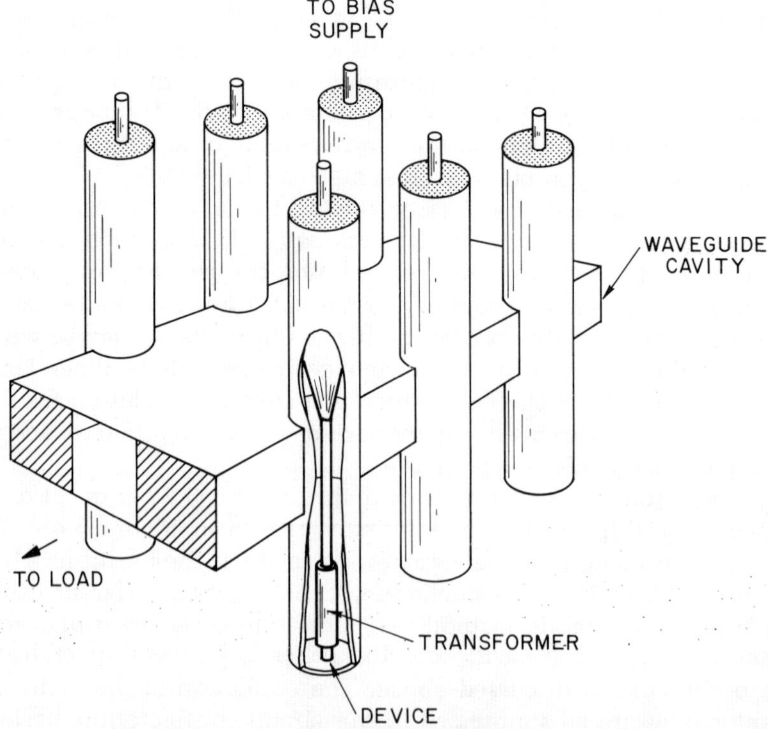

Figure 5.48 A single-cavity, multiple-device oscillator

the impedance locus is located away from the device line at those resonant frequencies different from the desired one. In addition, the coupling between the coaxial line and the cavity in these modes is generally weaker than in the desired mode, and hence the diameter of the impedance locus becomes smaller, making the intersection with the device line more unlikely. For the desired resonant mode of the cavity, there are N possible modes of oscillation, where N is the number of active devices. However, all but the 0th mode will be suppressed since the excitations of the cavity by the N devices in the ith mode $(i \neq 0)$ just cancel out because of the phase relation among the device currents, and consequently the devices see their respective coaxial terminations. On the other hand, for the 0th mode, each device sees N times the resonant impedance in series with the termination, making the intersection between the device line and the impedance locus possible. Twelve impatt diodes were originally combined using a rectangular waveguide cavity TE_{106}.[27] No spurious oscillations were observed during the circuit adjustment. More recently, a power combiner of 16 impatt diodes has been demonstrated using a TM_{010} cylindrical cavity at X-band.[28] It appears that the mode problem of multiple-device oscillators is finally understood and under control.

5.3 MEASUREMENTS

5.3.1 Device line

One of the most important parameters for the design of microwave oscillators is the device line, namely, the negative of the device impedance as a function of the r.f. current amplitude. The device impedance can be measured by a standing wave detector or a network analyser after mounting the device in an appropriate fixture and applying the bias supply. The measured impedance is the device impedance transformed by the fixture. So, depending on the application, it has to be transformed back to the device terminals. To do so, the fixture has to be thoroughly characterized as a two-terminal pair network connecting the transmission line and the device. This is usually done by connecting in turn three known inpedances in the place of the device and measuring the corresponding impedances from the transmission line. The three independent measurements are sufficient to characterize the two-terminal pair network provided that the network is reciprocal. A short, open and a 50 Ω resistor or a known capacitor are conveniently used as the three impedances. The depletion capacitance of a semiconductor diode can be also used for this purpose if the capacitance is assumed to be independent of frequency and measured at a convenient low frequency. Once the fixture is properly characterized, the measured impedance can be easily translated to the impedance seen at the device terminals.[29,30]

One drawback of the above method is that it requires a fairly powerful and hence expensive signal generator to drive the device into the useful operating point. An alternative method which does not require a signal generator is to utilize the oscillation condition, $Z(\omega) = \bar{Z}(A)$. A single-tuned oscillator circuit can present a variety of circuit impedance $Z(\omega)$ to the device and the circuit impedance can be measured separately by a passive measurement. Thus, the device impedance under various oscillating conditions and hence the device line can be obtained.[31] Strictly speaking, $\bar{Z}(A)$ is a function of frequency. Therefore the circuit must be adjusted so as to keep the oscillation frquency constant while varying the output power in the measurement. Furthermore, care must be taken to keep harmonic components sufficiently low during the circuit adjustment. If large subharmonic or higher harmonic components exist, the measured impedance will be considerably different from $\bar{Z}(A)$ without these harmonics.

5.3.2 Rieke diagram

A Smith chart representing the load impedance of an oscillator is called the Rieke diagram when the constant frequency and power contours are superimposed. The Rieke diagram indicates how the oscillation frequency and output power vary with the load impedance. The range of acceptable load impedance can be readily identified on the Rieke diagram.

To draw a Rieke diagram, a slide screw tuner may be conveniently used as schematically illustrated in Figure 5.49. By changing the insertion length and position of the screw tuner, the load impedance of the oscillator can be adjusted to any value in the entire Smith chart. So, the oscillation frequency and output power corresponding to each point on the Smith chart can be measured, and the constant frequency and constant power contours can be drawn on the Smith chart. The same insertion length and same position of the slide screw tuner present different impedances to the oscillator at different frequencies. However, the frequency dependence may be neglected if the frequency range involved is narrow, as is usually the case.

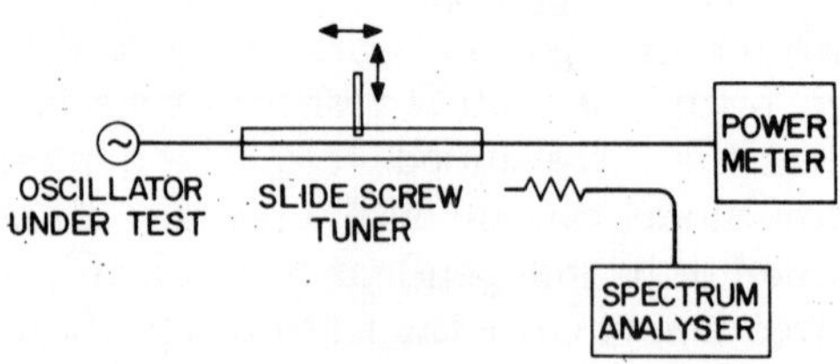

Figure 5.49 Measurements for a Rieke diagram

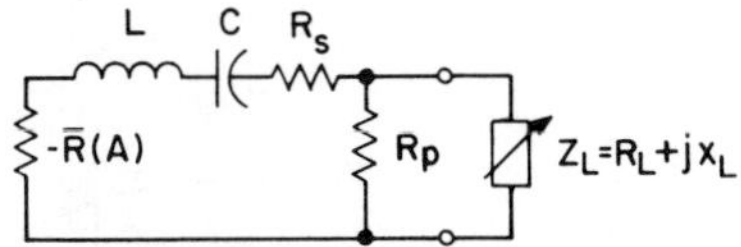

Figure 5.50 An equivalent circuit of practical single-tuned oscillators

In order to understand the general appearance of Rieke diagrams, let us consider the simple equivalent circuit of a cavity oscillator shown in Figure 5.50. In this diagram, $-\bar{R}(A)$ represents the device impedance, L,C and R_S represent the desired resonant mode and R_p the loss due to other resonant modes of the cavity. The variable load impedance is expressed by $Z_L = R_L + jX_L$. Note that R_p is usually large. When $|Z_L| \ll R_p$, the oscillation condition is given by

$$\left(\omega L - \frac{1}{\omega C}\right) + X_L = 0 \tag{5.110}$$

$$R_L - \bar{R}(A) + R_s = 0 \tag{5.111}$$

Equation 5.110 indicates that the constant frequency contours essentially coincide with the constant reactance loci. Notice that X_L decreases with increasing frequency. Since the power output is dependent primarily on $\bar{R}$ and A, Equation 5.111 indicates that the constant power contours essentially coincide with the constant resistance loci on the Smith chart. When R_L becomes large, (5.111) will not be satisfied regardless of the value of A. Consequently, there is an area on the Smith chart where constant frequency and power contours disappear. For Z_i inside this area, oscillation cannot occur. So far, $R_p \gg |Z_L|$ is assumed. When $|Z_L|$ becomes comparable to R_p, a substantial fraction of the generated power will be consumed in R_p. As a result, the constant power contours will deviate from the constant resistance loci in the region of large $|Z_L|$. The output power becomes smaller than that calculated in the absence of R_p. Consequently, the Rieke diagram may look like the one shown in Figure 5.51. In practice, the oscillator circuit will be far more complex than that shown in Figure 5.50, and the Rieke diagram will be distorted from the one shown in Figure 5.51. More often than not, constant power and constant frequency contours become discontinuous corresponding to a jump or jumps in the operating point. The same point on the Smith chart may correspond to two or more different values of frequency and power creating overlapping contours. Consequently, the interpretation of Rieke diagram of practical oscillators is not always simple and straightforward.

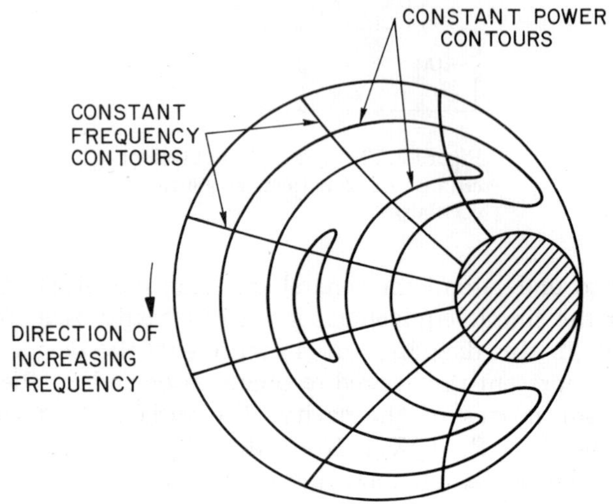

Figure 5.51 Rieke diagram of a single-tuned oscillator

Before closing the discussion of Rieke diagrams, it may be worth mentioning that the locking range discussed in Section 5.1.7, can be also obtained from the Rieke diagram. To show this, let us consider a small reflection of constant magnitude $|\Gamma|$ with its phase ψ, and hence the position of the reflection, being variable. Suppose the maximum frequency deviation due to this reflection is given by $\Delta\omega$. Since the load impedance varies with ψ as given by

$$Z_L = R_0(1+2\,|\,\Gamma\,|\,e^{i\psi}) \tag{5.112}$$

where R_0 is the characteristic impedance of the line and equal to the load resistance, the constant frequency contours corresponding to

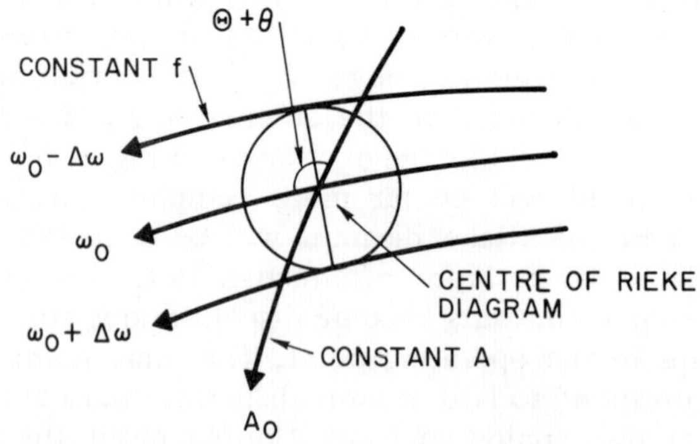

Figure 5.52 Explaining the calculation of injection locking range from the Rieke diagram

$\omega_0 \pm \Delta\omega$ must be tangential to the circle given by (5.112), as shown in Figure 5.52. Now, imagine the constant A contour corresponding to A_0 on this diagram. This contour does not necessarily coincide exactly with the constant power contour P_0 because of changing η. The angle between the constant frequency contour ω_0 and the constant A contour is equal to the angle between the device line and the impedance locus by (5.59). So the intersecting angle is given by $\Theta + \theta$ as shown in Figure 5.52. From this figure, we see that

$$\Delta\omega \left| \frac{\partial Z_{out}}{\partial \omega} \right| \sin(\Theta + \theta) = 2 \,|\, \Gamma \,|\, R_0 \tag{5.113}$$

Substituting (5.56) into (5.71) gives

$$\frac{\Delta\omega}{\omega_0} = \frac{2R_L}{\omega_0 \left| \frac{\partial Z_{out}}{\partial \omega} \right| \sin(\Theta + \theta)} \sqrt{\frac{P_i}{P_0}} \tag{5.114}$$

In our case, $R_L = R_0$. Thus the same $\Delta\omega$ is obtained if

$$|\, \Gamma \,| = \sqrt{\frac{P_i}{P_0}} \tag{5.115}$$

Consequently, if we draw a circle corresponding to a small reflection given by (5.115) on the Rieke diagram and read the maximum frequency deviation due to the reflection, it gives the locking range of the oscillator for injection power P_i. Whether the incident wave is due to the reflection or due to the injection signal from an external source, the oscillator gives the same maximum frequency deviation.

5.3.3 Injection locking test

The construction of a Rieke diagram is time-consuming. If one is interested only in the location of the operating point relative to the maximum power point along the device line and the angle between the device line and the impedance locus, these can be obtained by an injection locking experiment. When the device line is perpendicular to the impedance locus, the output power will vary symmetrically with the injection signal frequency about the free-running frequency. The increase of the output power above the free-running value upon the application of the locking signal indicates that the operating point is located at an amplitude below that for maximum power. If the device current is larger than the current for maximum power, that is, if the load is undercoupled, the output power will decrease below the free-running value. When the device line is not perpendicular to the impedance locus, the output power versus injection signal frequency becomes unsymmetrical with respect to the free-running frequency as shown in Figure 5.15. From the shape of the output curve and the motion of the injection vector indicated in Figure 5.14, one can easily visualize how the impedance locus intersects the device line. Furthermore, how fast the circuit impedance varies near the intersection can be easily estimated from the locking range.

5.3.4 AM noise[32,33]

A preferred method of measuring the AM noise of an oscillator is schematically shown in Figure 5.53. The output of the oscillator is attenuated and fed into a magic T. Two matched Schottky barrier detectors are connected to the two side arms of the magic T and the remaining port is terminated. The amplifier has a switch and, depending on the switch position, the sum or difference of the two input signals can be amplified. First, the Schottky barrier detectors must be calibrated. Suppose the detector output voltages measured by the voltmeters vary with the relative r.f. input voltage as shown in Figure 5.54. Set the r.f. amplitude to V_{RF} and amplify the difference of the two detected signals and let V_D be the wave analyser reading. Then, amplify the sum of the two detected signals and let V_s be the wave analyser reading. V_D is due to the detector noise and V_s is due to the combined effect of the detector noise and the AM noise of the oscillator. Thus, the contribution from the AM noise alone is given by

$$V_N = \sqrt{(V_s)^2 - (V_D)^2} \tag{5.116}$$

From the relation between the r.f. and detected voltages shown in Figure 5.54, the AM noise relative to the carrier is given by

$$\sqrt{\frac{N}{C}} = \frac{V_N/2}{V_0 GB} \frac{\tan\theta_1}{\tan\theta_2} \tag{5.117}$$

where V_0 is the d.c. voltage of each detector at V_{RF}; B is the bandwidth of the wave analyser (in Hz); G is the voltage gain of the amplifier; and N/C is the AM noise power in a 1 Hz bandwidth relative to the carrier.

The factor 1/2 in the numerator on the right-hand side of (5.117) comes from the fact that V_N is the sum contribution of the two detectors. N/C is, of course, a function of the baseband frequency which can be selected by the tuning of the wave analyser.

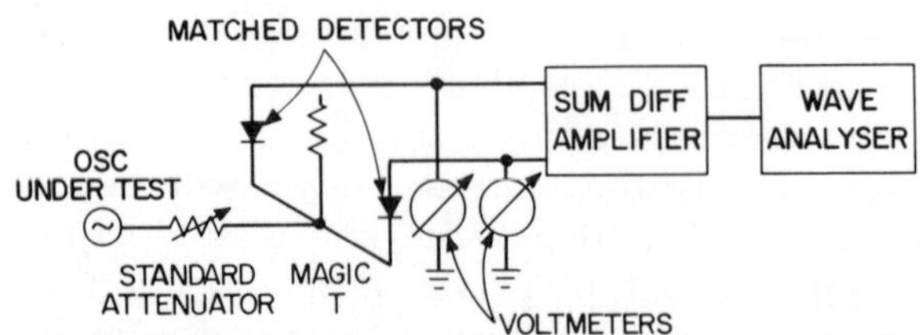

Figure 5.53 Schematic diagram of an AM noise set

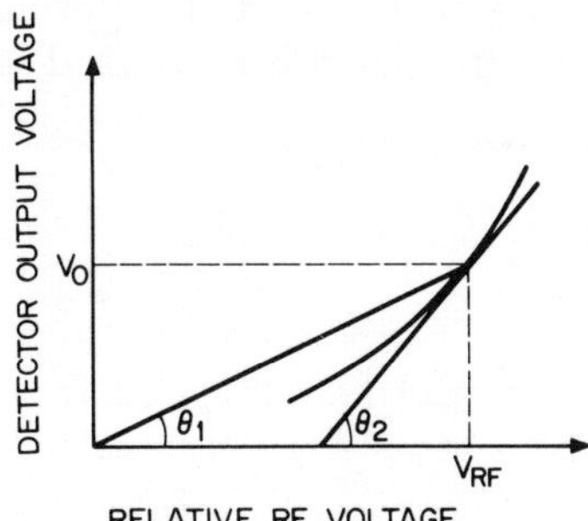

Figure 5.54 Calibration of the detectors in Figure 5.53

5.3.5 FM Noise[32,33]

The FM noise of an oscillator can be measured by a set-up schematically shown in Figure 5.55. The set-up works essentially as an FM discriminator. The oscillator output is split into two. One is fed into one port of the magic T after passing through a circulator-cavity arrangement. The other is fed into the conjugate port of the magic T through a phase shifter as a reference signal. The detection circuit is almost identical to the one used in AM noise measurement. The amplifier amplifies the difference of the two input signals. The coupling into the cavity is adjusted to be nearly critical to suppress the carrier: $Q_0 \simeq Q_{ext}$. The small reflection from the cavity near the resonant frequency is given by

$$r \simeq \frac{Z/Z_0 - 1}{2} = \frac{1}{2}\left(\frac{Q_{ext}}{Q_0} - 1\right) + jQ_{ext}\frac{\Delta\omega}{\omega_0} \tag{5.118}$$

where Z is the input impedance of the cavity and Z_0 is the characteristic impedance of the line, ω_0 is the resonant frequency

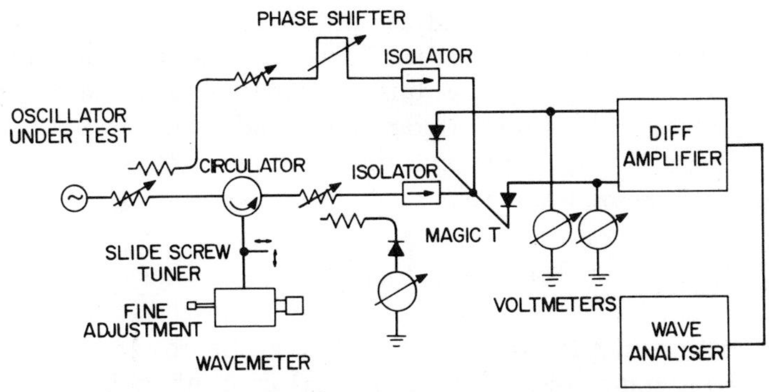

Figure 5.55 Schematic diagram of an FM noise set

which is adjusted to coincide with the oscillator frequency and $\Delta\omega = \omega - \omega_0$. If the incoming signal is expressed as

$$U(t) = A_0 [1+a(t)] \cos(\omega_0 + \Delta\omega)t \tag{5.119}$$

the reflected signal is given by $rU(t)$ and hence the input to the magic T is given by

$$U_s(t) = KA_0 \frac{1}{2}\left(\frac{Q_{ext}}{Q_0} - 1\right) [1 + a(t)] \cos(\omega_0 + \Delta\omega)t$$

$$-KA_0 Q_{ext} \frac{\Delta\omega}{\omega_0} [1 + a(t)] \sin(\omega_0 + \Delta\omega)t \tag{5.120}$$

where K is a constant which depends on the transmission loss in the signal path. The reference signal through the phase shifter can be expressed as

$$U_R(t) = A_R [1+a(t)] \cos[(\omega_0 + \Delta\omega)t + \varphi_R] \tag{5.121}$$

where φ_R is the phase which is adjustable by the phase shifter. If $A_R \gg KA_0$, the output of the balanced detector is proportional to the time average of

$$\left[\frac{U_s(t) + U_R(t)}{2}\right]^2 - \left[\frac{U_s(t) - U_R(t)}{2}\right]^2$$

to the first-order approximation and is given by

$$V = A_p \frac{1}{2}\left(\frac{Q_{ext}}{Q_0} - 1\right) [1 + 2a(t)] \cos\varphi_R$$

$$+ A_p Q_{ext} \frac{\Delta\omega}{\omega_0} [1 + 2a(t)] \sin\varphi_R \tag{5.122}$$

where A_p is a constant.

To make a measurement, first set the cavity coupling a few per cent off the critical coupling and adjust the phase shifter to minimize the output of the balanced detector. This adjustment brings about the condition $\cos\varphi_R = 0$. Then, readjust the cavity coupling as close to the critical coupling as possible. Under this condition, the output of the balanced detector is proportional to $\Delta\omega$. Since $a(t)$ is small, its effect on the output can be neglected:

$$V = A_p Q_{ext} \frac{\Delta\omega}{\omega_0} = (A_p Q_{ext}/f_0)\,\Delta f \tag{5.123}$$

The proportionality factor $(A_p Q_{ext}/f_0)$ can be obtained by changing the cavity resonant frequency by a small amount and measuring the corresponding change in the output of the balanced detector. The fine adjustment of the cavity frequency can be accomplished by inserting a small screw into the cavity. The calibration of the resonant frequency versus the insertion length must have been done prior to the experiment. Another method of calibrating the proportionality factor is to use a known FM signal. A reflex klystron can be used to provide the desired FM signal by applying a baseband sinusoidal signal to the repeller. The first null of the carrier takes place when the modulation index is 2.405. If the modulation frequency is f Hz, the corresponding peak deviation is given by $2.405\,f$ Hz, or equivalently, the r.m.s. deviation is given by $1.705\,f$ Hz. If the modulation voltage applied to the repeller is attenuated by 20 dB, the r.m.s. deviation becomes one-tenth of the original value. Consequently, any desired r.m.s. deviation can be produced by adjusting the modulation voltage. The wave analyser output can be calibrated using this known FM signal. Once the calibration is done, the set-up is ready for the FM noise measurement. It may be worth mentioning that the calibration must be made at the same input level as the test signal.

In Section 5.1.4, we have calculated the amplitude and frequency variations due to noise. They are the fluctuations of the device current. In the case of single-tuned oscillators, the device current is essentially the same as the output current except for a constant factor. However, in the case of more complex oscillators the device current may not be the same as the output current. Suppose that the transfer function of the current from the device terminals to the output port is given by $T(\omega)$, then the output current is given by

$$I_{\text{out}} = T\left(\omega_0 + \frac{\mathrm{d}\varphi}{\mathrm{d}t} - j\frac{1}{A}\frac{\mathrm{d}A}{\mathrm{d}t}\right) A\mathrm{e}^{j(\omega_0 t+\varphi)} \tag{5.124}$$

where $A\mathrm{e}^{j(\omega_0 t+\varphi)}$ represents the device current. Note that A as well as φ fluctuates with time owing to noise. The output amplitude and frequency fluctuations have to be calculated using (5.123) before comparison is made between the theory and experiment. Note that $T(\omega)$ is essentially constant for single-tuned oscillators.

REFERENCES

1. J. C. Slater, *Microwave Electronics*, New York: Van Nostrand, 1950.
2. K. Kurokawa, 'Some basic characteristics of broadband negative resistance oscillator circuits', *Bell Syst. Tech. J.*, 48, 1937–1955 (1969).
3. W. A. Edson, 'Noise in oscillators', *Proc. IRE*, 48, 1454–1466 (1960).
4. J. A. Mullen, 'Background noise in nonlinear oscillators', *Proc. IRE*, 48, 1467–1473 (1960).
5. M. Ohtomo, 'Experimental evaluation of noise parameters in Gunn

and avalanche oscillators', *IEEE Trans. MTT*, MTT-20, 425—435 (1972).

6. R. Adler, 'A study of locking phenomena in oscillators', *Proc. IRE*, 34, 351—357 (1946).
7. R. D. Huntoon and A. Weiss, 'Synchronization of oscillators', *Proc. IRE*, 35, 1415—1423 (1947).
8. K. Kurokawa, 'Injection locking of microwave solid state oscillators', *Proc. IEEE*, 61, 1386—1410 (1973).
9. T. Misawa and N. D. Kenyon, 'An oscillator circuit with cap structure for millimetre wave IMPATT diodes', *IEEE Trans. MTT*, MTT-18, 969 (1970).
10. N. D. Kenyon, 'Equivalent circuit and tuning characteristics of resonant-cap type IMPATT diode oscillators', *1973 European Microwave Conference Proceedings A.1.1.*
11. An early contribution was made by C. B. Swan at Bell Laboratories. 12.
12. F. M. Magalhaes and K. Kurokawa, 'A single-tuned oscillator for IMPATT characterizations', *Proc. IEEE*, 58, 831—832 (1970).
13. N. D. Kenyon, 'A circuit design for MM-wave IMPATT oscillators', *1970 G-MTT Symposium Digest*, pp. 300—303.
14. H. Tjassens, 'Circuit analysis of a stable and low noise IMPATT oscillator for X band', *1973 European Microwave Conference Proceedings A.1.2.*
15. J. K. Pulfer, 'Voltage tuning in tunnel diode oscillators', *Proc. IRE*, 48, 1155 (1960).
16. E. T. Harkless, *U.S. Patent 3,534,293*, October 13, 1970.
17. H. Komizo and coworkers, 'A 0.5 W CW IMPATT diode amplifier for high-capacity 11 GHz FM radio-relay equipment', *1972 ISSCC Digest of Technical Papers*, pp. 32—37.
18. K. Kokiyama and K. Momma. 'A new type of frequency-stabilized Gunn oscillator', *Proc. IEEE*, 59, 1532—1533 (1971).
19. M. Omori, 'The YIG-tuned Gunn oscillators; its potentials and problems', *1969 G-MTT Symposium Digest*, pp. 176—181.
20. D. C. Hanson, 'YIG-tuned transferred electron oscillator using thin-film microcircuits', *1969 ISSCC Digest of Technical Papers*, pp. 122—123.
21. P. M. Ollivier, 'Microwave YIT-tuned transistor oscillator amplifier design Application C band', *IEEE Jour. Solid State Circuits*, SC-7, 54—60 (1972).
22. J. J. Sie and W. J. Crowe, 'A one watt CW X-band avalanche diode source or power amplifier', *1969 G-MTT Symposium Digest*, pp. 266—272.
23. C. T. Rucker, 'A multiple-diode high-average-power avalanche-diode oscillator', *IEEE Trans. MTT*, MTT-17, 1156—1158 (1969).

24. K. Kurokawa, 'An analysis of Rucker's multidevice symmetrical oscillator', *IEEE Trans. MTT*, MTT-18, 967—969 (1970).
25. K. Kurokawa, 'The single-cavity multiple-device oscillator', *IEEE Trans. MTT*, MTT-19, 793—801 (1971).
26. R. H. Knerr and J. H. Murray, 'Microwave amplifier using several IMPATT diodes in parallel', *1974 ISSCC Digest of Technical Papers*, pp. 92—93.
27. K. Kurokawa and F. M. Magalhaes, 'An X-band 10-watt multiple-IMPATT oscillator', *Proc. IEEE*, 59, 102—103 (1971).
28. R. S. Harp and H. L. Stover, 'Power combining of X-band IMPATT circuit modules', *1973 ISSCC Digest of Technical Papers*, pp. 118—119.
29. C. N. Dunn and J. E. Dalley, 'Computer-aided small-signal characterization of IMPATT diodes', *IEEE Trans. MTT*, MTT-17, 691—695 (1969).
30. J. W. Gewartowski and J. E. Morris, 'Active IMPATT diode parameters obtained from computer reduction of experimental data', *IEEE Trans. MTT*, MTT-18, 157—161 (1970).
31. P. W. Dorman, 'Gunn diode impedance measurements using a single-tuned oscillator', *1974 G-MTT Symposium Digest*, pp. 150—151.
32. J. R. Ashley, C. B. Searles and F. M. Palka, 'The measurement of oscillator noise at microwave frequencies', *IEEE Trans. MTT*, MTT-16, 753—760 (1968).
33. J. G. Ondria, 'A microwave system for measurements of AM and FM noise spectra', *IEEE Trans. MTT*, MTT-16, 767—781 (1968).

CHAPTER 6

Microwave Amplifier Circuit Considerations

H. W. THIM and W. HAYDL

6.1 INTRODUCTION

The solid state amplifiers of today are well established devices and are widely used in numerous applications. Here, the term 'microwave' comprises the frequency range from roughly 1 GHz to the highest frequency which can be amplified (or generated) coherently, and this is presently somewhere around 300 GHz. The devices to be covered in this chapter — we restrict ourself to one-port devices — have already been described in previous chapters in their oscillating state. Amplifiers have in general identical structures and exhibit similar properties when used as large signal amplifiers.

One important difference becomes apparent in the absence of an externally applied signal: the device is stable in the sense that coherent radiation decays after the signal has been switched off. Only incoherent radiation (noise) is generated within the device. This stability is purely circuit-induced in the case of avalanche and baritt devices. The situation is much more complex in the case of transferred-electron devices as circuit stabilization fails under certain circumstances. It is therefore understandable that more space will be needed to describe transferred-electron amplifiers (Section 6.3) than baritt (Section 6.4) and avalanche devices (Section 6.5) for which we can refer to the descriptions in Chapter 3.

6.2 GENERAL CHARACTERISTICS OF HIGH FREQUENCY AMPLIFIERS

6.2.1 Definition of amplifier

An amplifier is a box with input and output terminals and is supposed to have a transducer gain over a certain frequency band. In the case of 'one-port' or 'two-terminal' devices, a circulator is commonly used to separate the incoming from the amplified signal. This type of amplifier circuit normally called 'a reflection type circuit' is shown in Figure 6.1. The advantage of this circuit is that the gain is unilateral which makes circuit stabilization much easier than in the case of a transmission amplifier without circulator. According to Figure 6.1 the diode sees some

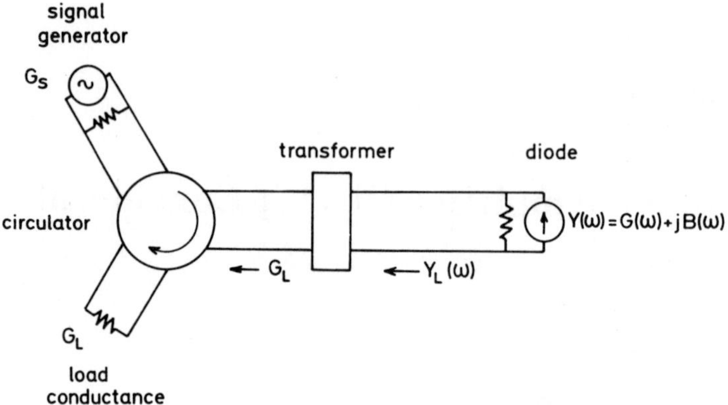

Figure 6.1 Reflection amplifier

complex admittance $Y_L(\omega)$ or, in the absence of the transformer, the load conductance G_L in a frequency range for which the circulator has been designed. A typical circulator bandwidth is one octave up to X-band and less at higher frequencies.

6.2.2 Small-signal gain—bandwidth and negative Q Factor

Since real devices are always shunted by some reactance (parallel plate capacitance, load inductance, package capacitance, etc.), the negative conductance exhibited over a certain frequency range cannot be exploited fully over that frequency range. The small-signal, or linear, gain of the reflection amplifier of Figure 6.1 is given by

$$G_{ss} = \left| \frac{Y(\omega) - Y_L(\omega)}{Y(\omega) + Y_L(\omega)} \right|^2 \tag{6.1}$$

Since the frequency appears in this expression, a gain—bandwidth product can easily be defined from (6.1) at the frequency band over which the average gain is 3 dB. It is clear from (6.1) that the gain—bandwidth product is smaller the larger the susceptances present in the circuit. The ratio of negative conductance to susceptance has been introduced as a very useful quality factor of negative conductance diodes. For a large gain—bandwidth product, a low negative Q factor is required.

6.2.3 Stability

The amplifier is small-signal stable if the gain given by (6.1) remains finite over the whole frequency band. This simply means that the a.c. power generated by the diode's negative conductance must be smaller than the power absorbed by the load, or

$$|V^2 G(\omega)| < V^2 G_L \tag{6.2}$$

A sometimes less elaborate method is the Nyquist stability analysis which either investigates the zeros and poles of the admittance (or impedance) function[1] or inspects these functions when plotted in the complex plane. When a device is found to be small-signal stable it is not necessarily large-signal stable as will be seen in the case of a transferred-electron (Gunn) diode. The opposite situation may also occur: a small-signal unstable device might eventually be large-signal stable. In general, however, a small-signal unstable device will oscillate in a large-signal mode.

6.2.4 Noise measure

It has been shown[2] that the noise measure is a more meaningful way of characterizing the noise performance of amplifiers than the more conventional noise figure. It is related to the noise figure via the small signal gain G_{ss} by the expression

$$M = \frac{F_{op} - 1}{1 - 1/G_{ss}} \tag{6.3}$$

The noise figure F_{op} of the reflection amplifier shown in Figure 6.1 reads

$$F_{op} = 1 + \frac{\sqrt{\overline{i}^2}}{k_T T_0} \frac{G_L}{[G_L - G(\omega)]^2} \tag{6.4}$$

The optimum noise figure is obtained when all reactances are tuned out. Substituting expressions (6.1) and (6.4) into (6.3) yields

$$M = \frac{\overline{i}^2}{k_T T_0} \frac{1}{[-4G(\omega)]} \tag{6.5}$$

which is independent of gain. It is clear from (6.3) that the noise measure is equal to the excess noise figure $F_{op}-1$, which is the ratio of noise introduced in the amplifying process to the amplified generator noise. For gains smaller than unity, M is negative.

6.2.5 Saturation power and efficiency

A quantity closely related to oscillator operation is the maximum available signal power or saturation power. It occurs at low gain and if associated with gain compression, the device is called a stable power amplifier (stability occurs in the absence of the signal). If the device is tuned to very high or even infinite gain, the device is said to be operated as locked oscillator (it oscillates even without external signal applied). The transition from one operation to the other is smooth.

The amplifier efficiency is conveniently defined as

$$\frac{\text{output signal power} - \text{input signal power}}{\text{d.c. power}}$$

or (6.6)

$$\eta = \frac{P_{out} - P_{in}}{P_{DC}}$$

6.2.6 Distortion and frequency conversion

If the input—output relation is nonlinear ('gain compression') the output signal will be distorted and power is generated at other frequencies. There are many applications for which such nonlinear effects cannot be tolerated. A simple example is amplitude modulation. For FM and PCM, distortion is sometimes harmless. This effect can be incorporated in form of a negative conductance decreasing with r.f. signal amplitude (see Chapter 5).

6.3 TRANSFERRED-ELECTRON (GUNN) AMPLIFIERS

6.3.1 Two-valley model

Almost 15 years ago, Ridley and Watkins[3] and, independently, Hilsum[4] predicted that negative differential mobility may be found in semiconductors that exhibit a conduction band consisting of a high mobility central valley and a low mobility energetically higher-lying satellite valley.

Current oscillations in the microwave range were shortly thereafter found in GaAs and in InP by Gunn.[5] Indium phosphide has a different band structure from GaAs,[6,7] with two satellite valleys occupied by electrons, and has only recently been considered as a serious competitor to GaAs. Negative differential mobility and associated current oscillations have been observed not only in the III—V compounds GaAs and InP, but also in some II—VI compounds such as CdTe[8,9] and ZnSe.[9,10]

It is very useful to describe the Gunn effect by the average drift velocity of the conduction electrons as a function of electric field.[1,11] To be able to do this, one must know how the electrons are distributed between the conduction band valleys as a function of electric field. Calculations for GaAs employing displaced Maxwellian distributions,[12] Monte Carlo procedures[13] and the Chambers method[14] are in good agreement with the measured velocity—field characteristic of Ruch and Kino.[15]

The average carrier drift velocity in GaAs has been measured by several methods in the past.[15–19] A number of theoretical calculations have been performed[12,13,20] which substantially agree with the experimental results. The most widely accepted theoretical and experimental values of the electron velocity versus drift field for GaAs

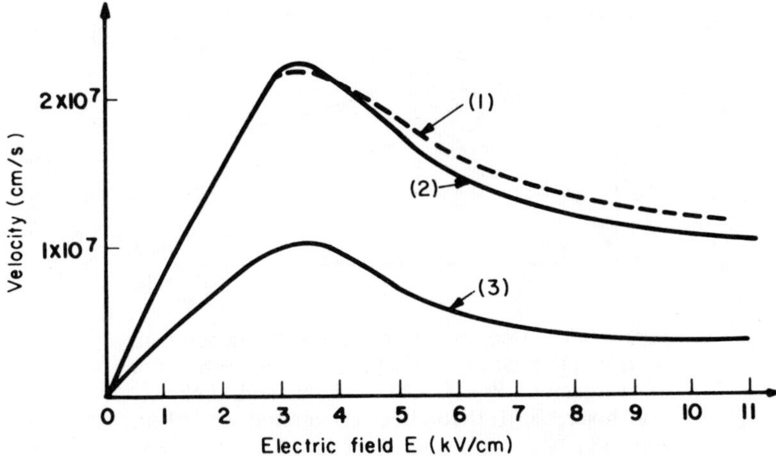

Figure 6.2a Theoretical (solid) and experimental (dashed) conduction-band drift velocity versus electric field for GaAs. ((1) after Ruch and Kino;[15] (2) after Butcher and Fawcett.[12])

are illustrated in Figure 6.2(a). These curves describe the steady state distribution and are therefore valid only for time intervals which are long compared to intervalley scattering and energy relaxation times. The instantaneous velocity—field curve of Figure 6.2(b) is very useful to study space charge dynamics provided the dielectric relaxation time is sufficiently long. The stability criteria derived in the next section are based on this 'isothermal' approximation. However, the isothermal

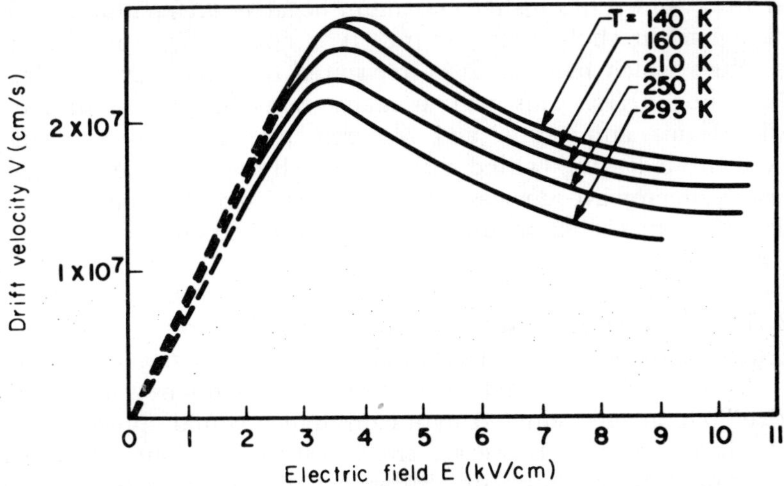

Figure 6.2b The conduction-band drift velocity versus electric field for six temperatures as measured by Ruch and Kino.[15]

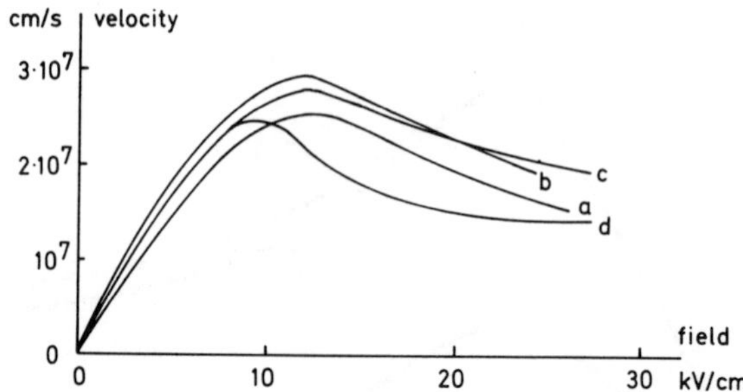

Figure 6.3 Measured velocity—field characteristic of InP. (a) Nielson,[29] (Reproduced by permission of Pergamon Press), (b) 'T Lam and Acket,[30] (c) Glover,[31] (d) theoretical calculation by Hilsum and Rees[28]

approximation is certainly inadequate for finding the upper frequency limit of the Gunn effect. For this, the time and space dependent Boltzmann equation has been solved in the small signal limit.[21–24]

The conduction band of InP is somewhat different from that of GaAs because two energetically higher-lying minima exist. There still exists some controversy as to how important the highest-lying minimum is in the transferred-electron process.

A great number of theoretical calculations also exist for InP.[25–28] Few experimental velocity—field measurements exist at this time.

The differences between the velocity—field characteristics of GaAs and InP, the v–E curve for which is illustrated in Figure 6.3, are a higher peak velocity, a higher maximum negative differential mobility, and a higher threshold field for InP. It is not readily apparent from these values which material is better suited for microwave oscillators, amplifiers or FETs. This will in many cases be dependent on the specific device application and requirements. The development of epitaxial InP is lagging behind that of GaAs by several years. Still, very impressive microwave devices have been realized. InP may prove to provide efficient oscillators and low noise amplifiers in the 30 GHz and 60 GHz bands.

6.3.2 Space charge waves and small-signal impedance

When a perfectly homogeneous crystal is biased in the negative mobility region of the velocity—field characteristic, a frequency independent negative conductance shunted by a capacitance would appear between two fictitious planes. However, small disturbances such as thermal noise, doping fluctuations or diffusion tails at contacts grow as a consequence of the negative differential mobility. The growth factor can be derived conveniently from an isothermal small signal as shown below for a one-dimensional structure. Combining the one-dimensional current density equation

$$J(t) = env(E) - e\frac{\partial(Dn)}{\partial x} + \epsilon\frac{\partial E}{\partial t} \tag{6.7}$$

and Poisson's equation

$$e(n - n_0) = \epsilon\frac{\partial E}{\partial x} \tag{6.8}$$

where n_0 is the doping density, one obtains the second-order differential equation

$$J(t) = \epsilon\frac{\partial E}{\partial x} v(E) + en_0 v(E) - \frac{e}{\epsilon}\frac{\partial}{\partial x}\left(D\frac{\partial E}{\partial x}\right) + \epsilon\frac{\partial E}{\partial t} \tag{6.9}$$

An interesting trial solution of (6.9) is

$$E(x, t) = E_0(x) + E_1 \exp[j(\omega t - kx)] + \frac{J_1}{jv\epsilon k}\exp(j\omega t) \tag{6.10}$$

$$J(t) = J_0 + J_1 \exp(j\omega t) \tag{6.11}$$

where 0 and 1 denote d.c. and a.c. quantities, respectively.

Substituting Equations 6.10 and 6.11 into 6.9 yields

$$J_0 = v(E_0)\left(en_0 + \epsilon\frac{dE_0}{dx}\right) - \epsilon\frac{d}{dx}\left(D\frac{dE_0}{dx}\right) \tag{6.12}$$

and the dispersion relation

$$kv = \omega - j(\omega_c + k^2 D) \tag{6.13}$$

where

$$\omega_c = \frac{en_0\mu}{\epsilon} \tag{6.14}$$

is the dielectric relaxation frequency, and v is given by

$$v = v_0 - \frac{en_0}{\epsilon}\frac{dD}{dE} \tag{6.15}$$

where $v_0 = v(E_0)$ is the d.c. electron drift velocity. The dispersion relation, was derived from a homogeneous d.c. electric field.

Equations (6.10) and (6.13) indicate that space charge waves grow if the differential mobility is negative. Without this growth the electric field would remain uniform throughout the diode and the diode would represent a frequency-independent negative conductance shunted by the parallel plate capacitance according to the term $J_1 \exp(j\omega t)/jv\,\epsilon\,k$

in (6.10). The diode would no longer be a transit time device with significant properties as will be shown in Section 6.3.4.

From (6.15) it follows that the phase and group velocity, both approximately equal to v, can exceed the electron drift velocity v_0 for high doping levels, since $\mathrm{d}D/\mathrm{d}E$ is negative above threshold.

As already pointed out space charge grows exponentially leading to traveling domain formation. During the initial growth period one can derive the small-signal impedance function by integrating (6.10) over the length of the diode. The result is

$$Z(\omega) = \frac{L^2}{Av\epsilon}\,\frac{(e^s - 1 - s)}{s^2(1 - Ds/vL)} \tag{6.16}$$

where A is the device cross-sectional area, and the complex dimensionless parameter s has been introduced to represent

$$s = -\frac{L}{v}\left(\omega_c + \frac{\omega^2 D}{v^2}\right) - j\frac{\omega L}{v} \tag{6.17}$$

The complex impedance, according to (6.16), for several different $n_0 L$ products is shown in Figure 6.4. It is interesting to note that the real part of the impedance is negative at the transit time frequency ($F = 1$) and harmonically related frequencies ($F = 2, 3, \ldots$) but is positive at d.c. ($F = 0$). The diode thus can amplify signals at those frequencies as a consequence of internal space charge growth.

6.3.3 Stable amplifiers

Since space charge disturbances grow in the absence of a signal it appears that instability occurs making stable amplification impossible. There are however conditions under which stabilization of the device can be achieved. There are basically three stabilization mechanisms.

6.3.3.1 $n_0 L$ stability

By inspection of Figure 6.4 one immediately finds that for $n_0 L < 10^{12}$ cm $^{-2}$ the impedance function does not encircle the origin counter-clockwise in the upper half plane. In this case the Nyquist theorem tells us that the system is stable. For $n_0 L = 1.5 \times 10^{12}$ cm $^{-2}$ the system is seen from Figure 6.4 to be unstable. A large enough series resistance, however, which shifts the $n_0 L = 1.5 \times 10^{12}$ cm $^{-2}$ curve to the right so that the origin is not encircled in the upper half plane the device is stable.

McCumber and Chynoweth[1] have derived an analytical expression for the critical $n_0 L$ product by investigating the zeros and poles of the impedance. They found that $Z(\omega)$ has no poles, which according to the Nyquist theorem indicates that the transferred-electron device is stable in a constant current circuit, or if it is connected to a large series

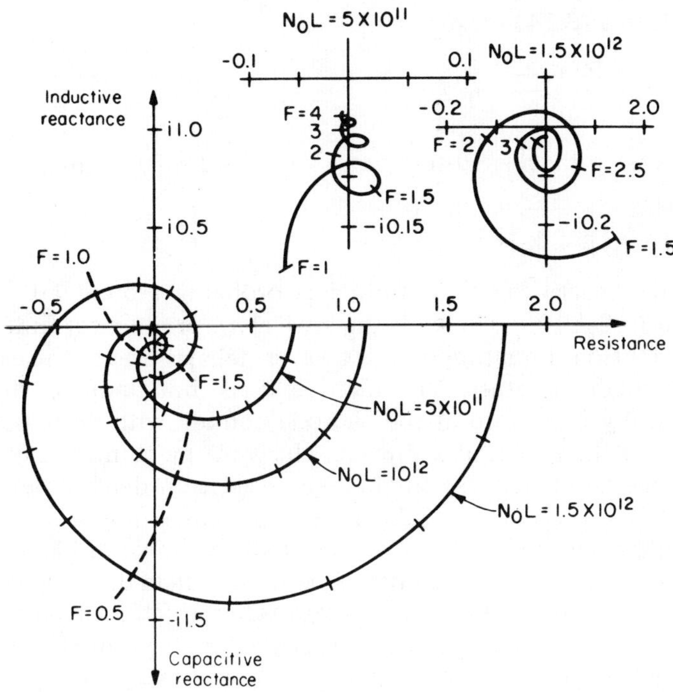

Figure 6.4 The complex impedance of a low n_0L-product amplifier diode as given by (6.16). The normalized frequency, F shown as a running parameter on each curve, is unity when the frequency is equal to the inverse transit time. The separate curves are for several values of the product of doping density and length, n_0L. It was assumed that the average negative mobility and drift velocity were such that the ratio of transit time to growth time was 2.0 when $n_0L = 10^{12}/\text{cm}^2$ (from Copeland and S. Knight,[33] reproduced by permission of Academic Press)

resistance. $Z(\omega)$ has, however, a innumerable number of zeros. One of the complex frequencies at which $Z(\omega)$ vanishes, lies in the upper half plane the system is unstable against small perturbations.

The smallest zero of s is $2.09 \pm j\,7.46$ which corresponds to ω via (6.17)

$$\omega = -j\left(+\,2.09\,\frac{v_0}{L} + \omega_c\right) \pm 7.46\,\frac{v_0}{L} \tag{6.18}$$

(diffusion has been neglected).

The ω_c decreases from its positive low field value to negative values, ω remains in the lower half plane as long as

$$\omega_c < 2.09\,\frac{v}{L} \tag{6.19}$$

or substituting (6.14) for ω_c

$$n_0 L < \frac{2.09\, v_0 \epsilon}{e\,|\,\mu\,|} = (n_0 L)_{crit} \tag{6.20}$$

For n-GaAs with $\mu = -1400$ cm^2/V s, $v = 2 \times 10^7$ cm/s, $\epsilon = 1.11 \times 10^{-12}$ F/cm,

$$(n_0 L)_{crit} \approx 2 \times 10^{11} \text{ cm}^{-2} \tag{6.21}$$

(For n-InP the critical $n_0 L$ product is probably around 10^{12} cm^{-2}.) The value of 2×10^{11} cm^{-2} is the critical nL product for uniform d.c. field distribution. Equation 6.12, however, tells us that in the case of an ohmic cathode contact the d.c. field is not uniform but rises monotonically from zero at the cathode contact with decreasing slope dE/dx up to threshold, above which electrons move more slowly. The consequence is an increase in the excess electron density beyond this point so that the electric field in turn increases monotonically with increasing gradient dE/dx towards the anode as shown in Figure 6.5.

As a consequence of the inhomogeneous d.c. field distribution $|\,\mu\,|$ in (6.20) has to be replaced by an average value which is smaller than the maximum value. In real devices the critical $n_0 L$ product is therefore larger than 2×10^{11} cm^{-2}.

The I—V characteristic of an electric field distribution as in Figure 6.5 does not show a static negative differential resistance. The reason for this has been described by Shockley[34] and is known as Shockley's positive conductance theorem: with increasing voltage, the entire electric field distribution is shifted upwards and the threshold field will be reached at a point closer to the cathode. As a result, the electric field

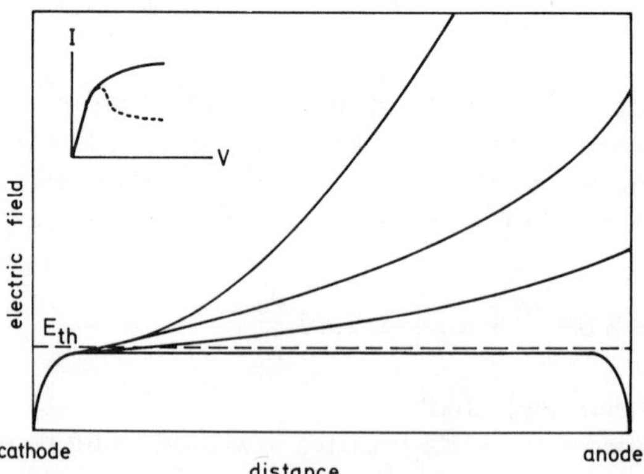

Figure 6.5 Stable field distribution in a diode with subcritical $n_0 L$ product and I—V characteristic (diffusion neglected)

gradient there will increase, and more electrons are travelling with peak velocity which means an increase in the current. The result is a positive differential resistance.

A similar field configuration has also been calculated by Kroemer[35] for the limiting case of zero doping.

The n_0L stabilized amplifier is stable only if a proper load impedance is connected to it (for example, a short or a positive resistor of any magnitude). If the capacitive component of the impedance is tuned out by an inductive load and a positive resistor of magnitude equal to the negative resistance the gain becomes infinite and oscillations result.

6.3.3.2 Circuit stabilization

A diode with overcritical n_0L products can still be stabilized by adding enough positive resistance to the device. This method can be applied successfully only if the n_0L product is slightly larger than the critical n_0L product simply because a 'pure resistive' resistor does not exist. There is always present some parasitic susceptance in a microwave circuit causing instability. Sterzer[36] has considered this possibility originally suggested by McCumber and Chynoweth[1] in greater detail.

6.3.3.3 Diffusion stabilization

If n_0L of a diode in a constant voltage circuit is greater than $(n_0L)_{crit}$ small space charge disturbances grow. This does not necessarily mean that the device remains unstable, after some time an entirely different field distribution could evolve which might be stable. Such a solution, has, in fact, been found by computer simulation[37,38] and experimentally by probing.[39,40] It is characterized by a stationary high field (accumulation) layer at the ohmic anode contact as illustrated in Figure 6.6. Magarshack and Mircea[37,38] have first pointed out that the diffusion might be the dominating stabilization mechanism. Later, Gueret[41,42], Thim[43] and Jeppesen and Jeppsson[44,45] have calculated a condition for the existence of a stationary anode domain which reads

$$n_0 \geqslant \frac{\epsilon v^2}{4e\,|\,\mu\,|\,D} \tag{6.22}$$

Gueret[41] derived this expression as a condition for the occurence of an absolute instability. Thim[43] derived it by requiring that the accumulation layer readjust more quickly than it moves into the anode. The physical mechanism is probably that diffusion balances the outflow of the accumulation layer.[46]

From the electric field distribution of Figure 6.6 it is apparent that part of the device is biased below threshold and represents a positive series resistance. The device current is below that required for threshold.

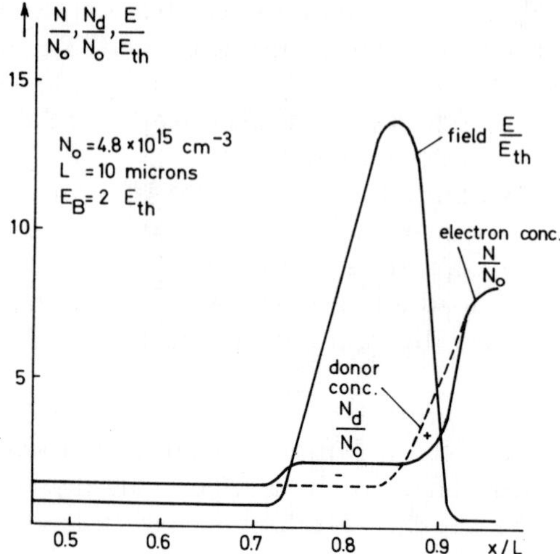

Figure 6.6 Stable regime of over-critically short Gunn diode (after Magarshak and Mircea[38])

Figures 6.7 and 6.8 illustrate calculated values for the anode domain width and the peak domain field as a function of the bias for two different $n_0 L$ products.

For a practical amplifier a wide domain is wanted, a peak field which is low enough such that as much of the device as possible is in a region where a high negative differential mobility exists. It is clear that a

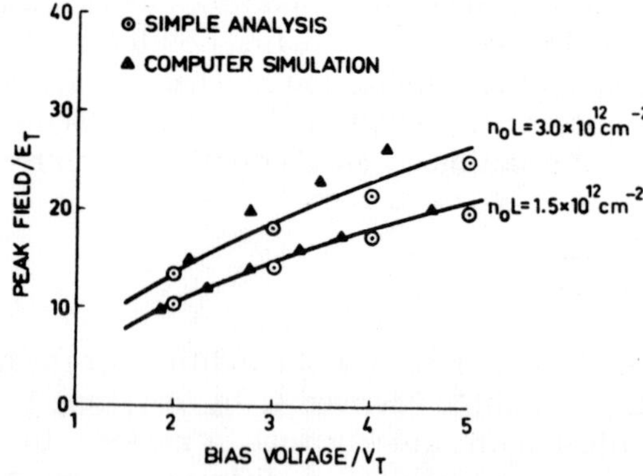

Figure 6.7 Normalized peak domain field versus normalized bias voltage for simple analysis and computer simulation considering two typical $n_0 L$-products (from Jeppesen and Jeppsson,[44] reproduced by permission of IEEE Inc.)

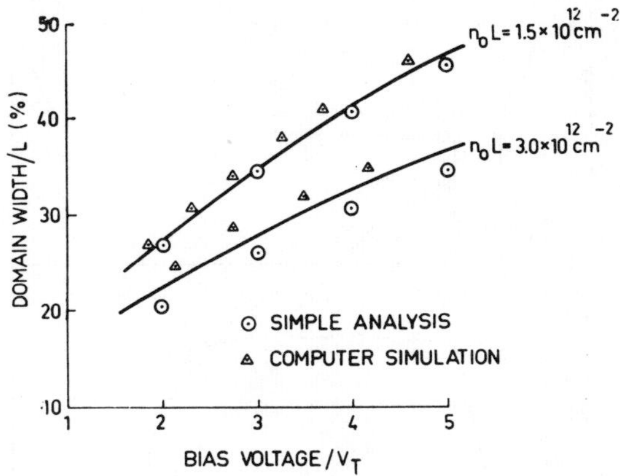

Figure 6.8 Normalized domain width versus normalized bias voltage for simple analysis and computer simulation considering two typical $n_0 L$-products (after Jeppesen and Jeppsson,[44] reproduced by permission of IEEE Inc.)

doping inhomogeneity large enough (10%) to lift the field above threshold somewhere in the low field region will cause domain formation at that spot which leads to cyclic domain formation ('Gunn oscillations'). This will also happen if a large signal is applied to this diode even in the absence of doping inhomogeneities. Therefore, it appears that this type of amplifier is not useful for large signal amplification.

6.3.3.4 Notch stabilized amplifiers

An apparently much more useful high $n_0 L$ product amplifier is a diode with a cathode doping notch. (A high $n_0 L$ product is desirable for high output power and low negative Q factor as discussed in 6.3.3 and 6.3.3.9.) This type of stabilization has been observed by McCumber and Chynoweth[1] performing computer simulations of TE diodes with overcritical $n_0 L$-product ($2 \cdot 10^{12}$ cm^{-2}) and a narrow region of reduced doping near the cathode. This 'notch' causes the electric field to rise sharply and to remain fairly constant throughout a major portion of the diode. In front of the anode the field decreases to a low value. Charlton and Hobson[47] have performed computer simulations and experiments on notch-stabilized TE diodes. Figure 6.9 shows an example of the calculated field profile of a device with a cathode notch and a slightly increasing doping profile towards the anode. The field distribution is stable because a composition of the low negative differential mobility in the high field leads to an effectively increased critical $n_0 L$-product (6.20). It is to be noted that the field is low at the anode. The point where the field abruptly decreases, moves towards the cathode as the device voltage is decreased[48] or the $n_0 L$-product is

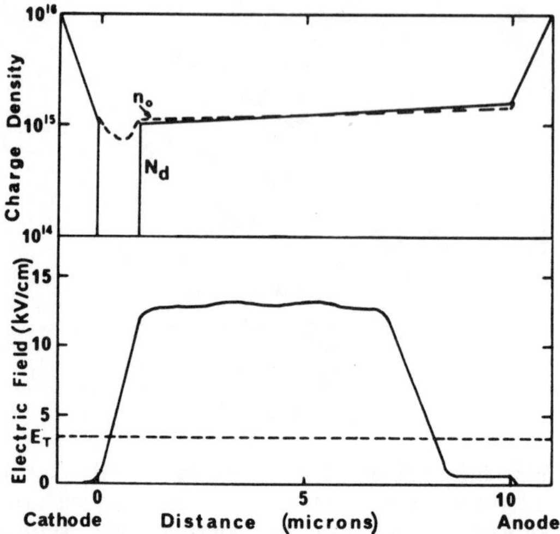

Figure 6.9 Doping and electric field profiles of a transferred-electron amplifier using cathode-notch stabilization (from Charlton and Hobson,[47] reproduced by permission of IEEE Inc.)

increased. The low field region represents a positive series resistance which must be avoided since it reduces the negative conductance of the device. In practice, a conductivity notch may be grown epitaxially either intentially or unintentially at the interface between the substrate and the active layer[49] or the active layer and a regrown n^+ contact layer. For optimum operation the field should be as uniform as possible which depends critically on the size of the notch.[49] For a 10 micrometre device the optimum ratio of the doping levels in the flat region and in the notch is of the order of 3.

6.3.3.5 Transverse stabilization

The one-dimensional approach breaks down for devices with very thin transverse dimensions. It was pointed out by Kataoka and coworkers[50] that the growth of Gunn domains is greatly reduced in a device which is thin in one lateral direction especially when the surface is loaded by a high permittivity layer. The effect is enhanced in the presence of a metal film on top of the dielectric layer which tends to short out the longitudinal r.f. field.

A two-dimensional theory which neglects diffusion developed by Kino and Robson[51] yields a dispersion relation which reads

$$v_0 k = \omega - j \frac{\omega_{cz}}{1 + \dfrac{\epsilon_b k_y}{\epsilon_a k_z} - \cot(k_y a)} \tag{6.23}$$

where k_z and ω_{cz} are the wave number and the dielectric relaxation frequency in the direction of d.c. current flow, respectively, k_y the wave number in transverse direction, ϵ_b/ϵ_a the ratio of the permittivities outside and inside the semiconducting layer and $d = 2a$ the thickness of the semiconducting layer. Equation 6.14 is difficult to solve in general. It is, however, possible to find simple solutions for the two limiting cases

$$\left|\cot(k_y a)\right| \begin{matrix}\gg \\ \ll\end{matrix} 1 \qquad (6.24)$$

If the upper sign is valid we can also assume that

$$|k_z a| \simeq \frac{\omega}{v_0} a \ll 1 \qquad (6.25)$$

In (6.16) k_z was replaced by ω/v_0, since we are looking for a stability criterion, that is we assume, that space charge growth is small over one transit length. For stability we obtain from (6.14) that

$$\left|\frac{\omega_{cz}\epsilon_a \omega a}{\epsilon_b v_0} \frac{L}{v_0}\right| < 1 \qquad (6.26)$$

Substituting (6.8) into (6.17) and keeping in mind that oscillations occur at the transit time frequency, that is,

$$\omega \frac{L}{v_0} = 2\pi \qquad (6.27)$$

we obtain, for stability to occur, the relation

$$n_0 d < \frac{v_0 \epsilon_b}{e|\mu|\pi} \simeq 10^{11}\ \text{cm}^{-2} \qquad (6.28)$$

for n-GaAs. The length has dropped out in this case, which because of (6.16) is valid for low frequencies.

For high frequencies ($k_z a \simeq \omega a/v_0 > 1$) the lower sign in (6.15) becomes valid. In this case, which has been treated by Hofmann[52–54] the layer thickness is equal to the transverse wavelength, that is the longitudinal r.f. field is shorted out at the boundaries. In the high frequency limit the ratio of k_y/k_z goes to zero. Thus the growth rate approaches that of the one-dimensional case.[55,56] At high frequencies, however, diffusion becomes effective. Thus the growth rate shows a distinct maximum at intermediate frequencies. As a consequence, stability occurs in samples which do not satisfy (6.16) only if L is limited (one-dimensional case) unless diffusion becomes effective at sufficiently low frequencies.

Becker and coworkers[57] recently derived the *nd* criterion by entirely different arguments. They considered a stripline with an active semiconducting strip and a dielectric (substrate) between the strip and a metal ground plate. The propagation velocity of quasi-TEM-waves was found to be smaller than the drift velocity of electrons if *nd* is smaller than the critical value. In this case no domain can form since it would overtake its induced current drop thereby collapsing.

Dielectric loading is only one possible way of stabilizing thin semiconducting layers. Resistive[58] or magnetic[59] layers have also been suggested as possible candidates.

The lateral stabilization has been used exclusively for building travelling wave amplifiers (see 6.3.6). It is not practicable for sandwich structures.

6.3.3.6 Small-signal gain, bandwidth and negative Q

The small-signal impedance and gain of a stabilized amplifier has been measured by many authours[49,60–62] and has also been calculated for both uniform and non-uniform d.c. field distributions.[1,36,37,47,63] A closed analytical expression for the impedance has been obtained only for the case of a uniform d.c. field and has the form given by (6.16). A plot of this expression is shown in Figure 6.4. The real part is negative at the transit time frequency and at integer multiple values of it. This of course is physically not admissible and indicates the importance of d.c. field non-uniformity diffusion and energy transport effects neglected in the derivation of (6.16). If for example field non-uniformity and diffusion in an *nL* stabilized diode

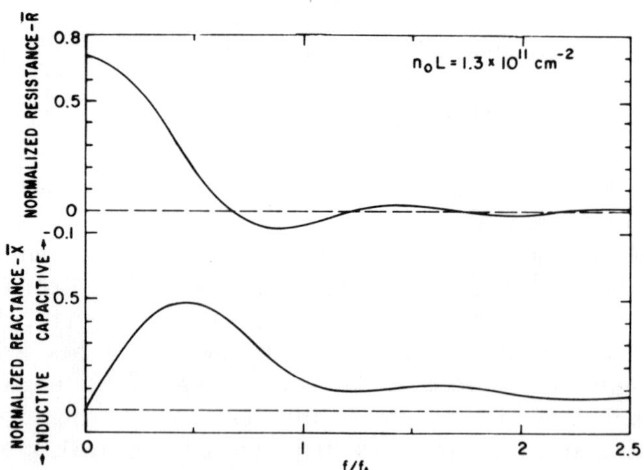

Figure 6.10 Real and imaginary parts of small-signal impedance of a bar of *n*-type GaAs biased at an average d.c. electric field of approximately 5000 V/cm as a function of the ratio of signal frequency to the transit time frequency (from Sterzer,[36] reproduced by permission of IEEE Inc.)

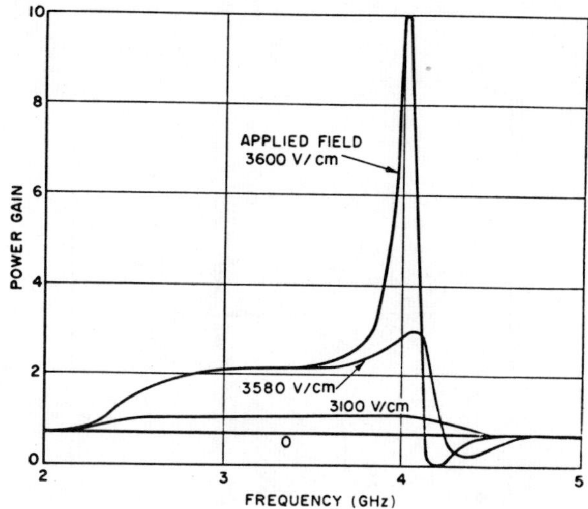

Figure 6.11 Gain versus frequency characteristics for different values of the average bias field of a sub-critically doped GaAs TE amplifier (from Thim and Barber,[60] reproduced by permission of IEEE Inc.)

($n_0 L < 10^{12}$ cm^{-2}) are taken into account, negative resistance occurs only at the fundamental (transit time) frequency and twice that frequency as shown in Figure 6.10.

Although negative resistance is exhibited over almost one octave at the fundamental frequency the electronic contribution to the reactance in addition to the plate capacitance reduces the band over which high gain can be achieved. A typically observed gain versus frequency characteristic measured on a 50 Ω coaxial reflection type circuit is shown in Figure 6.11. In this case the reactive part is dominating the overall impedance leading to a small gain times bandwidth product. It is therefore obvious that the ratio of reactance to negative resistance which is called the negative Q (quality) factor of the device should be kept as small as possible in order to obtain high gain—bandwidth products. Assuming a uniform d.c. field distribution the negative Q of the diode is given by the simple expression

$$Q = \frac{2\pi \epsilon v_0}{q \mid \mu \mid n_0 L} \tag{6.29}$$

It is thus inversely proportional to the $n_0 L$ product. This is an optimistic expression because only the parallel plate capacitance appears in (6.29) neglecting any electronic contribution arising from space charge waves. In terms of a parallel equivalent circuit the negative

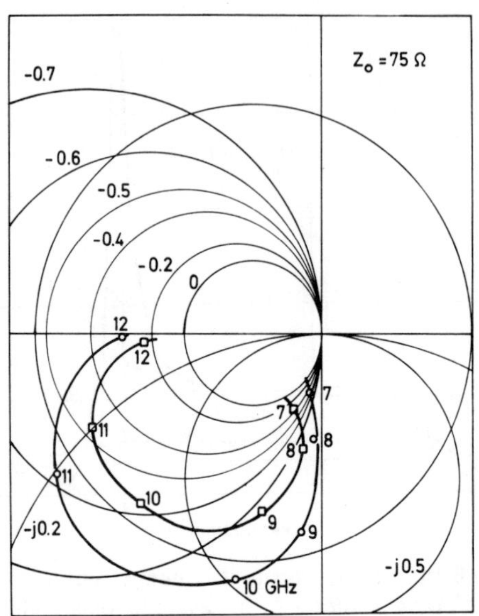

Figure 6.12 Impedance of a notch-stabilized amplifier (after Magarshak and coworkers,[49] reproduced by permission of IEEE Inc.)

conductance is largest in the case of a uniform field and is proportional to the negative differential mobility and $n_0 L$.

A uniform d.c. field distribution is exhibited in notch stabilized amplifiers with slightly over-critical $n_0 L$ product. Indeed, a large gain—bandwidth product has been measured by Magarshack and coworkers[49] using 'notch-stabilized' transferred-electron diodes. A typical diode exhibited negative resistance over almost one octave and 8 dB gain from 9—11 GHz as seen in Figure 6.12.

An exact expression for the negative Q of a diode with non-uniform field distribution and a large space charge component of the reactance has not yet been derived explicitly. Nevertheless, since space charge waves always limit the bandwidth over which negative conductance is exhibited ('transit time limitation') it is desirable to shoot for uniform field distribution which can occur in notch stabilized devices. For this, (6.29) holds indicating that for optimum gain bandwidth a high $n_0 L$-product should be used. However, too high a value of $n_0 L$ leads to a large growth factor (6.13) and hence, to large space charge for which case (6.29) does not hold. More work needs to be done in this direction to establish a Q versus $n_0 L$ relationship. Negative Q factors between —0.5 and —5 should be feasible.

6.3.3.7 Upper frequency limit

Since the transfer of energy from the externally applied field to the electrons and the scattering of electrons between the conduction band valleys take a finite time the negative resistance properties must disappear at frequencies comparable to the corresponding relaxation frequencies. The Chambers formula[14] and displaced Maxwellian approximations[22] have been employed to calculate the small-signal upper frequency limit for both uniform and non-uniform a.c. field distributions. The d.c. field was assumed uniform. The resultant upper limiting frequency as a function of bias field for a uniform a.c. field is represented by the dotted line in Figure 6.13. The highest frequency of operation is obtained for about 16 kV cm^{-1} because such high fields accelerate the electrons more rapidly through the weak scattering region causing the overall response speed to rise.[14]

The non-uniform a.c. field case with diffusion included has been treated by Grasl and Zimmerl.[22] They found that the upper limiting frequency for space charge growth increases with increasing carrier density as shown by the solid lines in Figure 6.13. Interestingly, diffusion causes an increase of limiting frequency at any bias field as the broken line representing the diffusion free case indicates. Slightly above threshold diffusion even leads to an overshoot with respect to the diffusion free uniform a.c. field case. This is perhaps due to the strong field dependence of the diffusion coefficient just above threshold[15] which leads to an increase of phase velocity as indicated by (6.15). In

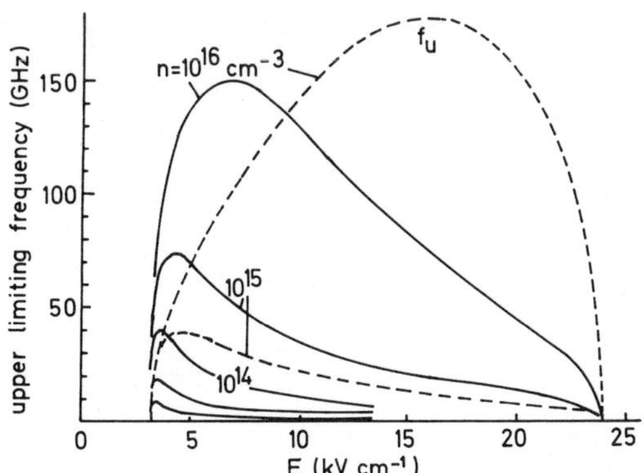

Figure 6.13 Limiting frequency against bias field for the negative differential conductivity (dotted curve) and for the TSCW amplifier (solid curves) for various dopings. The dashed curve corresponds to $N_d = 10^{15}$ with diffusion neglected (from Grasl and Zimmerl,[22] reproduced by permission of Akademie — Verlag Berlin)

other words, upper valley electrons have a smaller diffusion coefficient than lower valley electrons, which partly compensates for the delayed scattering into the lower valley.

The upper frequency limit in Figure 6.13 is due to the electron scattering processes. Space charge waves in the drift space are subject to this limit. However, actual devices will exhibit lower limits because of transit-time and injection effects. The travelling space charge wave amplifier (TSCWA) described in Section 6.3.6 has a frequency limit determined by the geometry of the coupling electrodes. For example the gap must be approximately half a space charge wavelength long in order to excite a wave efficiently (1 micrometre for 50 GHz). The two-terminal amplifier is a factor 2 better off as the drift space is one wavelength long (2 micrometres for 50 GHz). Here, electron heating sets the frequency limit because electrons injected from the cathode contact do not posess enough energy to be scattered into sub-bands. It has been shown theoretically[24,25,64,65] that electrons must travel a few tenths of a micrometre ('dead zone') before they gain that energy. Figure 6.14 shows that InP might be the favoured transferred-electron material because of its short 'dead zone'. Gunn amplifiers have been operated at 60 GHz using slightly over-critically doped diodes ($n_0 L = 2.5 \times 10^{12}$ cm^{-2}) but the highest frequency of operation is probably around 80 GHz.

6.3.3.8 Noise

Transferred-electron amplifiers are not particularly promising as low noise devices because hot electrons produce a significant amount of

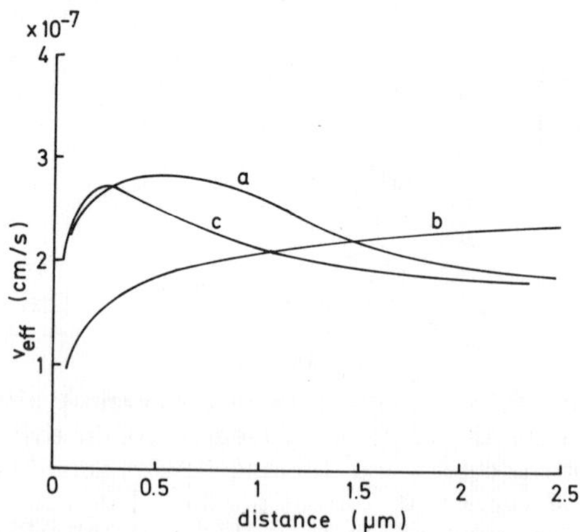

Figure 6.14 Effective velocity against distance for InP. (a) GaAs; 5 kV/cm (b) InP; 10 kV/cm (c) InP; 15 kV/cm (after Maloney and Frey[25])

noise power due to thermal velocity fluctuations. The total noise power generated by the amplifier can be calculated by integrating the local contributions over the length of the diode. Shockley and coworkers[67] have given a practical method ('impedance field method') for this calculation. Several authors[66,68,69,70] have applied this method to the TE amplifier with uniform d.c. field distribution. A generalized expression for arbitrary cathode boundary factors reads:[73]

$$M = \frac{2e^2 D n_0}{V_0 \epsilon k_T T} \cdot \frac{1 - \dfrac{1-e^{-jkL}}{jkL} - \dfrac{1-e^{-jk^*L}}{jk^*L} + \dfrac{1-e^{-j(k+k^*)L}}{j(k+k^*)L}}{\left| k + k^* - k^*\left(1 - \dfrac{k}{k_B}\right)\left(\dfrac{1-e^{-jkL}}{jkL}\right) + k\left(1 - \dfrac{k^*}{k_B^*}\right)\left(\dfrac{1-e^{-jk^*L}}{jk^*L}\right)\right|} \tag{6.30}$$

According to expression (6.30) the noise measure depends on the complex wave number k given by (6.13), the complex cathode wave number k_B which relates the cathode a.c. field to total a.c. current via

$$k_B = \frac{J_1 + j\omega\epsilon E_1}{v_0 \epsilon E_1} \tag{6.31}$$

on the diode length L and on the doping concentration n_0. For a device with a small growth constant ($k \approx \omega/v_0$) and with an ohmic cathode contact ($k_B \to \infty$), the noise measure expression has the simple form[68]

$$M = \frac{e}{kT} \frac{D}{|\mu|} \tag{6.32}$$

This is essentially the lower limit of M, but for certain cathode conductivities (real part of k_B) and for small values of (imaginary part of k) the noise measure can assume lower values as shown in Figure 6.15.

Expression (6.32) indicates that materials with a small diffusion constant and a high negative differential mobility are candidates for low noise amplifiers. Sitch and Robson[71] have performed uniform field calculations for various electric fields and n_0L-products for GaAs and InP, including diffusion damping of the forward travelling space charge wave. A minimum noise measure of 7 dB is predicted for GaAs ($n_0L = 4 \times 10^{10}$ cm^{-2}, $E_{av} = 5$ kV/cm) and 2 dB for InP ($n_0L = 5 \times 10^{10}$ cm^{-2}, $E_{av} = 20$ kV/cm). Figure 6.16 shows the results of these calculations.

Measured noise figures are several decibels higher. 'notch stabilized'

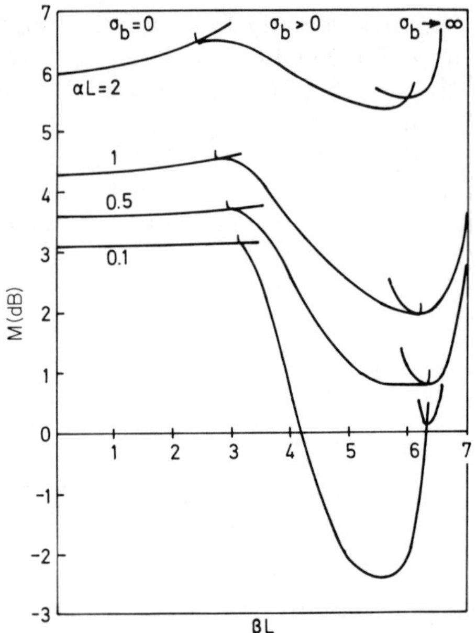

Figure 6.15 Minimum noise measure as a function of transit angle (after Johnson[70])

X-band GaAs amplifiers with presumably fairly uniform field distributions exhibited noise figures of 10 dB[49] which is only 2 dB above the calculated minimum, whereas InP amplifiers exhibited noise figures of the order of 11 dB.[72] Figure 6.17 shows the measured noise figure for various bias levels at X-band frequencies. 7.7 dB has been measured at Q-band frequencies.[73] A comparison with other two-terminal amplifiers (Table 6.1) shows that except for the baritt diode the TE amplifiers exhibits the lowest noise figures.

6.3.3.9 Saturation power and efficiency

If uniform d.c. and a.c. field distributions were present within the device the maximum available r.f. power would be 25% of the d.c. input power. In this hypothetical case, saturated output power levels of the order of 3 W could be obtained for a d.c. input power of 10 W which is a safe input level from the thermal point of view. In reality, the efficiency is only a few per cent owing to space charge non-uniformities. Perlman[74] has obtained the highest saturated output power which was close to one watt. Gain compression usually occurs several decibels below the maximum power level. Figure 6.18 shows output power versus input power measured on a 12 dB gain stage.

Besides gain compression, several other nonlinear effects exist in

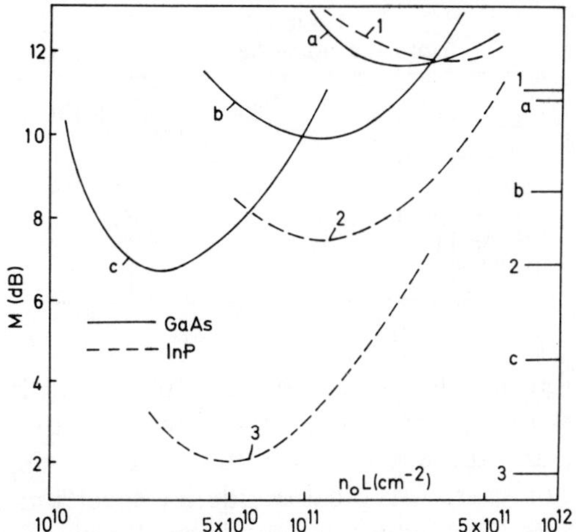

Figure 6.16 Variation of minimum noise measure M with n_0L-product for various uniform fields. GaAs (a) 15 kV/cm, (b) 11.5 kV/cm, (c) 5 kV/cm. InP (1) 10 kV/cm, (2) 30 kV/cm, (3) 20 kV/cm. Respective asymptotic values of M are shown on right-hand ordinates (after Sitch and Robson[71])

amplifiers at large r.f. signal levels: (1) Gain expansion, which occurs if under large r.f. signal conditions the average field is shifted into the high negative mobility range. (2) Intermodulation distortion, which occurs if two signals are present AM/FM(PM) conversion which is the change of frequency (phase) with drive level.

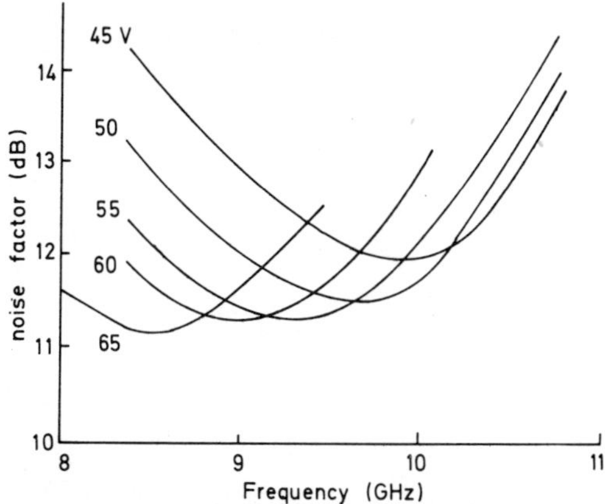

Figure 6.17 Measured noise figure of InP TE amplifier as function of frequency for various bias voltages (after Braddock and Gray[72])

Table 6.1
Amplifier noise figures (dB)

	Best	Typical
Silicon impatt	27	35
Silicon baritt	9	12
GaAs impatt	17	32
GaAs TE	10	18
InP TE	8	12

Intermodulation distortion of an amplifier can be measured by performing a two-frequency test. Two single frequencies, generally 1—5 MHz apart, are applied to the input and the amplifier output is observed on an analyser. A typical frequency spectrum of the output is illustrated in Figure 6.19, where f_1 and f_2 are the signal frequencies, f_{3+} and f_{3-} the upper and lower third order intermodulation products respectively, and f_{5+} and f_{5-} the upper and lower fifth order intermodulation products. Intermodulation products are defined as the ration of the main carrier to the sidebands caused by the intermodulation.

Figure 6.20 illustrates the intermodulation characteristics of stable amplifiers.

The third order intercept point for Gunn diodes is commonly 6—10 dB above the saturated power output level. For impatt devices, the intercept lies approximately at the saturated power output level.

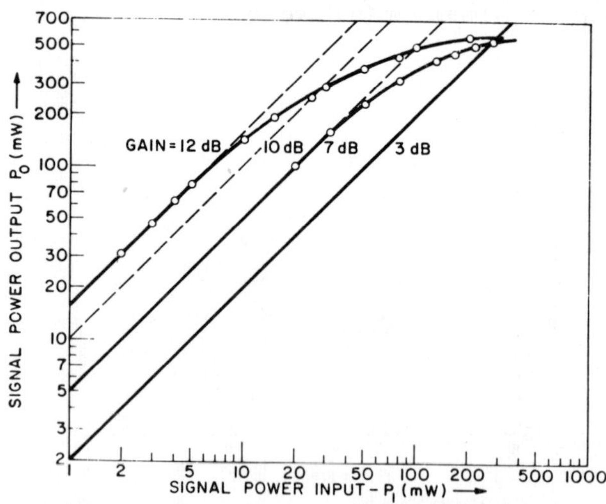

Figure 6.18 Power input-output characteristic for a single 12 dB stage high $n_0 L$ product GaAs amplifier (from Perlman,[74] reproduced by permission of RCA Corporation)

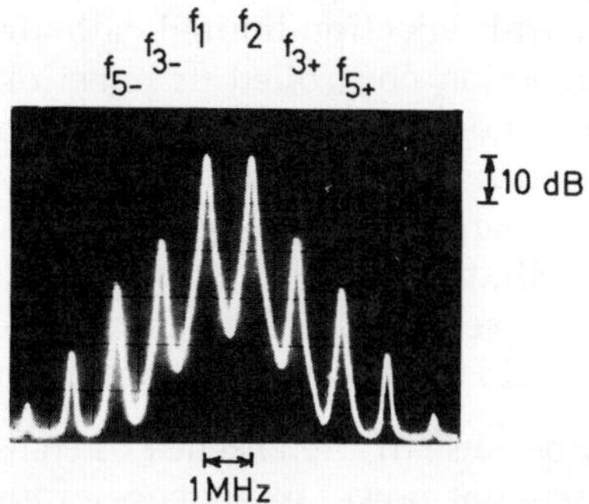

Figure 6.19 Typical frequency spectrum of two-tone intermodulation test

The saturated output power decreases with diode length. Since in transit time operation the diode length decreases as the frequency increases, TE diodes show the typical $1/f^2$ dependence of the power impedance product:

$$PZ \propto E_1{}^2 L^2 = E_1 v_0{}^2 \frac{1}{f^2} \tag{6.33}$$

E_1 is the average a.c. field swing and v_0 the electron velocity. The output power can be approximated by

$$P = E_1^2 |\mu| eA(n_0 L) \tag{6.34}$$

where A is the cross-sectional area. The power output is thus proportional to $n_0 L$.

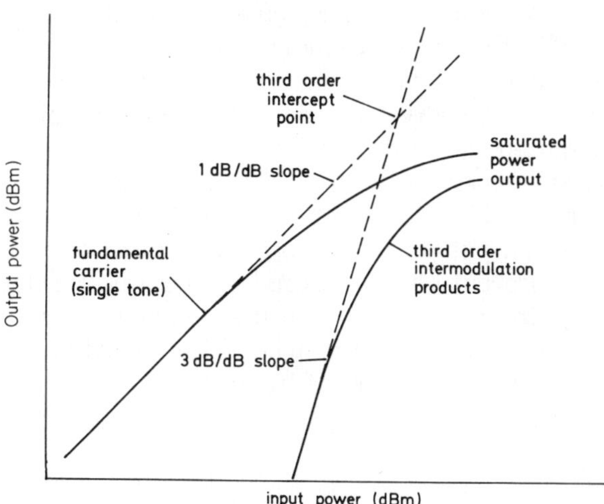

Figure 6.20 Typical intermodulation characteristics of stable amplifiers

6.3.4 Stable amplifier with injection limited cathode

Emitter controlled, injection controlled or injection limited amplifiers are transferred electron diodes with special provisions at the cathode to limit the number of carriers entering the cathode region thereby preventing formation of large space charge layers. As a consequence of this, the d.c. field distribution remains uniform. If the $n_0 L$-product is kept below the critical level the a.c. field distribution will also remain uniform. This device would exhibit the following features:

(a) a large bandwidth because of the absence of transit time effects
(b) a frequency independent power impedance product
(c) an efficiency of 25% for GaAs
(d) a small negative Q factor
(e) a low noise figure of about 8 dB for GaAs and perhaps 2 dB for InP

Until now, no device has been built which has the predicted microwave properties. One exception perhaps is the current limiting contact on InP[75] made with oxidized and alloyed Ag—In—Ga. These diodes which had supercritical $n_0 L$-product yielded high efficiency (20%) oscillations. No results on amplification have been reported so far.

There have been several proposals[76–78] on how to build injection limiting cathode contacts like a shallow Schottky barrier contact. Such low barriers are not in practice obtainable on GaAs because of the high density of surface states present. Atalla and Moll[77] have proposed a transistor (n—p) emitter which injects carriers depending on the base emitter voltage. Recently, Clark and coworkers[79] have been able to realize by liquid epitaxy a structure as proposed by Dumke and coworkers[80] which is basically a transistor structure consisting of an n^+ (GaAlAs)-p^+ (GaAs)-n(GaAs)-n^+ (GaAs). Although transistor action of this device has been observed, no microwave amplification or oscillation due to the Gunn effect has been obtained yet.

Hariu and coworkers[81] have, from a generalized expression of the diode admittance for arbitrary field dependent carrier injection from the cathode, computed the device conductance and found large bandwidth behaviour, as is illustrated in Figure 6.21. These results have been obtained for an assumed barrier height of 0.25 eV. One disadvantage of injection controlled amplifiers is the $n_0 L$ limitations imposed on the diode in order to prevent dipole formation within the device. This limitation may be circumvented by biasing the diode out in the flat portion of the $v(E)$ characteristic where the negative differential mobility is very low. If a large signal of frequency

$$\omega > v\left(\frac{en_0\,|\mu|}{\epsilon D}\right)^{1/2} \tag{6.35}$$

is applied, diffusion will suppress domain formation, and the device works as a class C amplifier or oscillator.

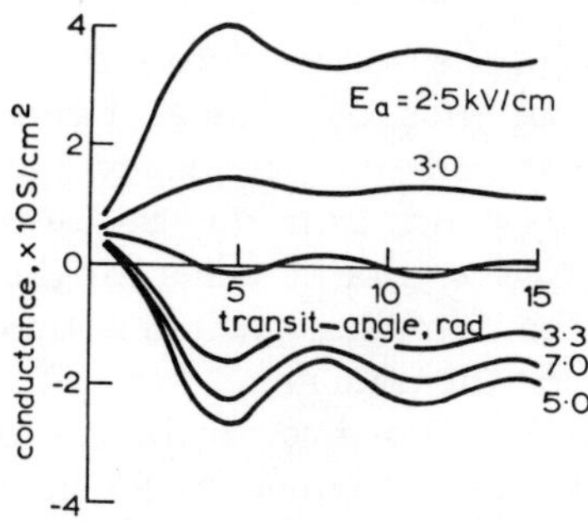

Figure 6.21 Conductance of a depleted Schottky barrier contact GaAs diode; active layer thickness = 0.5 μm, Schottky barrier height = 0.25 eV (after Hariu, Ono and Shibata[81])

6.3.5 Oscillating amplifier

Another novel type of TE amplifier[66] is the oscillating amplifier. It makes use of the negative branch of the I—V characteristic of a TE device oscillating in the travelling domain mode at the transit time frequency. The I—V characteristic of such a diode is shown in Figure 6.22. The negative resistance occurs because of the decreasing field

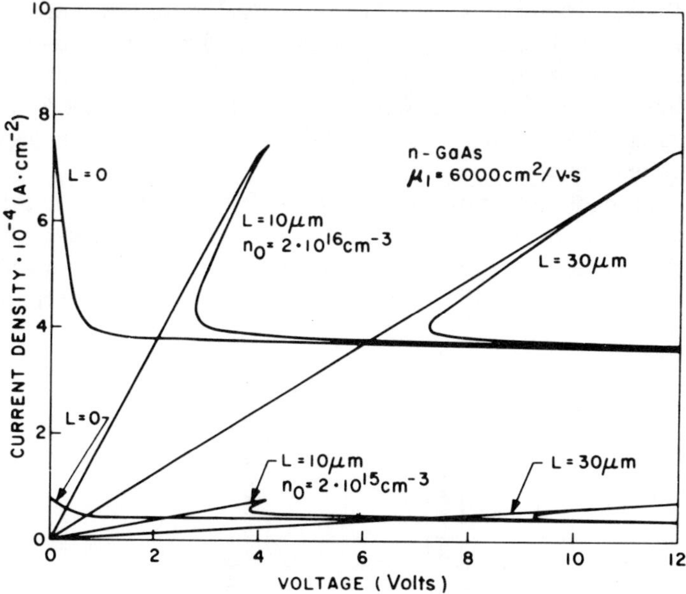

Figure 6.22 Theoretical I-V characteristics for Gunn oscillators with a range of lengths and doping densities (from Thim,[66] reproduced by permission of IEEE Inc.)

outside the domain when the voltage across the device is increased. An oscillating TE diode can thus be used as a reflection-type amplifier which oscillates at the transit time frequency and is capable of amplification over a very wide band, from d.c. on up to a frequency whose period is comparable to the time it takes for the domain to readjust. This time is given by the domain capacitance times the positive resistance of the region outside the domain.

Experimentally, amplification has not been observed at frequencies much above the transit time frequency. So far there has been little application for this type of TE amplifier since it is preferable to work with a non-oscillating device. The circuit requirements are more complex.

6.3.6 LSA amplifiers

Little has so far been reported on LSA amplifiers. Lidgey and Foulds[82] have measured a linear gain of 14—26 dB at 9.5 GHz for a reflection-type device, which is simultaneously oscillating at 4.15 GHz. The external circuit is very complicated because of the requirements of the LSA mode of operation.

6.3.7 Travelling wave amplifiers

Travelling space charge wave amplifiers invented by Robson and coworkers[83] are four-terminal devices which provide unidirectional gain without the use of ferrites. As illustrated by Figure 6.23, an r.f. signal applied to the input terminals is carried by the drifting space charge and is amplified above a threshold when negative differential mobility exists. The velocity and growth rate of the space charge waves is dependent on the applied voltage. This leads to a voltage dependent gain, phase and delay.[83]

Subcritically doped amplifiers ($n_0 L < 10^{11}$ cm^{-2}) can handle only low r.f. power levels and have generally low gain. For higher r.f. power levels and large gain, travelling space charge wave amplifiers are supercritically doped and stabilized by either a small transverse dimension or

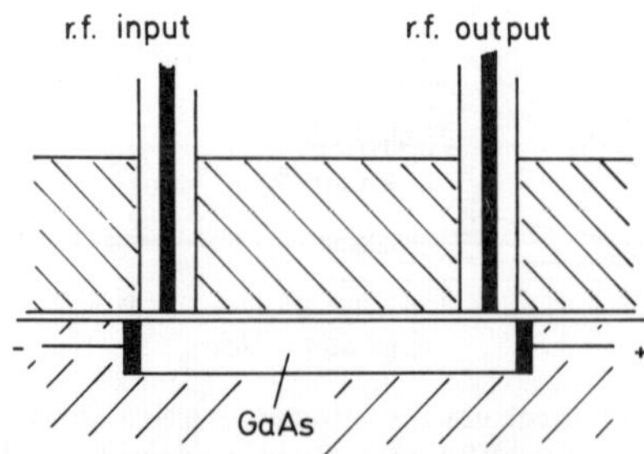

Figure 6.23 A typical configuration for the space charge wave amplifier

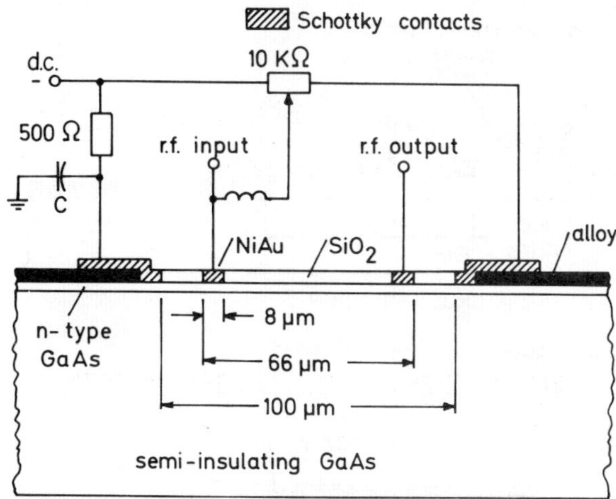

Figure 6.24 GaAs travelling wave amplifier (from Dean and coworkers,[84] reproduced by permission of IEEE Inc.)

by loading the surface with a material with high dielectric constant, as discussed in 6.3.3.5. This reduces the electric field inside the device and thus raises the critical n_0L-product. Dean and coworkers[84] have built laterally stabilized amplifiers for X-band operation using a Schottky barrier in the vicinity of the ohmic cathode contact in order to raise the cathode field. This resulted in low coupling losses but the overall gain was still low and only 0.5 mW saturated output power was obtained. The structure is shown in Figure 6.24.

Further improvement was obtained by combining the GaAs Schottky barrier field effect transistor with the TE travelling wave amplifier.[85–87] In this 'travelling wave transistor' the FET couples space charge waves into the biased GaAs epitaxial layer which serves as the amplifying drift space as illustrated in Figure 6.25. Very large bandwidth has been obtained with this type of amplifier as shown in Figure 6.26. Since the upper frequency limit of the travelling wave transistor is still determined by the length of the gate, no obvious reason exists why this type of amplifier should be superior to the Schottky barrier FET without amplifying drift region as the drift region undoubtedly must add noise.

6.3.8 Design procedures and practical realization of TE amplifiers

Since presently available TE amplifier diodes exhibit typical noise measures of 15 dB they will mainly be used as power amplifiers. Unfortunately, not all diodes commercially available can be used as amplifiers because many of these will turn out to be inherently unstable when embedded in the usual coaxial or wave guide circuits. This is usually due to doping fluctuations and to a high n_0L-product. As has

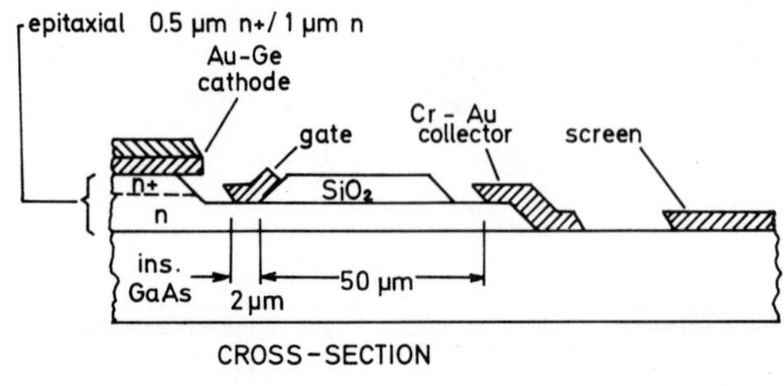

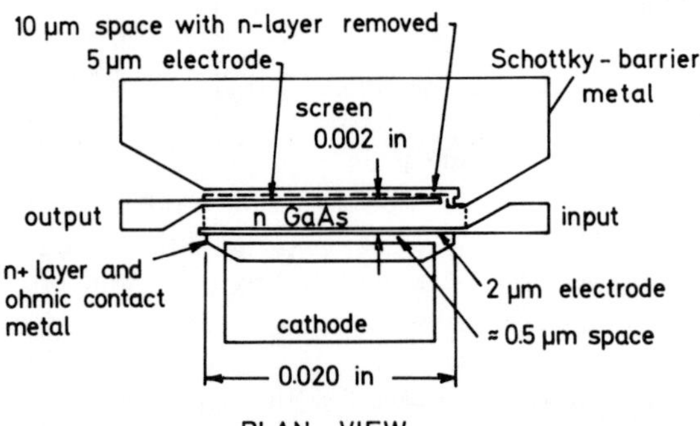

Figure 6.25 Schematic cross-section of travelling wave transistor (from Dean and coworkers,[85] reproduced by permission of IEEE Inc.)

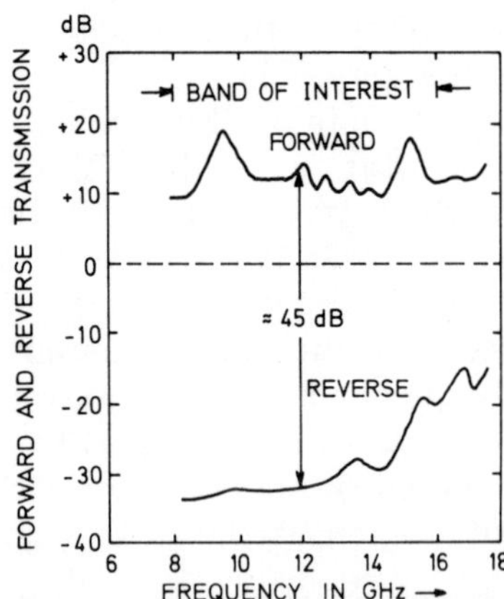

Figure 6.26 Instantaneous forward and reverse net gain for travelling wave transistor mounted in header package. Displayed results include effects of all (built-in) bias strays and blocking capacitors (from Dean and coworkers,[85] reproduced by permission of IEEE Inc.)

been pointed out previously any unstable impedance can be stabilized by adding a suitable stabilizing circuit (Nyquist theorem) like an infinite resistor. However, because of unavoidable reactances always present in practical circuits and packages one has to tailor the device to obtain stability.

In order to obtain high bandwidth (low negative Q) and high power one has to choose the n_0L-product as high as possible according to expressions (6.29) and (6.34). An n_0L-product around 10^{12} cm^{-2} for GaAs is probably the optimum choice because such a diode will be stable regardless of doping fluctuations. A doping notch at the cathode would be desirable but it is difficult with present day techonology to control the shape of the notch. Perhaps an incomplete alloying process might be a practicable way of realizing it.

The next step is an impedance measurement over the frequency range where negative conductance appears. The load conductance should be larger than the negative conductance of the device, otherwise oscillations might occur. At frequencies up to X-band the device can be measured after packaging since standard types of packages have been studied extensively.[88-91] At frequencies above X-band, but especially at millimetre wavelengths, packaged devices are practically unusable and quartz stand-offs are generally used with a short band going to the device. This configuration has been used quite successfully in the past.[92] Once the impedance of the device is known one can then design equalizer networks to tailor the gain—bandwidth characteristic to the desired shape. Figures 6.27 and 6.28 illustrate this design procedure. Amplifier gain and power output can be increased by using several cascaded stages[93] which may be accomplished by using circulators and hybrids.

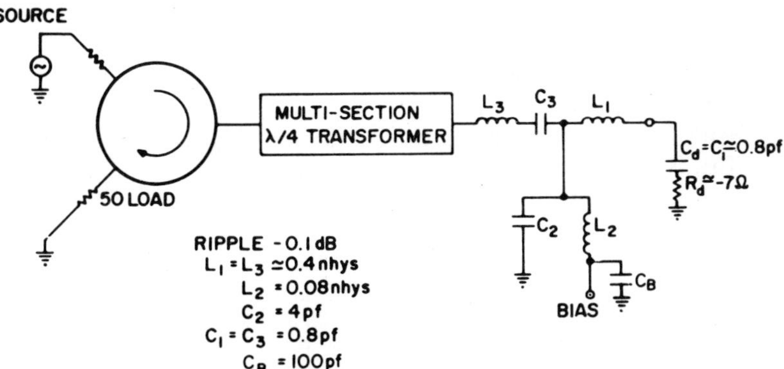

Figure 6.27 Circulator-coupled reflection-type amplifier network, comprising a multisection impedance transformer and a three-section filter-type equalization network. The active device equivalent circuit is included in the final filter resonator (from Perlman,[74] reproduced by permission of RCA Corporation)

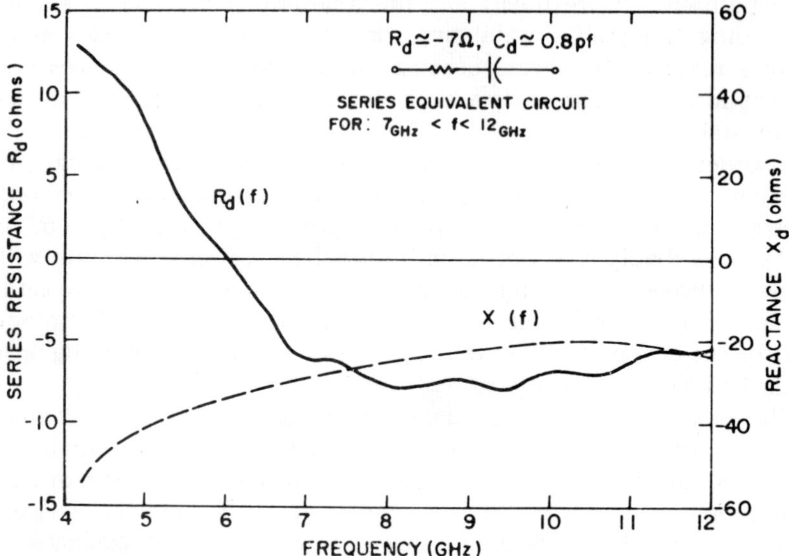

Figure 6.28 Equivalent series impedance components of the stabilized active transferred-electron device as a function of frequency. An equivalent RC network is also indicated (from Perlman,[74] reproduced by permission of RCA Corporation)

As can be seen from Figure 6.18 higher output powers are obtained when the gain is low. Because of this, the final output stage should be designed to have low gain. The interstage isolation must also be generally of the order of 15—20 dB or better to insure stability of the amplifier which requires a great number of isolators as shown in Figure 6.29. More will be said in Section 6.4 on multistage TE and impatt amplifiers.

6.4 BARITT AMPLIFIERS

6.4.1 Principle of operation

The possibility of obtaining high frequency negative resistance in a punched-through structure had been considered by Wright[94] and

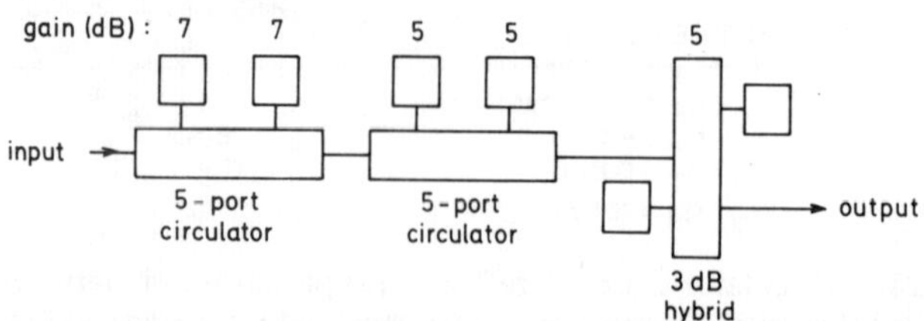

Figure 6.29 Block diagram of a multistage cascaded amplifier

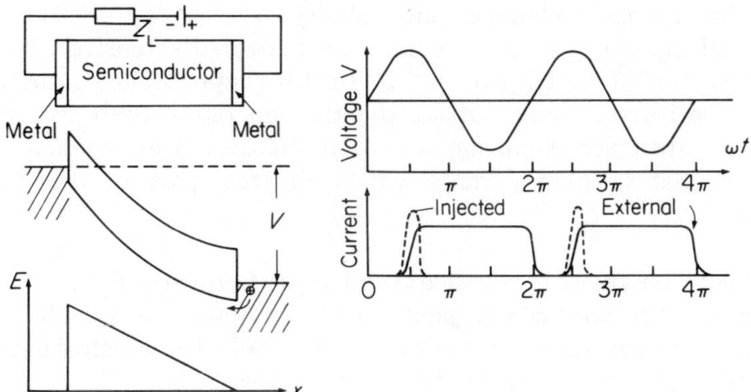

Figure 6.30 Baritt diode structure, potential and field distributions, voltage and current waveforms

Ruegg.[95] They had shown theoretically that injection of minority carriers over a barrier into a depleted drift region produces a current waveform as shown in Figure 6.30. Recently, microwave oscillations have been observed in metal—semiconductor—metal and in *p—n—p* structures. The main difference between baritt and impatt diodes is the injection mechanism. The principle of operation is also shown in Figure 6.30. When the bias applied to the two contacts exceeds the punch-through voltage, a hole current is injected from the positively biased contact into the drift region. This current rises exponentially with voltage (dotted curve) but is in phase with it. The necessary overall phase delay is established by making the length of the drift region equal to about three-quarters of the distance the carriers travel during one period. The frequency is therefore given by the expression

$$f = 0.75\, v_s/L \tag{6.36}$$

Since the phase delay between current and voltage is greater than 90° and less than 180° the real part of the impedance is negative; but a strong, capacitive reactance appears which reduces the bandwidth. The diode thus has a large negative Q factor ($Q < -20$). This is a severe drawback which limits the usefulness of the baritt diode.

6.4.2 Static field distribution

Since the baritt diode is a punch-through structure, the field is low or even negative at the cathode (right contact in Figure 6.30). This could lead to a favourable phase delay,[95] but this has not yet been established experimentally. Experimental results obtained with n^+—p—n—n^+ structures[96] have not shown the expected high efficiency of 15% (see Section 6.4.7).

6.4.3 Small-signal impedance and stability

The small signal r.f. properties have been covered extensively by Weissglas in Section 3.5.2 (Equation 3.59) of his chapter. The r.f. impedance has a negative real part and a capacitive reactance. Stability occurs if the load impedance is chosen such that the overall impedance remains finite in that frequency range where the real part of the device is negative.

6.4.4 Gain—bandwidth, negative Q and upper frequency limit

The bandwidth over which gain can be obtained is severly limited by the capacitive reactance of the device. The Q factor which is a measure of available gain times bandwidth can be expressed by

$$Q = \frac{1 - \cos\theta + \theta_B \sin\theta}{\theta(1 + \theta_B{}^2) + \theta_B(1 - \cos\theta) - \sin\theta} \tag{6.37}$$

The transit angle and the injection angle are given by

$$\theta = \frac{\omega L}{v_0} \tag{6.38}$$

$$\theta_B = \frac{\omega E}{\sigma_B} \tag{6.39}$$

Q assumes a minimum value of approximately —20 for $\theta = 13\pi/8$ and $\theta_B = 1.88$. Any small series resistance at the contacts or in the package can completely wipe out the negative resistance. This has not yet been achieved. For the same reason a low upper frequency limit is to be expected. This can be estimated to be of the order of 80 GHz for a device with 1 micrometre active length. So far only X-band devices have been operated successfully. No experimental results on gain—bandwidth products are available.

6.4.5 Noise

The expression for the small signal noise measure has already been derived by Weissglas in Section 3.5.4 of his chapter. Ordinary shot noise will be present but the main contribution will arise from thermal fluctuations in the low field and drift region. The general expression also applicable to the baritt diode is given by (6.30) in Section 6.3.3.8 of this chapter. Noise measures of the order of 10 dB have been calculated from this expression which agree fairly well with measured results (see Chapter 3).

6.4.6 Saturation power and efficiency

Although a large-signal calculation using the waveforms shown in Figure 6.30 yields 15% efficiency, actually observed (oscillator) efficiencies lie considerably lower. 2% have been measured on X-band devices with power output in the 100 mW range. Perhaps inefficient carrier injection into the low field region is responsible for this discrepancy but more data are necessary in order to draw a correct conclusion.

6.4.7 Design procedure

In order to design a baritt amplifier circuit much the same procedures can be applied as in the case of stable TE amplifiers. Since no experimental results are available on baritt amplifiers we refer the reader to the oscillator section described by Weissglas in Section 3.5.5 of his chapter.

6.5 AVALANCHE INJECTION AMPLIFIERS

There are basically two devices which fall into this category: the impatt diode and the trapatt diode. Since the trapatt diode is a large-signal device which oscillates at the impatt frequency while oscillating or amplifying at a lower frequency or more correctly, is phase locked at a lower frequency, the reader is referred to the large-signal trapatt theory covered by Weissglas in Section 3.6 of his chapter. It should be added here that trapatt amplifiers exhibit very high noise levels (60 dB) which is probably due to time jitter of the avalanche 'ignition' process.

6.5.1 Impatt amplifier principle

In 1958 Read[97] proposed a power generating diode consisting of a *p—n* junction biased into breakdown, at one end of a relatively high resistance region serving as the drift space for the avalanche-generated carriers. This structure is shown in Figure 6.31. Also shown is the steady state field distribution under d.c.-biased condition. The narrow high field region at the left (reversed-biased *p—n* junction) is generating carriers during that quarter of the period following the maximum of the

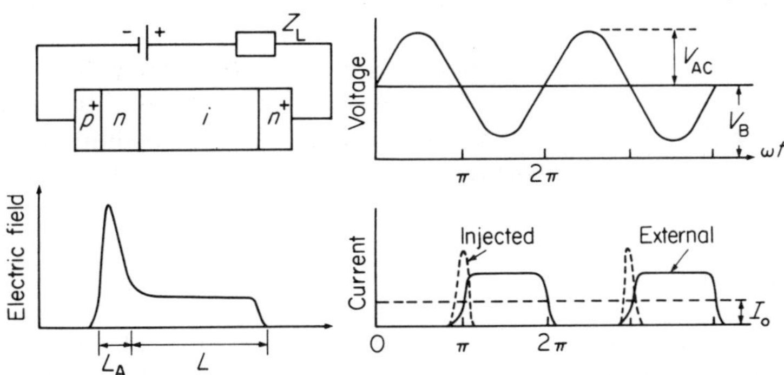

Figure 6.31 Read-type impatt diode structure, field distribution, current and voltage waveforms

a.c. voltage shown on the right. Since the avalanche is an exponentially rising process it turns off when the a.c. voltage decreases fastest. Thus the injected current reaches its maximum right at the point marked by π. This 90° phase lag is very significant since it yields a terminal current which is 180° out of phase if the drift length L is chosen such that the carriers reach the anode contact right at the instant when the a.c. voltage rises through zero. Under these conditions the diode produces maximum negative terminal resistance at half the transit time frequency f_t given by

$$f_t = v_s/2L \tag{6.40}$$

6.5.2 Static field distribution

The static field distribution is shown in Figure 6.31. When an a.c. voltage is applied to the amplifier this field is moved up and down. Since the carriers in the drift region must not drop out of saturation and, on the other hand, must not reach the energy level necessary for breakdown, the a.c. voltage swing must not exceed a certain level which limits the power and efficiency of the amplifier in large-signal operation.

Essentially identical field profiles can be obtained with GaAs Schottky barrier diodes with a 'hi-lo' or 'lo-hi-lo' profile as discussed by Weissglas in Section 3.4.3 of his chapter.

6.5.3 Small-signal impedance and stability

The small-signal impedance of impatt diodes has been derived by Weissglas in Section 3.4.1 of his chapter and need not be repeated here. Above a certain frequency (avalanche frequency) the impedance exhibits a negative real part and a capacitive reactance. When the diode is connected to a load impedance which leads to a positive over all impedance within that frequency range at which the real part is negative the amplifier will be stable.

6.5.4 Gain—bandwidth, negative Q and upper frequency limit

According to Figure 3.12 in Weissglas's chapter the negative resistance decreases above the resonant frequency ω_a so that the diode works well only over a frequency band of one octave. The negative Q factor is approximately given by

$$Q = \frac{\pi}{2}(1 - \omega^2/\omega_a{}^2) \tag{6.41}$$

According to expression (6.41) and also to experimental results, the Q factor lies between -1 and -10 which makes circuit matching much easier than in the case of the baritt diode. However, small-signal gain—bandwidth products are reaching those of transferred-electron devices.

Although amplifiers have not yet been operated above X-band it is obvious that the results obtained with impatt oscillators can be extrapolated to impatt amplifiers. Hence, impatt amplifiers should be operable well above 100 GHz and perhaps 300 GHz, which is so far the highest frequency of operation of impatt oscillators.

6.5.5 Noise

The noise level of impatt amplifiers is very high corresponding to noise measures between 20 dB and 40 dB. GaAs impatt amplifiers are 10 dB quieter than Si impatt diodes owing to equal electron and hole ionization rates. More on noise can be found in Section 3.4.5 of Chapter 3.

6.5.6 Saturation power and efficiency

Impatt amplifiers are high power, high efficiency devices exhibiting high gain and adequate bandwidth for the system's need. Under large-signal conditions a number of special problems are commonly observed. These include gain compression or expansion, parametric or triggered oscillations, large intermodulation effects and excess noise. Some of these effects are associated or caused by instabilities in the bias circuit due to low frequency negative resistance. To avoid these instabilities one must carefully design the bias circuit as will be discussed in the next section.

6.5.7 Design procedures and practical realization

6.5.7.1 Single-stage amplifiers

Because of the high noise level impatt amplifiers are exclusively used as power amplifiers. As usual the diode is selected for a certain frequency range by means of (6.40). The Smith chart plot of the packaged diode is then used for tailoring the desired gain—bandwidth characteristic. However, for large-signal operation the small-signal impedance is not the only characteristic one needs. As mentioned in Section 6.5.6, the large-signal impedance at low frequencies (usually in the MHz range) must be known in order to design a bias circuit with proper loading to avoid low frequency instabilities. Brackett[98] has shown that these instabilities cause excessive noise, bias oscillations and possibly diode burn-out. An example of a practical circuit is shown in Figure 6.32.

6.5.7.2 TE/Impatt-multistage amplifiers

Transferred-electron and impatt amplifiers have been combined to yield high power, high gain at fairly low noise levels. Their lower noise property makes TE amplifiers suitable for the input stages. Compared with impatt amplifiers, a wider bandwidth is obtained at a given r.f.

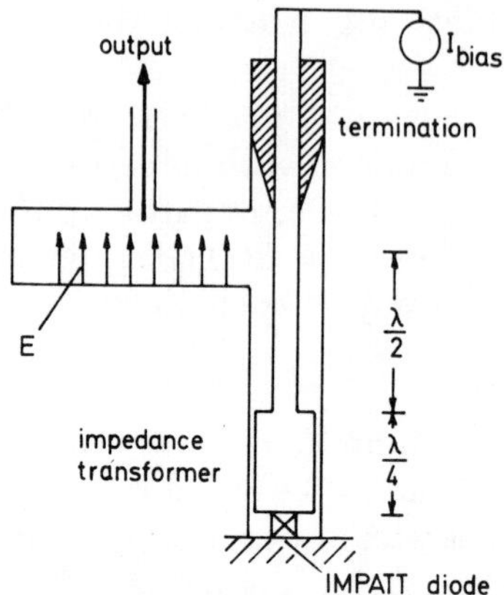

Figure 6.32 Impatt amplifier circuit. Coaxial module magnetically coupled to a microwave cavity (after Harp and Russel,[99] reproduced by permission of IEEE Inc.)

input level owing to the fact that the negative resistance of a TE diode is much larger than that of an impatt diode. In order to obtain a large over all bandwidth the impatt stage is tuned for low gain over a large bandwidth. Besides the lower FM/AM noise of TE devices, their non-linear distortion is also less as Figure 6.33 indicates. Impatt devices are generally used in the driver and high power output stages.

Multistage amplifiers are characterized by the high degree of isolation which for stability is required between the individual stages, implemented in general by two ferrite isolators per stage. Stable operation

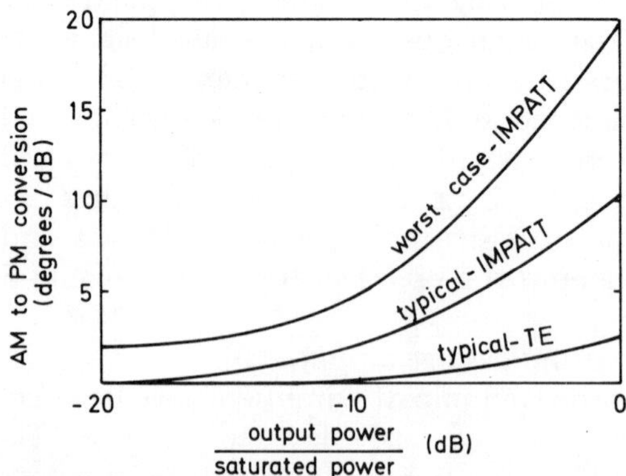

Figure 6.33 AM to PM conversion for multi-stage stable amplifiers

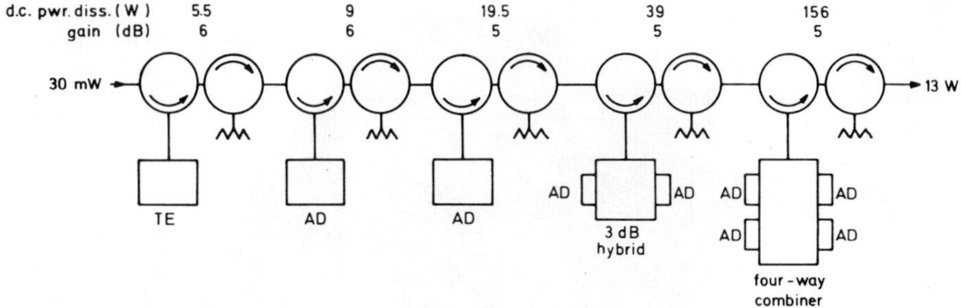

Figure 6.34 A five-stage hybrid type amplifier (after Baines[100])

requires that the total return loss of the amplifier—isolator chain is always greater than the forward gain. Figures 6.34 and 6.35 show the circuit of a five-stage hybrid-type amplifier and its characteristics, respectively.

Hanson and Heinz[101] have developed an X-band power amplifier for FM and PM communications systems applications having 1 W output and 30 dB gain, utilizing two electrically tuned injection—locked oscillator stages operated as amplifiers which are tunable over a 500 MHz range with 250 MHz locking bandwidth. The amplifier combines the low noise properties of GaAs TE diodes and the high-power, high-efficiency properties of impatt devices. A varactor-tuned Gunn diode is utilized in the first stage to achieve low FM noise and a varactor tuned silicon impatt diode in the second-stage oscillator. The noise figure is about 30.5 dB with a 0 dBm input signal level. An overall view and an interior view of this amplifier are shown in Figure 6.36.

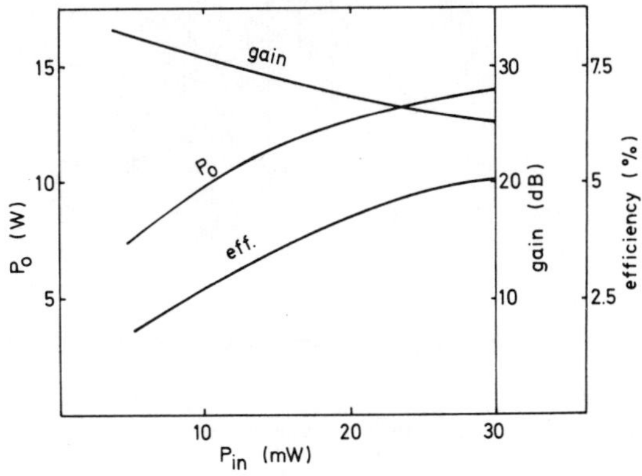

Figure 6.35 Characteristics of a 13 W five-stage amplifier (after Baines[100])

Figure 6.36a Overall view of integrated X-band power amplifier

Figure 6.36b Interior view of integrated X-band power amplifier. The first stage varactor-tuned Gunn oscillator is on the bottom right, the second stage varactor-tuned impatt diode oscillator at the bottom left (after Hanson and Heinz,[101] reproduced by permission of IEEE Inc.)

6.6 COMPARISON OF TE, BARITT AND IMPATT AMPLIFIERS

1. impedance level	:	baritt	$<$	impatt	$<$	TE
2. Q	:	TE	$<$	impatt	$<$	baritt
3. power	:	baritt	$<$	TE	$<$	impatt
4. noise	:	baritt	$\leqslant$	TE	$<$	impatt
5. amplifier gain x bandwidth	:	baritt	$<$	impatt	$<$	TE
6. amplifier gain—saturation	:	baritt	$<$	TE	$<$	impatt
7. bias voltage	:	TE	$<$	baritt	$<$	impatt
8. efficiency	:	baritt	$<$	TE	$<$	impatt
9. current density	:	baritt	$<$	impatt	$\leqslant$	TE

REFERENCES

1. D. E. McCumber and A. G. Chynoweth, 'Theory of negative-conductance amplification and of Gunn instabilities in two-valley semiconductors', *IEEE Trans, Electon Dev.*, ED-13, 4—21 (1966).
2. W. A. Haus and R. B. Adler, *Circuit Theory of Linear Noisy Networks*, Wiley, New York, 1959.
3. B. K. Ridley and T. B. Watkins, 'The possibility of negative resistance effects in semiconductors', *Proc. Phys. Soc. London*, 78, 293—304 (1961).
4. C. Hilsum, 'Transferred electron amplifiers and oscillators', *Proc. IRE*, 50, 185—189 (1962).
5. J. B. Gunn, 'Instabilities of current and of potential distribution in GaAs and InP, *Proc. Conf. Physics of Semiconductors (Paris), 1964.*
6. C. Hilsum, J. B. Mullin, B. A. Prew, H. D. Rees and B. W. Straughan, 'Instabilities of InP 3-level transferred electron oscillators', *Electron. Lett.*, 6, 307—308 (1970).
7. C. Hilsum and H. D. Rees, 'Three level oscillators: A new form of transferred electron device', *Electron. Lett.*, 6, 277—278 (1970).
8. A. G. Foyt and A. L. McWorther, 'The Gunn effect in polar semiconductors', *IEEE Trans. Electon Dev.*, ED-13, 79—87 (1966).
9. G. W. Ludwig, R. E. Halsted and M. Aren, 'Current saturation and instability in CdTe and ZnSe', *IEEE Trans. Electron Dev.*, ED-13, 671 (1966).
10. G. W. Ludwig and M. Aren, 'Gunn effect in ZnSe', *J. Appl. Phys.*, 38, 5326—5331 (1967).
11. H. Kroemer, 'Theory of the Gunn effect', *Proc. IEEE*, 52, 1736 (1964).
12. P. N. Butcher and W. Fawcett, 'Calculation of the velocity—field characteristics for gallium arsenide', *Phys. Lett.*, 21, 489—490 (1966).

13. A. D. Boardman, W. Fawcett and H. D. Rees, 'Monte Carlo calculation of the velocity—field relationship for gallium arsenide', *Solid State Comm.*, 6, 305—307 (1968).
14. H. D. Rees, 'Calculation of distribution functions by exploiting the stability of the steady state', *J. Phys. & Chem. Solids*, **30**, 643—655 (1969).
15. J. G. Ruch, and G. S. Kino, 'Transport properties of GaAs', *Phys. Rev.*, 174, 921—931 (1968).
16. J. B. Gunn, 'Measurement of the negative differential mobility of electrons in GaAs', *Phys. Lett.*, 22, 369—371 (1966).
17. G. A. Acket, 'Determination of the negative differential mobility of *n*-type gallium arsenide using 8 mm-microwaves', *Phys. Lett.*, 24A, 200—202 (1967).
18. D. M. Chang and J. L. Moll, 'Direct observation of the drift velocity as a function of the electric field in gallium arsenide', *Appl. Phys. Lett.*, 9, 283—285 (1966).
19. J. G. Ruch and G. S. Kino, 'Measurement of the velocity—field characteristics of GaAs', *Appl. Phys. Lett.*, **10**, 40 (1967).
20. E. M. Conwell and M. O. Vassell, 'High-field transport in *n*-type GaAs', *Phys. Rev.*, **166**, 797—821 (1968).
21. H. D. Rees, 'Time response of the high field electron distribution function in GaAs', *IBM Journ. Res. Develop.*, 13, 537—542 (1969).
22. M. L. Grasl and O. F. Zimmerl, 'Frequency behaviour of space-charge-wave amplifiers in GaAs', *Phys. Stat. Solid*, 2, 391—405 (1970).
23. H. W. Pötzl and O. F. Zimmerl, 'Limiting frequency of GaAs travelling wave amplifiers', *Moga*, 16.29—16.33 (1970).
24. R. Bosch and H. W. Thim, 'Computer simulation of transferred electron devices using the displaced Maxwellian approach', *IEEE Trans. Electron Dev.*, ED-21, 16—25 (1974).
25. T. J. Maloney and J. Frey, 'Electron dynamics in short channel InP field-effect transistors', *Electron. Lett.*, 10, 115—116 (1974).
26. H. Hillbrand, H. D. Rees and D. Jones, 'Theoretical characteristics of transferred-electron amplifiers', *Electron. Lett.*, 10, 87—89 (1974).
27. W. Fawcett and D. C. Herbert, 'High field transport in indium phosphide', *Electron. Lett.*, 9, 308—309 (1973).
28. C. Hilsum and H. D. Rees, 'Three-level transferred-electron effects in InP', *Electron. Lett.*, 7, 437—438 (1971).
29. L. D. Nielsen, 'Measurement of the velocity/field characteristic for electrons in indium phosphide', *Solid-State Commun.*, **10**, 169—171 (1972).
30. H. 'T Lam and G. A. Acket, 'Comparison of the microwave veloc-

ity/field characteristics of *n* type InP and *n* type GaAs', *Electron. Lett.*, 7, 722—723 (1971).

31. G. H. Glover, 'Microwave measurement of the velocity field characteristic of *n*-type InP', *Appl. Phys. Lett.*, 20, 224—225 (1972).
32. R. E. Hayes 'Measurement of the velocity—field characteristic of indium phosphide by microwave absorption technique', *IEEE Trans. Electron Dev.*, ED-21, 233—235 (1974).
33. J. A. Copeland and S. Knight, *Semiconductors and Semimetals*, *7A*, Willardson and A. C. Beer Ed., New York, Academic Press, 1971, pp. 3—72.
34. W. Shockley, 'Negative resistance arising from transit time in semiconductor diodes', *Bell Syst. Tech. J.*, 33, 799—826 (1954).
35. H. Kroemer, 'Detailed theory of the negative conductance of bulk negative mobility amplifiers, in the limit of zero ion density', *IEEE Trans. Electron Dev.*, ED-14, 476—492 (1967).
36. F. Sterzer, 'Stabilization of supercritical transferred electron amplifiers', *Proc. IEEE*, 57, 1781—1783 (1969).
37. J. Magarshack and A. Mircea, 'Wideband CW amplification in X-band with Gunn Diodes', *ISSC Conf. Dig. Technical Papers*, 1970, pp. 134—135.
38. J. Magarshak and A. Mircea, 'Stabilization and wide-band amplification using overcritically doped transferred electron diodes', *Proc. Int. Conf. Microwave and Optical Generation and Amplification, Amsterdam, The Netherlands*, 1970, pp. 16.19—16.23.
39. H. W. Thim and S. Knight, 'Carrier generation and switching phenomena in *n*-GaAs devices', *Appl. Phys. Lett.*, 11, 85—87 (1967).
40. M. P. Shaw, P. R. Solomon and H. L. Grubin, 'The influence of boundary conditions on current instabilities in GaAs', *IBM Journal Res. Dev.*, 13, 587—590 (1969).
41. P. Gueret, 'Connective and absolute instabilities in semiconductors exhibiting negative differential mobility', *Phys. Rev. Lett.*, 27, 256 (1971).
42. P. Gueret and M. Reiser, 'Switching behaviour of over-critically doped Gunn diodes', *Appl. Phys. Lett.*, 20, 60—62 (1972).
43. H. Thim, 'Stability and switching in overcritically doped Gunn diodes', *Proc. IEEE*, 59, 1285—1286 (1971).
44. P. Jeppesen and B. I. Jeppsson, 'A simple analysis of the stable field profile in the supercritical TEA', *IEEE Trans. Electron Dev.*, ED-20, 371—379 (1973).
45. P. Jeppesen and B. I. Jeppsson, 'The influence of diffusion on the stability of the supercritical transferred electron amplifier', *Proc. IEEE*, 60, 452 (1972).

46. H. W. Thim, J. V. DiLorenzo, W. Gramann, W. Haydl and R. Bosch, 'Bistable switching in high quality GaAs Gunn diodes with overcritical n_0L product', *1972 Symposium on GaAs*, 267—274.
47. R. Charlton and G. S. Hobson, 'The effect of cathode notch doping profiles on supercritical transferred-electron amplifiers', *IEEE Trans. Electron Dev.*, ED-20, 812—817 (1973).
48. R. Charlton, K. R. Freeman and G. S. Hobson, 'A stabilization mechanism for supercritical transferred electron amplifiers', *Electron. Lett.*, 7, 575—577 (1971).
49. J. Magarshak, A. Rabier and R. Spitalnik, 'Optimum design of transferred-electron amplifier devices in GaAs', *IEEE Trans. Electron Dev.*, ED-21, 652 (1974).
50. S. Kataoka, H. Tateno and M. Kawashima, 'Suppression of travelling high-field domain mode oscillations in GaAs by dielectric surface loading', *Electron. Lett.*, 5, 48—50 (1969).
51. G. S. Kino and P. N. Robson, 'The effect of small transverse dimensions on the operation of Gunn devices', *Proc. IEEE.*, 56, 2056—2057 (1968).
52. K. R. Hofmann and H. 'T Lam, 'Suppression of Gunn domain oscillations in thin GaAs diodes with dielectric surface loading', *Electron. Lett.*, 8, 122—124 (1972).
53. K. R. Hofmann, 'Stability theory for thin Gunn diodes with dielectric surface loading', *Electron. Lett.*, 8, 124—125 (1972).
54. K. R. Hofmann, 'Stability criterion for Gunn oscillators with heavy surface loading', *Electron. Lett.*, 5, 469—470 (1969).
55. W. Frey, R. W. H. Engelmann and B. G. Bosch, *Arch. der Elektr. Übertr.*, 25, 1—8 (1971).
56. R. W. H. Engelmann, 'Plane wave approximation of carrier waves in semiconductor plates with nonisotropic mobility', *IEEE Trans. Electron Dev.*, ED-18, 587—591 (1971).
57. R. Becker, B. G. Bosch and R. W. Engelman, 'Domains and guided electromagnetic waves in GaAs stripline', *Electron. Lett.*, 6, 604—605 (1970).
58. P. Gueret, 'Stabilization of Gunn oscillations in layered semiconductor structures', *Electron. Lett.*, 6, 637—638 (1970).
59. M. Masuda, N. S. Chang and Y. Matsuo, 'Suppression of Gunn-effect domain formation by ferrimagnetic materials', *Electron. Lett.*, 6, 605—606 (1970).
60. H. W. Thim and M. R. Barber. 'Microwave amplification in a GaAs bulk semiconductor', *IEEE Trans. Electron Dev.*, ED-13, 110—114 (1966).
61. A. G. Foyt and T. M. Quist, 'Bulk GaAs microwave amplifiers', *IEEE Trans. Electron Dev.*, ED-13, 199 (1966).
62. B. S. Perlman, 'CW microwave amplification from circuit stabi-

lized epitaxial GaAs transferred electron devices', *1970 Digest Tech. Papers, ISSC Conf.*, pp. 136—137.

63. D. Holstrom. 'Small signal behaviour of Gunn diodes', *IEEE Trans. Electron Dev.*, ED-14, 464 (1967).
64. D. Jones and H. D. Rees, 'Electron-relaxation effects in transferred-electron devices revealed by new stimulation method', *Electron. Lett.*, 8, 363—364 (1972).
65. D. Jones and H. D. Rees, 'Theoretical characteristics of transferred-electron amplifiers', *Electron Lett.*, **10**, 87 (1974).
66. H. W. Thim, 'Linear negative conductance amplification with Gunn oscillators', *Proc. IEEE*, 55, 446—447 (1967).
67. W. Shockley, J. A. Copeland and R. P. James, 'The impedance field method of noise calculation in active semiconductor devices', *Quantum theory of atoms, molecules and the solid state*, Academic Press, 1966.
68. H. W. Thim, 'Noise reduction in bulk negative resistance amplifiers', *Electron. Lett.*, 7, 106—108 (1971).
69. F. Maloborti and V. Svelto, *Alta Frequ.*, 40, 667 (1971).
70. H. Johnson, 'A unified small signal theory of uniform-carrier-velocity semiconductor transit time diodes', *IEEE Trans. Electron Dev.*, ED-19, 1156 (1972).
71. J. E. Sitch and P. N. Robson, 'Noise measure of GaAs and InP transferred electron amplifiers', *4th European Microwave Conf. Proceedings*, pp. 232—236, 1974.
72. P. W. Braddock and K. W. Gray, 'Low noise wideband indium phosphide transferred electron amplifiers', *Electron. Lett.*, 9, 36—37 (1973).
73. S. Baskaran and P. N. Robson, 'Noise performance of InP reflection amplifiers in *Q* band', *Electron. Lett.*, 8, 137—138 (1972).
74. B. S. Perlman, 'Microwave amplifications using transferred electron devices in prototype filter equalization networks', *RCA Rev.*, 32, 3—23 (1971).
75. D. J. Colliver, L. D. Irving, J. E. Pattison and H. D. Rees, 'High efficiency InP transferred electron oscillators', *Electron. Lett.*, **10**, 221—222 (1974).
76. H. Kroemer, 'The Gunn effect under imperfect cathode boundary conditions', *IEEE Trans. Electron Dev.*, **ED-15**, 819—837 (1968).
77. M. M. Atalla and J. L. Moll, 'Emitter controlled negative resistance in GaAs', *Solid State Electron.*, 12, 619—129 (1969).
78. H. W. Thim, *U.S. Patent 3,537,021*, 1970.
79. B. W. Clark, H. G. B. Hicks and J. S. Heeks, 'An electronically controlled injection limited cathode for GaAs transferred electron devices', *ESSDERC 1974, Nottingham, England, paper A6.2.*
80. W. P. Dumke, J. M. Woodall and V. L. Rideout, 'GaAs—GaAlAs

heterojunction transistor for high frequency operation', *Solid State Electronics*, 15, 1339—1343 (1972).

81. T. Hariu, S. Ono and Y. Shibata, 'Wideband performance of the injection-limited Gunn diode', *Electron. Lett.*, 6, 666 (1970).
82. F. J. Lidgey and K. W. H. Foulds, 'An X band LSA amplifier', *IEEE Trans. Microwave Theory Tech.*, MTT-21, 736—738 (1973).
83. P. N. Robson, G. S. Kino and B. Fay, 'Two-port microwave amplification in long samples of gallium arsenide', *IEEE Trans. Electron. Dev.*, ED-14, 612 (1967).
84. R. H. Dean, A. B. Dreeben, J. F. Kaminski and A. Triano, 'Travelling wave amplifier using thin epitaxial GaAs layer', *Electron. Lett.*, 6, 775—776 (1970).
85. R. H. Dean and R. J. Matarese, 'The GaAs travelling wave amplifier as a new kind of microwave transistor', *Proc. IEEE*, **60**, 1486—1502 (1972).
86. R. H. Dean, A. B. Dreeben, J. J. Hughes, R. J. Matarese and L. S. Napoli, 'Broad-band microwave measurements on GaAs "Travelling wave" transistors', *IEEE Trans. Microwave Theory Tech.*, MTT-21, 805—809 (1973).
87. R. H. Dean, R. E. De Brecht, A. B. Dreeben, J. J. Hughes, R. J. Matarese and L. S. Napoli, GaAs travelling wave transistor', 1973 IEEE-GMTT International Microwave Symp., *IEEE cat. No. 73 CHO 736—9 MTT*, pp. 250—251.
88. I. W. Pence and P. J. Khan, 'Broad-band equivalent-circuit determination of Gunn diodes', *IEEE Trans. Microwave Theory Techn.*, MTT-18, 784—790 (1970).
89. R. P. Owens and D. Cawsey, 'Microwave equivalent-circuit parameters of Gunn effect device package', *IEEE Trans. Microwave Theory Tech.*, MTT-18, 790—798 (1970).
90. W. J. Getsinger, 'The packaged and mounted diode as a microwave circuit', *IEEE Trans. Microwave Theory Techn.*, MTT-14, 58—69 (1966).
91. C. L. Upadhyayula and B. S. Perlman, 'Design and performance of transferred electron amplifiers using distributed equalizer networks', *IEEE J. Solid State Circuits*, SC-8, 29—36 (1973).
92. H. J. Kuno, D. L. English and R. S. Ying, 'High power millimetre wave IMPATT amplifiers', *1973 IEEE ISSCC Digest of Technical Papers*, pp. 50—51.
93. D. H. Steinbrecher, 'Optimum efficiency of a cascade of low-gain amplifiers', *IEEE Trans. Microwave Theory Techn.*, MTT-18, 951—963 (1970).
94. G. Wright, 'Punch through transit-times oscillator', *Electronic Lett.*, 4, 543 (1968).
95. H. Ruegg, 'A proposed punch-through microwave negative-resistance diode', *IEEE Trans. Electron Dev.*, ED-15, 577 (1968).

96. D. Delagebeaudeuf, 'Punch through injection structures for low voltage oscillation and low noise amplification', *Proc. 4th European Microwave Conf., Montreux*, 178, 1974.
97. W. T. Read, 'Negative resistance arising from transit time in semiconductor devices', *Bell System Tech. J.*, 37, 401 (1958).
98. C. A. Brackett, 'The elimination of tuning-induced burnout and bias oscillations in IMPATT oscillators', *Int. Solid State Circuits Conf. Proc. 1973*, pp. 114—115, (1973).
99. R. S. Harp and K. J. Russel, 'Improvements in bandwidth and frequency capability of microwave power combinatorial techniques', *1974 ISSCC Digest of Technical Papers*, pp. 94—95.
100. A. S. Baines, 'A multi-stage solid state amplifier', *Proceedings 4th European Microwave Conf., Montreux*, pp. 113—117, 1974.
101. D. C. Hanson and W. W. Heinz, 'Integrated electrically tuned X-band power amplifier utilizing Gunn and IMPATT diodes', *IEEE J. Solid-State Circuits*, SC-8, 3—13 (1973).

CHAPTER 7

Applications of Microwave Solid State Devices

I. W. MACKINTOSH

7.1 INTRODUCTION

Most microwave systems utilize the generation, transmission and detection of microwave power. For example, a communications system may carry speech, data or television pictures between two widely separated stations. A radar system transmits a modulated carrier which, on reflection off a distant target and detection by the radar receiver, provides range, velocity or other information about the target.

Semiconductor devices have been used in microwave receivers throughout the history of microwave exploitation, starting with the point contact diode as a detector or mixer diode. With increased understanding of semiconductors, improved technology and the investigation of new semiconductor materials new devices have been developed. The gallium arsenide Schottky barrier diode has emerged as superior to the point contact diode. The tunnel diode made available a device capable of the direct generation of microwave power, albeit at a low power level. The varactor diode provided the parametric amplifier, upconverters and multipliers. It was the advent of transferred-electron devices and avalanche diodes that ushered in a new era of solid state power devices, and has led to the widespread development of all-solid-state transmitters and receivers.

Although the achievable power levels of solid state devices are modest by microwave tube standards, nevertheless, as this chapter will show, there are a large number of applications where all-solid-state systems are proving viable. Modest power levels imply small systems, and solid state devices bring the advantages of small size, low weight, lower voltage operation and the potential of increased reliability over microwave tubes. However these claims must be examined with care for each system to ensure that real advantages are being offered by the solid state device, especially where the competitor may be a lightweight compact travelling wave tube or a cheap magnetron.

Microwave systems are built up from sub-systems and components which consist of devices and their circuits built into modules. Each module is designed to have properties, viewed from the input and

Table 7.1
Microwave components and solid state devices used in their realization

Component	Devices
Multipliers	Varactor diodes, step-recovery diodes
Upconverters	Varactor diodes, Schottky barrier diodes
Modulators	Schottky barrier diodes, *p-i-n* diodes
Driver stages	TEDS, impatts, bipolar transistors, FETs
Power stages	TEDs, LSA diodes, impatts, trapatts, bipolar transistors, multipliers
Phase shifters	*p-i-n* diodes, varactor diodes
Limiters	*p-i-n* diodes, varactor diodes
Low noise amplifiers	Varactor diodes (parametric amplifiers), FETs, bipolar transistors, tunnel diodes, TEAs
Switches	*p-i-n* diodes, Schottky barrier diodes
Downconverters, detectors	Schottky barrier diodes, back diodes, point contact diodes
Local oscillators	TEDs, impatts, bipolar transistors, crystal controlled oscillators/multipliers, baritts
Microwave IF amplifiers	Bipolar transistors
Demodulators	Schottky barrier diodes
Pulse generators	Trapatt diodes, step recovery diodes, avalanche transistors
Logic elements	Transistors, TEDs, tunnel diodes

output ports, which are appropriate to the role of the module in the system. The number of devices and circuit complexity inside these modules are increasing as techniques for miniaturization advance.

The application of solid state microwave devices may be broadly classified on the basis of their contribution to the performance of system components. Table 7.1 shows a list of components together with the devices that have been used in their practical realization, or for which devices are under active development. Those in the first half of the table are to be found mainly in transmitters, and those in the second half mainly in receivers.

A prime requirement of a transmitter is that it generates sufficient power, and that it does so with an acceptable conversion efficiency. A prime requirement of a receiver is that it has a low noise figure in order to detect weak signals in the presence of noise. Power and efficiency are referred to as primary parameters of transmitting devices, and noise figure as a primary parameter of receiving devices. The first two sections after this introduction review the primary parameter performance of transmitting and receiving solid state devices.

Each system demands additionally that its components and devices satisfy a detailed specification in regard of other parameters, referred to as secondary parameters. Examples of secondary parameters are bandwidth, frequency stability with temperature, phase linearity with frequency, etc. The secondary parameters whose control is of crucial importance to the success of the system depend upon the details of the particular system and are best discussed in the context of the system. In the remaining sections of the chapter, therefore, a number of systems are described, showing how their requirements on device performance arise and how solid state devices can, or may, satisfy these requirements. The emphasis is on the application of the newer active devices; and the applications considered include their use in local oscillators, as parametric amplifier pumps, in communications systems, in radar systems, in phased array antennas, in microwave control systems and in measurement.

We end this introduction with a note on nomenclature. The transit time gallium arsenide transferred electron device is on occasion referred to as the Gunn diode, in line with common usage. It is necessary on occasion to distinguish the use of Schottky barrier or p—n junction diodes as varistors (variable resistances under forward bias) in contrast to their use as varactors (variable capacitances under reverse bias). Also where the necessary distinction is required we shall distinguish between upconverters and downconverters, but otherwise the term mixer will refer to a downconverter.

7.2 POWER DEVICES

Figures 7.1 and 7.2 show laboratory c.w. and peak power outputs of solid state microwave sources as functions of frequency. These plots are

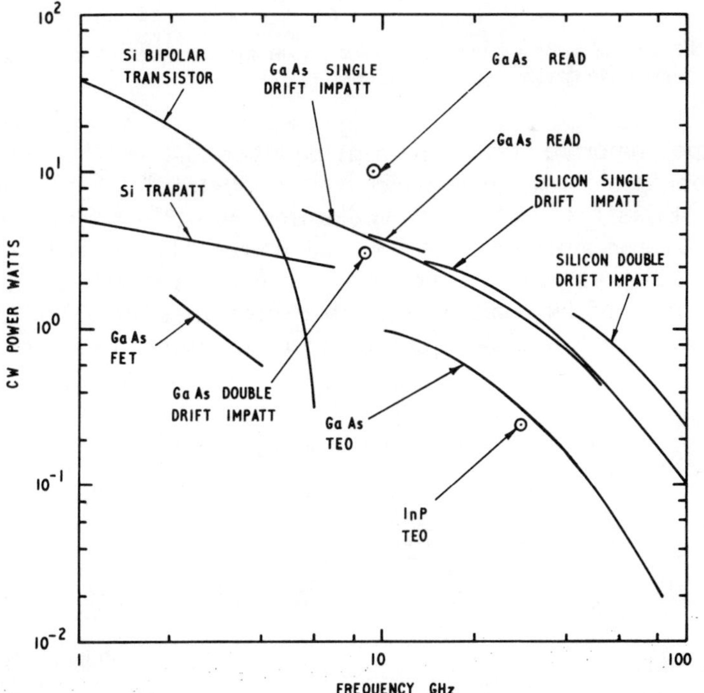

Figure 7.1 Best laboratory c.w. power performance of solid state microwave devices versus frequency (1974)

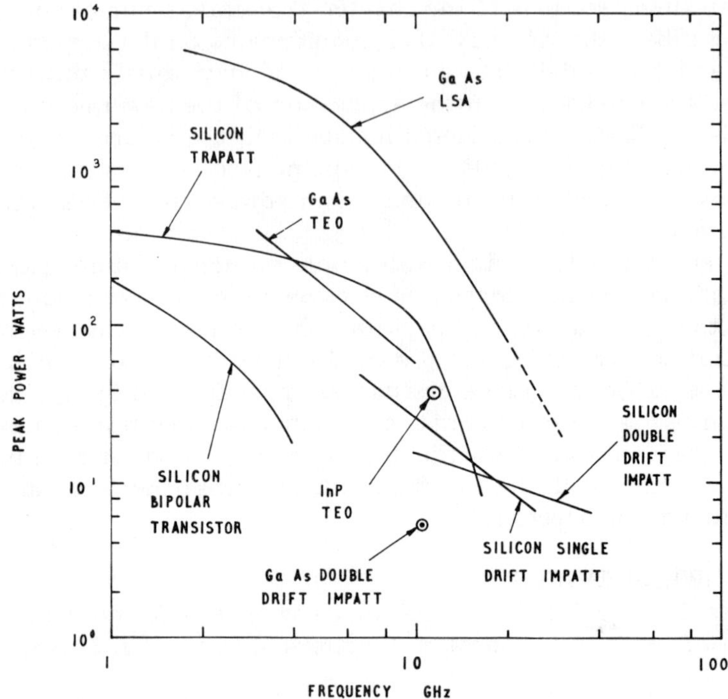

Figure 7.2 Best laboratory peak power performance of solid state microwave devices versus frequency (1974)

based upon reported best results with impatt oscillators,[1-12,273] trapatt oscillators,[13–18] transferred-electron oscillators,[19–31,39] gallium arsenide FETs[32,33] and silicon bipolar transistors.[34–36] It can be seen that LSA devices offer the greatest peak power, whereas transistors and impatt oscillators offer the best c.w. powers. Improved output power may be obtained by combining numbers of devices and Table 7.2 gives examples of performance reported in the literature. An active phased

Table 7.2
Performance of combinations of power devices

Device	Number combined	Frequency (GHz)	Power output (W)	Efficiency (%)	Pulsed or c.w.	Reference
Impatt	32	9.3	23	6.2	c.w.	37
Trapatt	5	1.9	1200	24	Pulsed	38
TEO	4	15.8	45		Pulsed	39
Step recovery diode multiplier	4	16	5		Pulsed	40

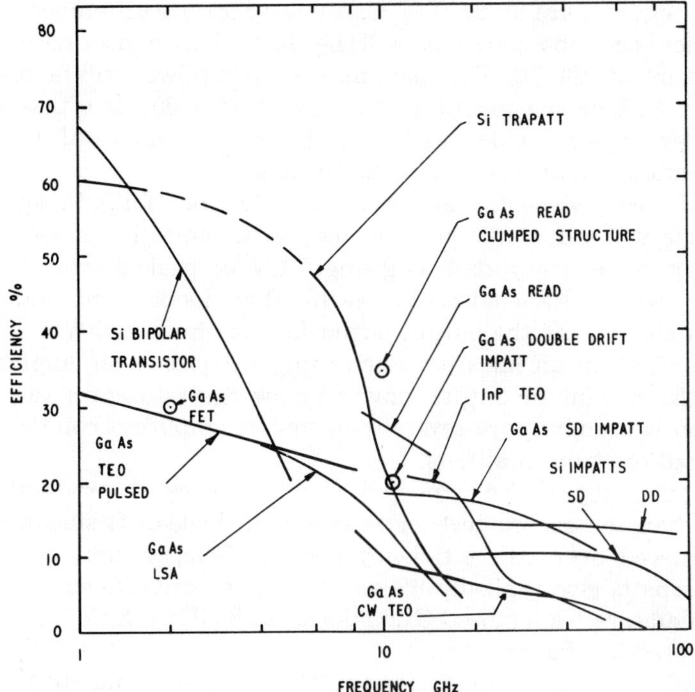

Figure 7.3 Best laboratory efficiencies of solid state microwave devices versus frequency (1974)

array antenna offers a method for combining a large number of devices and this is discussed in a later section.

Best laboratory efficiency performance is shown in Figure 7.3. Efficiency is an important system parameter for a number of reasons. Increased efficiency leads to reduced power input for the same power output, and for battery-powered equipments this means a longer operational time between battery renewals. Better efficiency can be exploited to reduce dissipated power and this eases problems with cooling. With pulsed devices, a smaller modulator is required on account of a reduced peak current and/or peak voltage. With improved efficiency for the same output power, the device temperature may be reduced and the system reliability is improved as a result of longer device life.

It must be emphasized that the results shown in Figures 7.1 and 7.2 are record results. For a device to be suitable for a particular application the output will be significantly less. For example, consider the case where c.w. output power has been measured at burn-out with the device on a water-cooled heat sink. Suppose the device temperature to have been 400°C. For a specification that requires useful life with an ambient temperature up to 100 °C, the temperature difference between the device and the ambient will be 100 °C at a maximum device temperature of 200 °C. The maximum output power will be restricted to about 25% of the record figure (assuming a similar efficiency and similar device area). A degradation in thermal impedance of the devices in production would further reduce this figure.

With pulsed devices, the additional specification of a pulse length and duty cycle will reduce the peak power performance. For example, LSA devices have been reported as giving 2 kW of peak power at 7 GHz. However, with pulse lengths of a few hundred nanoseconds and at duty cycles of about 0.2% the output power falls to about 200 W.

A specification on linearity when using a device as an amplifier will reduce the maximum output power below the saturation value. This reduction is usually more severe with impatt amplifiers compared with transferred-electron amplifiers.

The efficiency to be expected from a device in production will be less than the record level. For example, single-drift silicon impatts have achieved over 10% efficiency near 14 GHz, whereas commercial silicon impatts give typically 6% to 7%. The record efficiency reported in the literature for a pulsed Gunn device at 8 GHz is 22%,[21] whereas a realistic practical figure is 5%.

Typical c.w. Gunn efficiencies are 2% to 4% for commercial devices.

7.2.1 Reliability

It is clear that, in order for the system designer to be assured of obtaining a reliable device, the trade-off between device temperature and life must be known. This information is obtained by systematic life testing of a large number of devices at known temperatures.

It is usual to fit the time-to-failure data to a cumulative lognormal distribution, $\Delta(t)$, where

$$\frac{d\Delta}{dt} = \frac{1}{\sqrt{2\pi}\,\sigma t} \exp\left\{-\frac{1}{2\sigma^2}(\ln t - \mu)^2\right\} \tag{7.1}$$

where t is the time to failure and μ and σ are the mean and standard deviation of the normal distribution. It is then the standard practice to obtain an activation energy associated with a single predominant failure mechanism by fitting the median-time-to-failure (MTF) variation with temperature to an Arrhenius model using

$$\text{MTF} = A \exp(E_a/kT) \tag{7.2}$$

and noting that $\ln(\text{MTF}) = \mu$.

Staeker[41] has investigated the failure of silicon impatt diodes intended for use in a sattelite communication system in the frequency range 36—40 GHz. The diodes were diffused p^+—n—n^+ devices with chrome—platinum—gold metallization. Devices were screened for metallization defects, and were given a 100 hour burn-in at 280 °C. The activation energy was found to be 1.25 eV and the predicted MTF at 200 °C was over 4×10^5 hours. Murphy and coworkers[42] report an activation energy of 1.8 eV and a predicted MTF of over 4×10^4 hours obtained with 38 GHz GaAs impatts with Ti—Pt—Au metallization at 275 °C. (We note that 1 year = 8.8×10^3 hours.)

With microwave silicon bipolar transistors the MTF is given by[43,44]

$$\mathrm{MTF} = (Wt/CJ^2)\exp(E_a/kT) \tag{7.3}$$

where W is the metallization stripe width, t the stripe thickness, J the current density and C is a constant. E_a is typically 0.6 eV.[44] The MTF decreases as the current density increases, in contrast to the behaviour of silicon impatt diodes where no correlation with current density has been observed.[41]

The lognormal failure density function is an example of a wear-out distribution.[45] The mean wear-out life, M, for a lognormal distribution is related to the MTF by[46,47]

$$M = \mathrm{MTF}\exp(\sigma^2/2) \tag{7.4}$$

where σ is the lognormal standard deviation. Staeker[41] reports a value for σ of 1.12 for silicon impatt diodes, for which M = 1.9 MTF.

The overall mean-time-between-failures, MTBF, for a complex system depends upon the failure density function for each system component and the system maintenance procedure. For example, if components are replaced on failure, the equilibrium reliability function for the whole system becomes an exponential one.[45] In this case the system MTBF, m, is given by

$$\frac{1}{m} = \sum_{i=1}^{n} \frac{1}{M_i} \tag{7.5}$$

where the system has n components and M_i is the wear-out life of the ith component. Clearly, use of combinations of devices to obtain higher output powers will result in a reduction of MTBF compared to that obtained from a single device at the same temperature.

7.2.2 Environmental specifications

A device intended for incorporation in a system must satisfy requirements in regard of power output, efficiency and reliability, together with bandwidth and other secondary parameters. It may well be that there are requirements on weight, volume and shape. The device must also be designed so that the required performance is achieved in the environment in which the system is required to operate. Environmental factors which may have to be taken into account include a range in ambient operating temperature, a range in storage temperature, thermal and mechanical shock, thermal cycling, vibration, constant acceleration, humidity, moisture, salt atmosphere and nuclear radiation.

A high ambient operating temperature will limit the available output power by limiting the maximum device temperature in order to maintain the required reliability. A low operating temperature can cause problems with the self-starting of oscillators. Vibration can degrade the noise performance of an oscillator by introducing spectral components which dominate the intrinsic sideband spectrum. Moisture protection requires the hermetic sealing of device encapsulations or of modules.

Environmental specifications are laid down for military operational roles.[48,49] For example, portable ground equipment may be required to work over a temperature range of —40 °C to +70 °C.[48] A requirement for airborne equipment may be operation up to 70,000 ft altitude, and operation at sea level over the temperature range —54 °C to +71 °C.[49] Components in missiles are subject to severe accelerations. Environmental specifications and the methods for testing of components and the simulation of environmental conditions involves a large body of literature[48–50] to which the reader is referred for details.

7.3 MICROWAVE RECEIVERS

Figure 7.4 compares the best laboratory noise figure performance of silicon point contact diode mixers, Schottky barrier diode mixers, parametric amplifiers, tunnel diode amplifiers, gallium arsenide field effect transistors, silicon bipolar transistors, gallium arsenide and indium phosphide transferred-electron amplifiers.[51–59] Silicon impatt amplifiers have achieved noise figures of 31 dB at around 10 GHz,[60] and 41 dB to 46 dB at low power output levels at around 55 GHz.[61]

The noise figure likely to be achieved with components developed for system use will be higher than those shown for a number of reasons. With transistors, for example, the noise figures for amplifier modules will be greater than those shown in Figure 7.4 for individual devices. With a field effect transistor, the bias level required for maximum gain is different from that required for minimum noise figure.

In assessing the overall noise figure, the system designer must take into account other components, such as TR cells and limiters, which are placed between the antenna and the mixer or low noise amplifier. The overall noise figure also includes the effects of subsequent stages. It is standard practice to quote mixer noise figures which include an IF noise figure of 1.5 and this convention has been followed in Figure 7.4.

The dynamic range of a front end component is determined by the minimum detectable signal level and by the maximum tolerable signal level. The acceptability at the top end of the range depends upon system requirements in regard of output saturation and its effect on linearity. A possible measure is the level of the third-order inter-

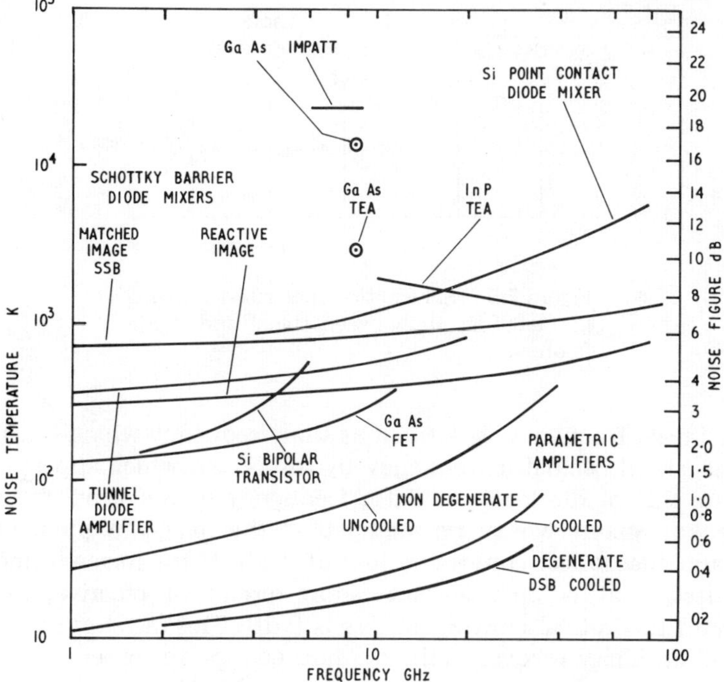

Figure 7.4 Best laboratory noise figure performance of solid state microwave devices (1974)

modulation intercept point. Typical values are in the range 0 dBm to 10 dBm for matched and reactive image mixers,[51] 16 dBm to 25 dBm for double balanced mixers,[62] −20 dBm to 0 dBm for parametric amplifiers,[51] −30 dBm to −10 dBm for tunnel diode amplifiers[51] and about 20 dBm for gallium arsenide FET amplifiers[63] and silicon bipolar low noise amplifiers.

7.3.1 Mixers

The use of semiconductor diodes in mixer downconverters is a well-established field and the reader is referred elsewhere for detailed treatments.[64,65] The purpose of this section is to note some points which are of consequence and interest to the system designer.

Mixers utilize a single diode in a single-ended mixer, two diodes in a balanced mixer and four diodes in a double balanced mixer. Balanced mixers are favoured for suppression of the transfer of AM noise from the local oscillator to the IF signal. Double balanced mixers are favoured for wide bandwidth operation and they give good intermodulation and harmonic performance with good isolation between ports.

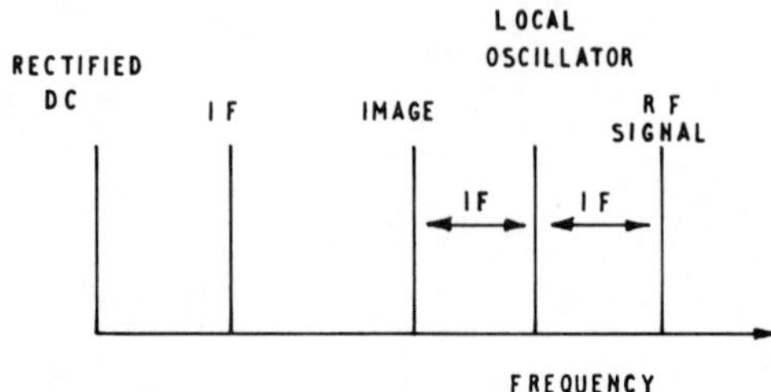

Figure 7.5 Schematic spectrum showing the relationship between signal and image frequencies

The image frequency is defined as that frequency which is displaced from the local oscillator frequency by the IF and which appears on the opposite side of the local oscillator frequency to the signal frequency, Figure 7.5. Mixers which are matched at the image frequency have a minimum theoretical conversion loss of 3 dB. If the image termination is lossless, that is open circuit, short circuit or reactive, then the minimum theoretical conversion loss is 0 dB. This latter type of mixer is called an image recovery mixer. Another type of mixer is the image rejection mixer; this type of mixer is designed to be insensitive to image frequency signals which otherwise would be downconverted and appear as unwanted IF signals.

The noise figure of a mixer varies with the IF. The noise figure is independent of the IF at high frequencies, but below a corner frequency, it rises, because of flicker noise, in a manner shown schematically in Figure 7.6. This effect is important in determining the sensitivity of 'zero IF' Doppler radars when detecting slowly moving targets.

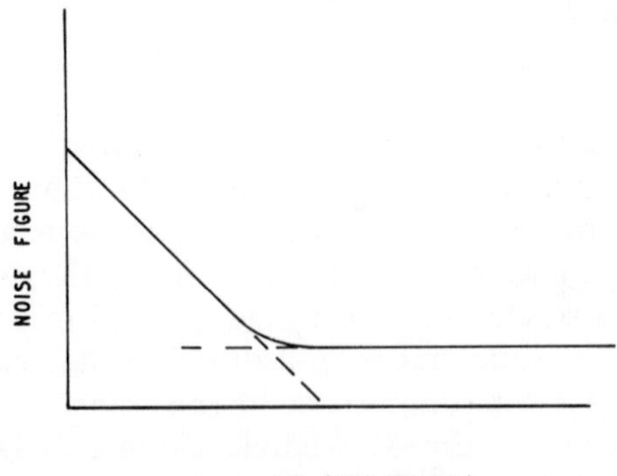

Figure 7.6 The variation of mixer noise figure with intermediate frequency

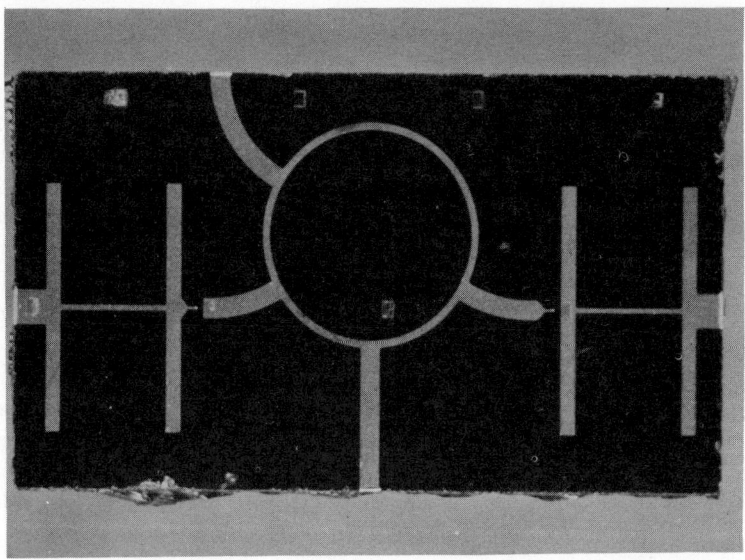

Figure 7.7 a 70 GHz monolithic integrated balanced mixer on gallium arsenide. (Reproduced by permission of Standard Telecommunication Laboratories Limited)

Germanium backward diodes have corner frequencies as low as 30 kHz, gallium arsenide Schottky barrier diodes typically around 100 kHz, germanium point contact diodes around 1 MHz and silicon point contact diodes around 3 MHz.[64,66] Schottky barrier diodes with corner frequencies less than 1 kHz are now commercially available.[121]

Receiver front ends incorporating mixers have received considerable attention with the aim of reducing their size and weight and cost by miniaturization. Triplate and microstrip techniques have been employed, and the reader is referred to bibliographies on microwave integrated circuits,[67,68] reviews [69] and examples of integrated receivers[70–72] for further details.

Monolithic integration of mixers in semi-insulating gallium arsenide has also been investigated. Figure 7.7 shows a gallium arsenide monolithic mixer described by Allen and Antell.[73] The chip measures 2.5 mm by 1.4 mm. The centre frequency is 70 GHz with a bandwidth of 10 GHz. The overall conversion loss is 7.5 dB with 2 mW of local oscillator power and an IF of 2.5 GHz.

7.3.2 Local oscillators

Of major concern is the noise performance and stability of an oscillator. The requirements in respect of AM and FM noise spectra and frequency and phase stability vary considerably from system to system. For example, the performance of a parametric amplifier is affected by AM noise on the pump. AM noise on the transmitter is responsible for limiting performance in c.w. and FM c.w. radar systems with monostatic antennas, and in c.w. Doppler radars using self-detecting oscillators. FM noise on the transmitter and local oscillator, resulting in phase instability on the received downconverted signal, can limit the performance of moving target indicator MTI radars.

Doppler radars specify limits on the combined effects of the spectra of FM noise on the transmitter and local oscillator, and they also specify the level of AM sidebands on clutter and strong target returns. Digital communication systems using phase comparison detection of differential phase-shift-key modulated signals require a specified level of phase stability over one bit period in order to avoid degradation of signal-to-noise ratio. FDM/FM communication systems require low FM noise local oscillators for driving upconverters and downconverters. Discussion on these requirements is to be found later in the chapter.

A carrier at frequency ω_0 whose amplitude and phase are subject to noise fluctuations may be described by

$$v(t) = A(t) \cos [\omega_0 t + \phi(t)] \tag{7.6}$$

where $v(t)$ is a voltage or a current, $A(t)$ and $\phi(t)$ are slowly varying real functions of time, and $\omega_0 (=2\pi f_0)$ is a constant. The instantaneous angular frequency is defined as $\omega_0 + \dot{\phi}(t)$. The time function $\dot{\phi}(t)$ has an associated 'power' spectral density function, $S_{\dot{\phi}}(\omega_m)$, where $\omega_m = 2\pi f_m$ and f_m is the modulation frequency.[74] $S_{\dot{\phi}}(\omega_m)$ is proportional to $[\Delta f(f_m)]^2$ where $\Delta f(f_m)$ is the r.m.s. value of the FM noise frequency deviation in a 1 Hz bandwidth at frequency f_m off the carrier. For small modulation indices $(\Delta f(f_m)/f_m \ll 1)$ for which the higher order sidebands may be neglected, the FM noise double sideband power P_s at modulation frequency f_m in a 1 Hz bandwidth is given by

$$\frac{P_s}{P_c} = \left[\frac{\Delta f(f_m)}{f_m}\right]^2 \tag{7.7}$$

where P_c is the carrier power.

Figures 7.8 and 7.9 compare the AM and FM noise performance of silicon impatts,[75–80] gallium arsenide impatts,[75,80,81] silicon trapatts,[14,82] Gunn oscillators,[25,75,80,83,84] baritts,[85,271] crystal-controlled oscillators followed by amplification and frequency multiplication,[83,84,86] and transistor oscillators followed by multiplication.[83,86,87]

In order to allow comparison, the FM noise data for impatts and Gunn oscillators has been selected or adjusted to be appropriate to a power output of 100 mW at a frequency of 10 GHz and an external Q of 100. The data for the silicon trapatt is for a power output of 1 W at a frequency of 5 GHz and a Q of 100, and that for the baritt is for

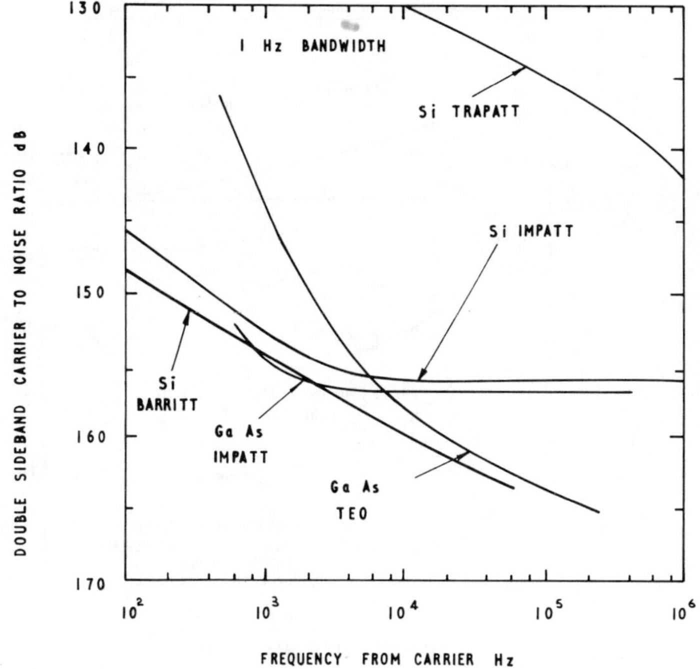

Figure 7.8 The AM noise performance of solid state microwave oscillators. The data for the baritt oscillator is for 6 GHz, that for the trapatt oscillator is for 5 GHz, and that for the other oscillators is 10 GHz

10 mW at 10 GHz with a Q of 100. The noise measure is defined by

$$M = \frac{Q^2 [\Delta f(f_m)]^2 P_0}{f_0{}^2 kTB} \tag{7.8}$$

where P_0 is the output power. The main noise measure scale on Figure 7.9 applies to impatt and Gunn oscillators, and the subsidiary scales to trapatt and baritt oscillators. The bandwidth B is 1 Hz.

It can be seen that the quartz-crystal-controlled oscillator followed by a multiplier offers the lowest FM noise. Baritt and Gunn oscillators offer similar levels of performance to each other above a few kHz off carrier. Silicon impatt FM and AM noise levels are worse than those of baritt and Gunn oscillators for modulation frequencies above a few kilohertz off the carrier, but below this frequency the silicon impatt offers a comparable (or possibly better[76]) performance. Silicon impatt devices which incorporate an intrinsic avalanche region show a significant improvement in both AM and FM noise performance.[88] Complementary silicon impatts also show an improved performance

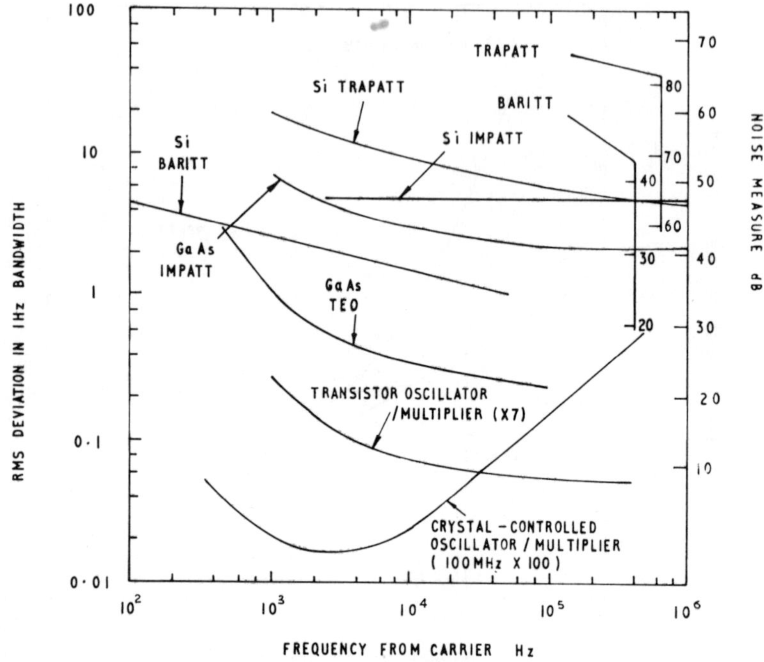

Figure 7.9 The FM noise performance of solid state microwave oscillators

compared to the conventional structure.[89] Data on gallium arsenide impatt noise performance is not extensive, but it indicates superior performance for AM noise below about 8 kHz off the carrier compared to Gunn oscillators. Impatt AM noise has a flat spectrum at frequencies above about 10 kHz, eventually falling at frequencies of tens of megahertz.[78] Levine and Chan[81] report AM noise ratios for this flat region of −157 dB in a 1 Hz bandwidth for double vapour epitaxial silicon impatts and −160 dB for gallium arsenide double vapour epitaxial impatts. Impatt AM noise is sensitive to the impedance of the bias, and with low bias impedances increases of 15 dB can occur.[78] Impatt FM noise increases with output power, and a noise measure as high as 90 dB has been observed.[80]

FM noise may be reduced by increasing the cavity Q, or by using an additional stabilizing cavity. Reduced AM noise has been observed on combining devices.[90] FM noise may be reduced by locking the oscillator to a source with lower FM noise. FM noise of the locked oscillator is reduced for modulation frequencies within the locking bandwidth. The noise is reduced by a factor given by the following formula derived from Hines,[86]

$$\frac{[\Delta f_L(f_m)]^2}{[\Delta f(f_m)]^2} = \left\{ \frac{[\Delta f_R(f_m)]^2}{[\Delta f(f_m)]^2} + \frac{f_m^2}{B^2} \right\} \Big/ \left(1 + \frac{f_m^2}{B^2}\right) \tag{7.9}$$

where $\Delta f_L(f_m)$ is the r.m.s. deviation of the locked oscillator, $\Delta f_R(f_m)$ is the deviation of the locking oscillator and B is the locking bandwidth. Ashley and Palka[77,83] have observed that for impatt and Gunn oscillators AM noise was not affected by locking.

Impatt and Gunn oscillators may be sub-harmonically locked, in which the locking signal is at f_0/n where n is an integer. For impatt oscillators, the results reported by Chien and Dalman[91] for $n = 2$ are in agreement with a modified form of Adler's formula,

$$\frac{B}{f_0} = \eta \left(\frac{P_i}{P_0}\right)^{n/2} \tag{7.10}$$

where B is the total locking bandwidth, η is a locking figure of merit and P_i is the sub-harmonic input power. Sub-harmonic locking provides a little-explored alternative to frequency multilplication.

Crystal-controlled oscillators followed by multipliers are a well-established form of local oscillator. The r.m.s. frequency deviation eventually rises with increasing modulation frequency (Figure 7.9) due to white noise introduced by transistor amplification at a level determined by the amplifier noise figure.[87] On multiplication, the frequency deviation is multiplied by n and the FM power spectrum by n^2 where n is the frequency multiplication factor. Multipliers have a reputation for time-consuming adjustment and require skilled maintenance. With correct adjustment, no significant noise additional to the multiplication effect should be introduced.

A microwave oscillator may also be stabilized by locking to a lower frequency source via a phase locked loop.[272] FM noise reduction of the locked oscillator occurs within a bandwidth determined by the frequency response of the phase lock servo loop. FM noise multiplication occurs as with a simple multiplier.[92]

The discussion so far has centred on consideration of short-term frequency stability. Medium-term frequency stability, corresponding to changes over periods of about an hour, is largely determined by temperature changes. Table 7.3 gives a comparison of quartz-crystal-controlled oscillator performance with that of the recently invented surface acoustic wave oscillator[93,94] and examples of impatt, Gunn and transistor oscillators. Fundamental microwave oscillators only approach the temperature stability of the lower frequency oscillators with difficulty.

Table 7.3
Oscillator frequency stability with temperature

Oscillator	Frequency (Hz)	Temperature range (°C)	$\left\lvert \frac{1}{f_0} \frac{df_0}{d\theta} \right\rvert$	Reference
Quartz crystal	$\leqslant 10^8$	−30 to +70	10^{-6}/°C	93
Surface acoustic wave oscillator	10^7 to 2×10^9	−30 to +70	10^{-6}/°C	93
TEO	7.64×10^9	0 to +50	0.4×10^{-6}/°C	95
Impatt with stabilising cavity	1.1×10^{10}		2.7×10^{-6}/°C	96
Silicon bipolar transistor oscillator (hybrid thin film with stabilizing cavity)	1.7×10^9	−70 to +70	17×10^{-6}/°C	97

Long-term stability (over periods of about a year) is determined by ageing effects. Quartz crystal oscillators have stabilities of a few parts in 10^6 whereas SAW oscillators have problems with ageing and their long-term stability is an order of magnitude worse.[94]

7.3.3 Upconverters

Although discussion on upconverters strictly belongs to discussion on transmitters, it is natural to complete the sections on mixer down-converters and local oscillators with a note on upconverters. In radar systems, in waveguide communication repeaters or in any system where the received signal and the transmitted signal are at the same frequency, the local oscillator can be used to drive both an upconverter and a downconverter.

Most commercial suppliers of balanced mixer downconverters recommend them for use as upconverters. Output powers are typically at the tens of milliwatts level. Higher output powers are obtainable from the varactor Upper Sideband UpConverter (USUC). The varactor upconverter is also a low noise device and has found application in the 100 MHz to 2 GHz part of the spectrum in this role.[51]

A number of other devices have been used as upconverters. Table 7.4 gives laboratory performance data for a variety of upconverters which have been reported on recently in the literature.

7.3.4 Parametric amplifiers

Parametric amplifiers, especially when cooled, offer a way of achieving a low system noise temperature. The expense of a parametric amplifier, and the complexity introduced by cooling, can be tolerated in systems where sensitivity is at a premium, as in earth receiving stations with satellite communication links, or in radar systems where the required

Table 7.4
Examples of upconverter performance

Type	Input signal frequency (GHz)	Local oscillator frequency (GHz)	Local oscillator power (mW)	Output power (mW)	Bandwidth (MHz)	Reference
MIS varactor	0.07	2.3	300	210	31.5	98
Si varactor	0.07	6.0	320	200	20	99
GaAs varactor	0.6	7.4	2000	870	200	100
Si impatt	0.07	7.46	17	16		101
Ge avalanche	4.0	43	26	1.7	800	102
GaAs varactor	1.25	50	100	8.7	500	103
Si varistor	1.5	60	35	3.5	600	104
Si varactor	1.5	60	100	20—25	600	104

sensitivity could only otherwise be achieved with a penalizing increase in transmitter power. Although a number of systems could benefit from increased sensitivity, a parametric amplifier is often a prohibitively expensive solution. Also, in order to obtain the benefit of low noise performance, the cooled parametric amplifier should be placed close to the receiving antenna to avoid intervening loss, and this may be an unacceptable design constraint.

The solid state parametric amplifier uses the nonlinear capacitance variation with voltage of the reverse-biased varactor diode. The varactor is pumped at a frequency f_p and in the appropriate circuit the varactor diode will amplify signals at the signal frequency f_s. The circuit must allow circulation of current at the idler frequency $f_i = f_p - f_s$. In the Pearson circuit[105,106] two varactor diodes are used, and the idler circuit is formed by the pair of diodes and associated parasitics.

A complete solid state parametric amplifier has had to await the arrival of solid state pumping sources. The requirement for a pump frequency which is much greater than the signal frequency has had the consequence that the first solid state pumps used were transistor, impatt or transferred-electron oscillators followed by multipliers. As performance at millimetre wavelengths has improved, transferred electron oscillators, and to a lesser extent impatt oscillators, have established themselves as fundamental pump sources. The multiplier solution has reappeared at frequencies of around and above 100 GHz for pumps for millimetre wave paramps. Table 7.5 gives data for some examples of laboratory paramps with solid state pumps, with amplifier gains typically in the range 10—14 dB.

The pump power required may be determined from the following equation[111,112]

$$P = \pi(f_p^2/f_c)C_0(V_B + \phi)^2 \tag{7.11}$$

where C_0 is the capacitance of the varactor diode at the maximum

Table 7.5
Examples of parametric amplifier performance

Signal frequency (GHz)	Bandwidth (MHz)	Noise temperature (K)	Pump frequency (GHz)	Pump power (mW)	Pump type	Reference
1.8	50	160	8.5	10	Transistor/ multiplier (4×)	107
4	500	62	45	100	Impatt	108
7.5	500	120	38	100	Impatt	109
14	300	300	40	50	Gunn	110
60	670	820	110	30	Gunn/multi-plier (3×)	111

reverse voltage excursion V_B, and ϕ is the varactor built-in voltage. This result applies to abrupt junction varactors with sinusoidal pumping current, and it is assumed that the minimum varactor elastance may be taken to be zero.

For this case, the modulation ratio, m, is 0.25 where in general m is given by

$$m = \frac{|S_1|}{S_{max} - S_{min}} \tag{7.12}$$

where $S_{max}(=1/C_0)$ is the maximum elastance of the varactor, S_{min} is the minimum elastance and S_1 is the Fourier component at the pump frequency of the time-varying elastance. The cut-off frequency, f_c, is given by

$$f_c = \frac{S_{max} - S_{min}}{R_s} \tag{7.13}$$

where R_s is the diode's series resistance.

The theoretical noise temperature of a non-degenerate parametric amplifier is given by the expression[113]

$$T = T_D\left(\frac{f_s}{f_i}\right)(m^2 f_c{}^2 + f_i{}^2)/(m^2 f_c{}^2 - f_s f_i) \tag{7.14}$$

where T_D is the varactor diode temperature. The noise temperature is at a minimum when $f_p = (m^2 f_c^2 + f_s^2)$.[112] When $mf_c \gg f_i$, the noise temperature is approximately $T_D(f_s/f_i)$.

Noise temperatures achieved in practice approach the theoretical values.

The effect of noise sidebands on the pump signal is to transfer noise sidebands onto the amplifier signal output. This effect is most trouble-some for a weak signal in the presence of a strong signal, because the

Table 7.6
Parametric amplifier noise sideband characteristics

Pump noise	Signal noise	X	Position of maximum noise
AM	AM	$g_n G$	Mid-band
AM	FM	$g_n G/2$	Band edge
FM	FM	$2f/B$	Mid-band
FM	AM	f/B	Band edge

noise sidebands on the strong signal may mask the weak signal. Clarke[114] has analysed the problem in terms of four coupling coefficients which describe the level of AM and FM sidebands appearing on the signal output due to AM and FM sidebands on the pump. The four coupling coefficients have maximum values given by an equation of the form

$$M = N + 20 \log X \tag{7.15}$$

where M(dB) is the double sideband power ratio of noise to output signal and N(dB) is the corresponding ratio for the pump. The factor X is dependent upon the type of coupling and on amplifier parameters as shown in Table 7.6. g_n is the normalized negative conductance, G is the mid-band gain, f is the sideband frequency offset and B is the effective bandwidth of the idler circuit referred to the signal circuit. Since $f \ll B$, the noise sidebands on the output signal due to FM noise on the pump are considerably reduced below the FM noise level on the pump. However, the noise sideband level on the signal output due to AM noise on the pump is above the AM noise level on the pump by a factor of the order of the amplifier gain. The results of this analysis have been confirmed experimentally by Dean and Clarke.[115]

The implications of these results are that systems which require the simultaneous detection of strong and weak signals will make stringent demands on the noise performance (especially the AM noise) of the pump. It would appear that impatt pumps are suitable for many applications which are not too demanding in this respect,[108,109,116] providing care is taken to remove spurious outputs by filtering. Gallium arsenide transferred-electron oscillators are a popular choice; but it remains to be seen whether the more stringent noise performance can be satisfied by present pumps, or whether alternatives, such as indium phosphide pumps, have a role.

7.3.5 Receiver protection

Receivers may be subject to overloads, either under normal operation or as a result of faulty operation. Examples with radars are the break-through of transmitter power in a pulsed radar, the shining of one radar transmitter into a nearby radar's receiver or the accidental reflection of large amounts of transmitted power back into the receiver.

Protection against burn-out involves the reduction of energy or power reaching the receiver front end by the use of TR cells and limiters, and by the increase of the burn-out resistance of front end components.

The transient response of a gas TR cell consists of an initial high power leakage spike followed by limiting action at a lower power level. Front end components need to be resistant to burn-out from spike leakage and, for pulse lengths greater than about 1 μs, they should be able to withstand the input power continuously.

Figure 7.10 shows the variation of noise figure with increasing microwave spike energy observed by Swallow and coworkers[117] for a gallium arsenide Schottky barrier mixer diode and a silicon point contact mixer diode. A measure of the burn-out resistance is to take the energy at which the noise figure increases by 1 dB. Anand and Howell[118] report that a degradation in flicker noise precedes the change in the 30 MHz IF noise figure. The energy for burn-out is also a function of the duration of the spike.[119] For example, for a spike duration varying from 1 ns to 100 ns, the energy for burn-out of a gallium arsenide Schottky barrier diode varied from 10 nJ to 100 nJ.[64] Typical c.w. burn-out powers are in the range 0.5 W to 1 W,[64,121] but powers as high as 12 W have been reported for detector diodes.[64]

Solid state limiters use *p—i—n* or varactor diodes. The reader is referred to Barber and coworkers[122] for a review. Typical incident powers for *p—i—n* limiters are from 100 W peak, 1 W mean[123] to 1 kW peak and up to 6 W mean[124] with a 1 μs at 10 GHz. Typical output powers are around 100 mW. Wideband detectors incorporating limiter

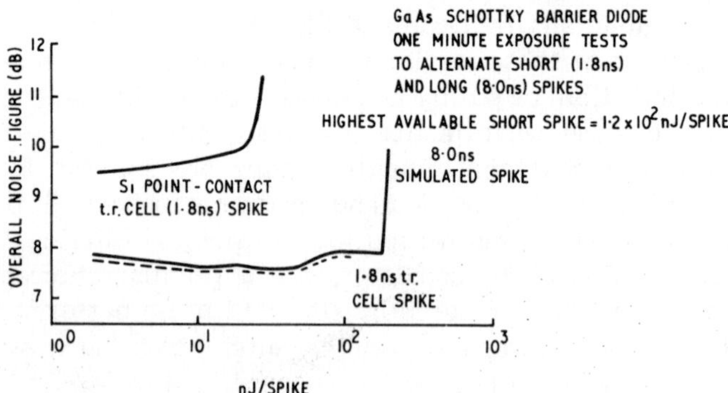

Figure 7.10 The measured variation of noise figure with microwave spike energy for a silicon point contact diode and for a gallium arsenide Schottky barrier diode. (From Swallow, Oxley and Hansom,[117] reproduced by permission of the 1973 European Microwave Conference)

protection are capable of withstanding peak powers of 100 W.[125] p—i—n diode switches may be used as active limiters in which the diode is switched to a reflecting state prior to the arrival of the high power microwave pulse. Multiple diode configurations are capable of withstanding kilowatts of incident power up to 10 GHz.

7.4 SOLID STATE MICROWAVE DEVICES IN COMMUNICATION SYSTEMS

In this section we consider the use of solid state microwave devices in communication systems. Terrestrial line-of-sight systems are well-established at microwave frequencies up to around 13 GHz. These systems use frequency division multiplexing in order to combine a number of telephone channels into a single baseband signal, and the baseband signal frequency modulates a microwave carrier (FDM/FM). Satellite communications have been established in similar frequency bands.

The drive for more capacity has led to the examination of systems at higher frequencies. The choice of frequency and the system configuration is constrained by the propagation properties of the atmosphere. Major features of propagation in clear air at sea level are the attenuation peaks due to water vapour at 24 GHz and oxygen at 60 GHz. At 24 GHz, for example, the attenuation is 0.17 dB/km and at 60 GHz it is 15 dB/km.[126] Attenuation due to rain is also severe. For example, for rainfall of 5 mm/hr measured values were 2 dB/km at 35 GHz,[127] and 2—3 dB/km at 48 GHz and 70 GHz.[126] This has the consequence that line-of-sight systems must incorporate link diversity in order to reduce the effects of severe fading due to rain.

Extremely large capacity systems are under development at millimetric frequencies which avoid problems with atmospheric attenuation by using a circular waveguide filled with dry nitrogen. In contrast, highly secure interference-free systems are proposed using the strong oxygen absorption at 60 GHz to limit the range of propagation.

Considerable attention is now being paid to the use of time division multiplexing using pulse code modulation (TDM/PCM), especially for millimetric waveguide systems, and for line-of-sight systems at frequencies around 20 GHz.

In this section we consider the use of solid state microwave devices in FDM/FM systems and television links and, with digital communication, in millimetric waveguide and free-space systems.

7.4.1 FDM/FM systems

Most existing line-of-sight relay systems use frequency division multiplexing with frequency modulated carriers (FDM/FM).[128,129] They are used for simultaneous transmission of large numbers of telephone channels on the same microwave carrier, or for television transmission,

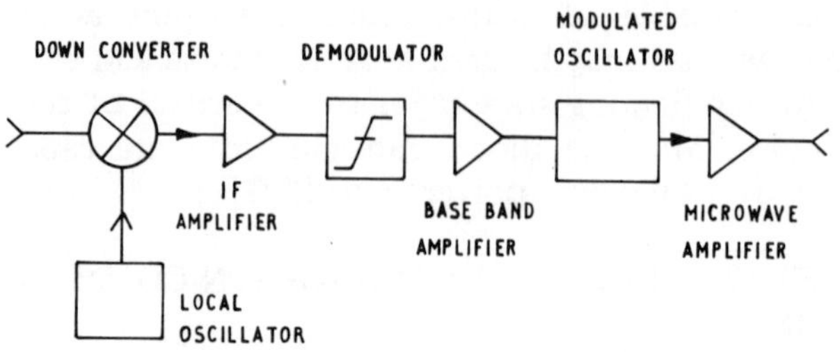

Figure 7.11 A baseband repeater

and they operate in allocated bands up to 13.2 GHz. Each telephone channel occupies a bandwidth of 300 Hz to 3.4 kHz, and, using single sideband suppressed carrier modulation, telephone channels are placed at 4 kHz intervals in the baseband. A 600-channel system occupies a baseband from 60 Hz to 2.54 MHz. The baseband signal is used to frequency modulate the microwave carrier. High capacity routes, and satellite links, may use a number of microwave carriers simultaneously on the same antennas.

FDM/FM repeaters are of two types: heterodyne repeaters and baseband repeaters. A radio-relay link has baseband repeaters at the ends, with hops in between through heterodyne repeaters.

The form of a baseband repeater is shown in Figure 7.11.[128–130] The incoming signal is downconverted to an IF, amplified, limited and FM demodulated. The baseband signal is amplified and remodulates a microwave carrier via a modulated oscillator. The signal may be amplified by a microwave amplifier and retransmitted. Baseband repeaters allow the addition or removal of telephone channels to or from the baseband, or frequency translation within the baseband.

Heterodyne repeaters[128,130] are of the form shown in Figure 7.12. The incoming signal is downconverted to IF, amplified, upconverted, amplified and retransmitted. The shift oscillator and shift converter are

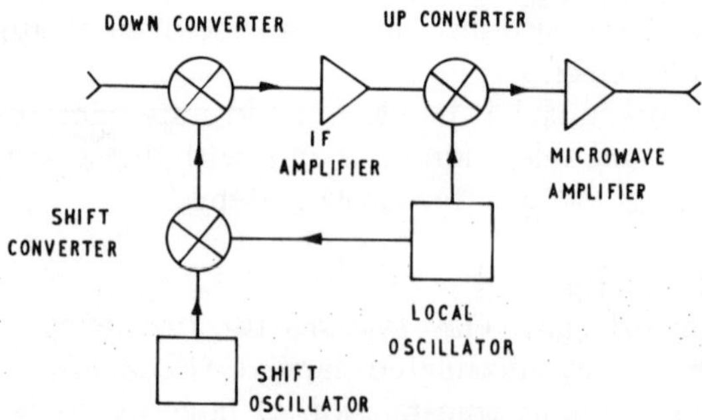

Figure 7.12 A heterodyne repeater

required in order that the transmitted signal is on a different frequency to that of the received signal.

Local oscillators at present are usually crystal controlled transistor oscillators followed by multiplier chains; but it is to be expected that Gunn diode oscillators will replace them.[130] Downconverters commonly use silicon Schottky barrier diodes in single-ended or double balanced mixers. Upconverters have used varactors for high power output without subsequent amplification, or low power Schottky barrier diodes when followed by an amplifier.[130]

Solid state power sources are increasingly being considered for use in the final power stages of the transmitter, on account of reduced size, reduced power consumption and improved reliability compared with tubes. For example, an all-solid-state repeater reduced repeater volume by a factor of two, reduced the d.c. power consumption by a factor of six and more than doubled the MTBF of the repeater by elimination of a TWT and its associated power supply.[131] Considerable development is going ahead on oscillators and amplifiers for FDM/FM systems using impatt diodes, Gunn diodes and transistors. We now turn our attention to these power sources and discuss the requirements on microwave output power and noise performance, indicating how well these devices perform under laboratory test conditions and on working links.

Transmitter output power

Consider a single hop between two repeaters. The power at the input to the receiver at the distant station is given by

$$P_R = \frac{P_T G_T A_R}{4\pi R^2 L} \tag{7.16}$$

where P_T = transmitter power
G_T = gain of transmitting antenna
A_R = effective area of the receiving antenna
R = hop length
L = loss.

The loss factor arises from partial obstruction of the beam, terrain reflection causing destructive interference at the receiving antenna and fading due to inverse bending of the rays or due to rain attenuation. In addition, losses in the feeders to and from the antennas at both ends of the link have to be included.

The transmitter power is determined by the requirement that each link perform to internationally agreed standards in regard to the maximum noise power present in any one telephone channel. The CCIR recommendations[132] require that the sum of thermal noise power and intermodulation noise power does not exceed prescribed values. Thermal noise is understood to include the effects of Johnson noise, shot noise, flicker noise etc; intermodulation noise is caused by the effects of nonlinearity in the system.

We now outline the procedure for determining the required transmitter power. The reader is referred to standard texts (e.g. Panter[128]) for more detailed discussion. Performance standards are set by specifying the maximum permissible noise power per channel at the receiving end of a radio-relay circuit. The noise level is expressed as an absolute power in picowatts referred to a reference point in the system known as the zero transmission level reference point. When the signal level is 1 mW at the zero transmission level reference point (expressed in decibel notation as a signal level of 0dBm0) and the noise power is 10 pW, the channel signal-to-noise ratio is 80 dB. Design proceeds by allocating a maximum noise contribution for each hop in the circuit and dividing this equally into thermal and intermodulation noise contributions. The thermal noise is then further divided into contributions from a number of sources including the transmitter, local oscillator and receiver. The channel signal-to-receiver-thermal-noise ratio is then converted into a carrier-to-noise ratio at the receiver output, and then a required received carrier power is obtained by taking into account the receiver noise figure. The required transmitter power then follows from Equation 7.16.

Table 7.7 shows data presented by Standal and Målsnes[133] on the required output powers in dBm as a function of frequency and channel capacity for a hop length of 46.5 km. The receiver noise figure has been taken as 6 dB, the antenna diameter is 3 m, the loss due to feeders and branching filters is 5.5 dB and the thermal noise level is 20 pW. It can be seen from the table that the required transmitter power increases with the number of telephone channels. A capacity of 2700 channels or multiple carrier operation demand even higher powers. For example, a 2700 channel link at 7 GHz requires 8—16 W (39—42 dBm).[130]

The requirements of Table 7.7 may be compared with performances shown in Figure 7.1. Varactor multiplier chains are established power sources for the smaller capacity links. Transistor sources are being

Table 7.7
Transmitter powers (dBm) for microwave links

Number of channels	Frequency (GHz) 2	4	6	7.5	8
300	25	19	15	13.5	13
960	35	29	25	23.5	23
1800	44	38	34	32.5	32

From Standal and Målsnes,[133] reproduced by permission of Microwave Exhibitions and Publishers Limited.

developed for the lower frequency systems with up to at least 960 channels,[133] and impatt and Gunn amplifiers and oscillators are being developed for the higher frequency systems. There is considerable interest in higher power, higher efficiency GaAs impatt diodes, and combinations of impatt diodes, for possible eventual replacement of TWTs with the higher capacity links.

Transmitter thermal noise

Thermal noise can be introduced by the power stages of repeater transmitters as well as by receivers. The signal-to-thermal-noise ratio for a channel at frequency f_m in the baseband at the output of a microwave power amplifier is given by Sweet[134] as

$$\frac{S}{N} = \frac{2P_{in}}{FkTB}\left(\frac{\Delta f_s}{f_m}\right)^2 \tag{7.17}$$

where F is the amplifier noise figure, P_{in} is the input power, B is the measurement bandwidth and Δf_s is the maximum single channel frequency deviation. Noise figures are around 20 dB for a Gunn amplifier and can be up to 60 dB[135] for a high power level impatt amplifier. Application of Equation 7.17 for a specified minimum signal-to-noise ratio of 80 dB leads to the conclusion that Gunn amplifiers are suitable for use with systems up to 1800 channels capacity, but that silicon impatt amplifiers, assuming they have a noise figure of 50 dB, require low noise pre-amplification to several hundreds of milliwatts to handle 1800 channels.[134]

Wood[136] has considered the use of an injection-locked silicon impatt oscillator and concluded, theoretically, that the gain per stage is limited by the thermal noise requirement, with a maximum single-stage gain of 8.2 dB for 960 channels, and a two-stage gain of 10.4 dB for 1800 channels.

Intermodulation

Intermodulation noise arises from the effects of nonlinear amplitude transfer characteristics, from the variation of gain with frequency (differential gain) and from the nonlinear variation of phase with frequency (group delay variation or differential phase). Intermodulation noise can also arise from the combined effects of AM noise introduced by a previous stage and AM/PM conversion. Intermodulation noise can be introduced by downconverters, upconverters, modulators, demodulators, IF and microwave amplifiers.

The standard approach for dealing analytically with the effects of amplitude nonlinearity is to express the output amplitude (voltage or current) as a polynomial of the input amplitude.[137-139] When two signals are present simultaneously, at frequencies f_1 and f_2 respectively, intermodulation signals are generated at frequencies $nf_1 \pm mf_2$ where n

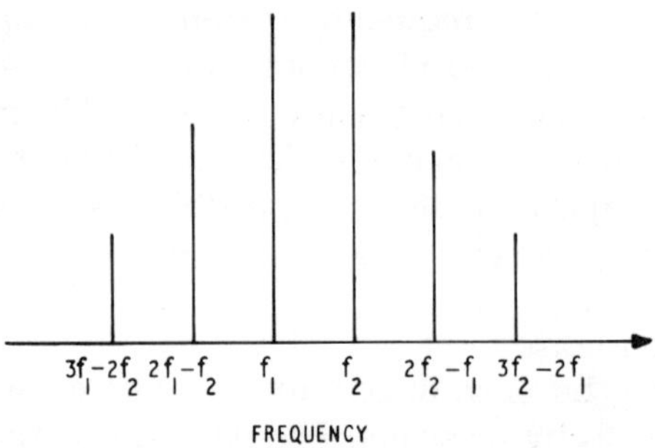

Figure 7.13 A schematic spectrum showing two signals at frequencies f_1 and f_2 and their associated intermodulation products

and m are positive integers. The order of an intermodulation product is defined by $n + m$. Figure 7.13 illustrates schematically the type of spectrum produced by signals at neighbouring frequencies f_1 and f_2. Third-order intermodulation products appear at $2f_1 - f_2$ and $2f_2 - f_1$.

For a cubic nonlinearity and two equal power inputs, the power contained in the third-order intermodulation products increases as the cube of the input power. Hence, whereas the output power increases 1 dB for each 1 dB increase in input power, the third-order intermodulation power increases by 3 dB/dB. This gives rise to a way of specifying the level of intermodulation by reference to the third-order intercept point which is defined by the intersection of the asymptotes to the output power and third-order intermodulation curves as shown on Figure 7.14.

The third-order intercept for an impatt amplifier is at a level nearly equal to the saturated output power,[140] whereas that for a transferred-electron amplifier is typically 6 dB above saturation.[141] A detailed investigation of intermodulation products for an impatt amplifier has been reported on by Trew and coworkers,[142] Komizo and coworkers[143] have obtained an improvement in intermodulation performance of a two-stage impatt amplifier by using a diode bias-current compensation technique; third-order intermodulation products were 40 dB down on an output of 21 dBm at 13 GHz with a gain of 11 dB. Improvement in intermodulation performance is possible with a feed-forward network; Lubell and coworkers[144] report a 15 dB improvement using this technique on a 1—2 GHz amplifier.

An AM/PM conversion of 3.5°/dB (comparable to that of a TWT) at an output power of about 0.5 W is observed from a silicon impatt amplifier at 11.25 GHz.[145] The increase in noise due to this level of AM/PM conversion following a Gunn diode modulator with a modulation linearity of 0.1 dB/±10 MHz and a delay characteristic of

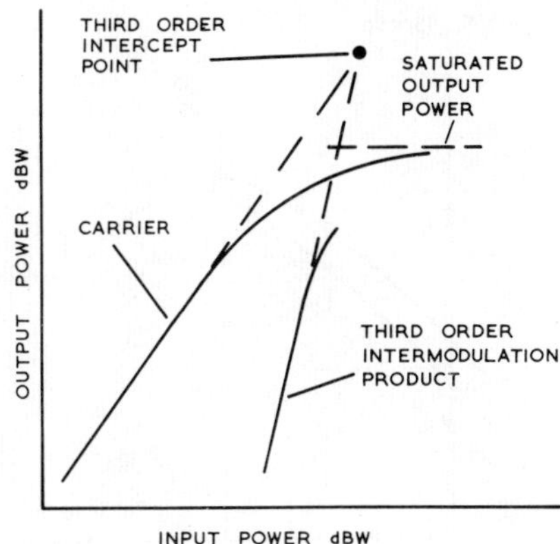

Figure 7.14 The variation of output power and third-order intermodulation product with input power for a non-linear component, and the identification of the third-order intercept point

1 ns/±10 MHz is calculated to be less than 5 pW. A two-stage injection-locked 11/GHz oscillator, with a Gunn diode first stage and a silicon impatt second stage had an AM/PM conversion of less than 3°/dB.[146]

Noise loading

The noise loading method for testing multichannel systems is one which closely simulates the traffic carried by a link under operating conditions by applying a wideband white noise signal in the baseband. The wideband noise loading is set to comply with the CCIR recommendation at a level of $(-15 + 10 \log_{10} N)$ dBm0 for $N \geqslant 240$ where N is the number of channels.

A band stop filter is placed to remove a selected frequency slot from the baseband noise signal. The noise power appearing in this slot at the output is then measured. At lower loading levels thermal noise predominates and at higher levels intermodulation noise predominates. On tests for a 960-channel system, Komizo and coworkers[145] observed a 2—5 pW increase in thermal noise, when using a 0.5 W, 7.5 dB gain, 11 GHz silicon impatt amplifier, and up to 10 pW due to intermodulation noise; no increase in either thermal or intermodulation noise was observed for a gallium aresnide amplifier. Endersz[147] has demonstrated that an X-band 250 mW FM modulated Gunn oscillator, utilizing varactor tuning and a linearization technique, can satisfy CCIR requirements for a 960-channel system. Figure 7.15 shows noise loading results for this oscillator, demonstrating a signal-to-noise ratio better than 76 dB at the zero reference point (corresponding to 25 pW).

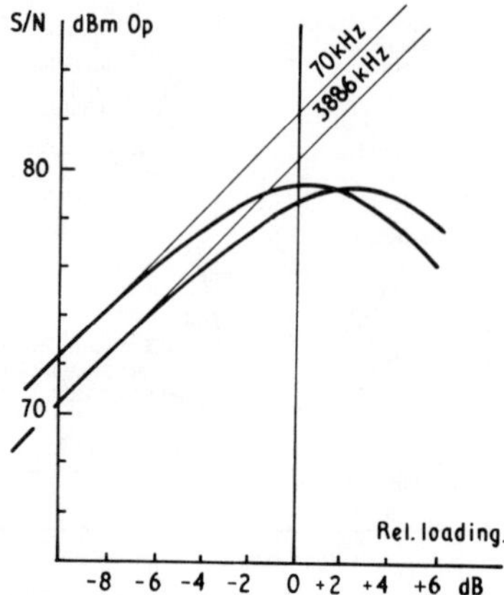

Figure 7.15 The results of noise loading tests on an X-band 50 mW FM Gunn oscillator. (From Endersz,[147] reproduced by permission of the 1973 European Microwave Conference)

A Gunn diode FM modulator followed by a silicon impatt amplifier has been used in equipment for the NTT 11 GHz 600-channel short haul system.[148,149] Noise loading tests have demonstrated a performance meeting the design objectives with an 11 dB margin. The estimated MTBF of the repeater is greater than 50,000 hours compared to 7000 hours with a repeater using klystrons.

Television transmission

Impatt amplifiers are establishing themselves in the transmitters of outside broadcast microwave links.[150] Hickin and Teesdale have described the use of an impatt diode amplifier in a 12 GHz outside broadcast transmitter head unit[151] manufactured by Microwave Associates Limited. Figure 7.16 shows a photograph of the unit. The impatt amplifier delivers 300—350 mW to the antenna. Ranges of up to 25—30 km are possible.

7.4.2 Digital communication

Established microwave links are predominantly based upon the use of FDM/FM. Future systems however will make extensive use of time division multiplexing (TDM) with pulse code modulation (PCM). PCM brings a number of advantages over conventional frequency modulation. At each repeater it is possible to restore the signal to a well-defined digital sequence, and so regenerate the original signal. Transmission distortions, therefore, do not accumulate from repeater to repeater. Since PCM is compatible with time division multiplexing, this

Figure 7.16 A 12 GHz outside broadcast television transmitter head unit containing a silicon impatt amplifier. (Reproduced by permission of Microwave Associates Limited)

allows considerable flexibility, in the selection of channels to be transmitted, in the interfacing with switching equipment and in allowing a number of different types of signal — television, telephony or data — to be transmitted on the same carrier simultaneously.

PCM systems are under development for satellite communication,[152] for terrestrial line-of-sight links, using small repeaters located on the tops of poles, towers or buildings,[131,153,154] and for millimetric waveguide systems. These systems use either two-level or four-level phase shift key modulation.

The overall performance of a digital communications link is measured by the probability of occurrence of a bit error. Errors are introduced into the digital stream on account of noise fluctuations at the input to the receiver demodulator. The reader is referred to standard texts[155–158] or to the review by Torrieri and O'Connor[159] for detailed accounts of the dependence of error rates on signal-to-noise ratio. Coherent detection gives the least error for a given signal-to-noise ratio. However, a relatively small increase in signal with incoherent or differential phase detection will recoup the difference in the two respective cases. Although QPSK does not give the least error, it is particularly favoured for high capacity systems on account of a higher information rate for the same bandwidth compared with the other types of modulation. With ASK, the average signal-to-noise ratio is required to be 3 dB greater than with PSK for the same error rate. However ASK systems are easy to implement and are therefore attractive for portable systems. A portable solid state system using incoherent ASK is described in the next section.

7.4.3 Portable systems

Systems are under development which consist of easily deployable transmit-receive units which can be aligned visually over short distances. In order to reduce the size of the equipment, it is advantageous to operate at millimetre wavelengths with a small antenna size. Rain attenuation limits the range at 30 GHz to an all-weather value of about 5 km. At 60 GHz, the strong oxygen absorption limits the range to about 1 km, with a rapid fall in signal strength beyond the receiving terminal. This strong attenuation could be exploited to provide communication links which are free from interference from neighbouring links[160] or to provide secure communication systems.[161,162] Interest is being shown in the use of low-cost millimetre wave links for ship-to-ship communication.[163]

The powers required for line-of-sight communication over a few kilometres at 30 GHz are readily met by presently available impatt sources. Figure 7.17 shows a block diagram of an all-solid-state terminal of a portable system based upon a 100 mW impatt oscillator described by Volckman, Gibbs and Denniss.[164] Each terminal of the link is contained in a volume of approximately $(30\ \text{cm})^3$ and weighs 9 kg.

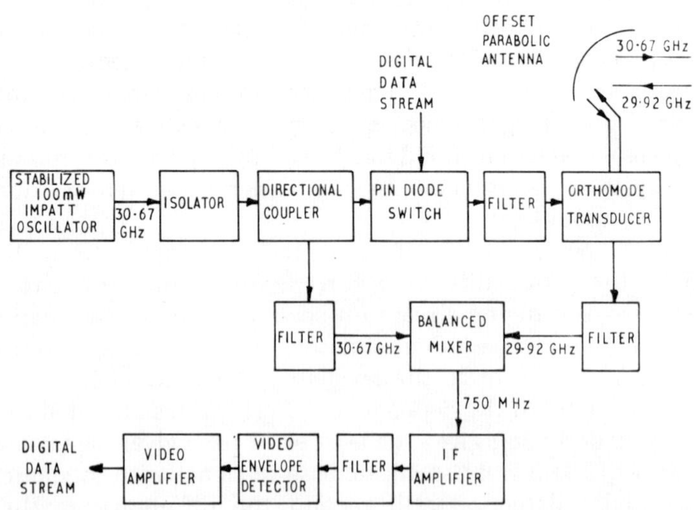

Figure 7.17 One terminal of a 30 GHz two-way link. (Copyright © Controller HMSO, London, 1975)

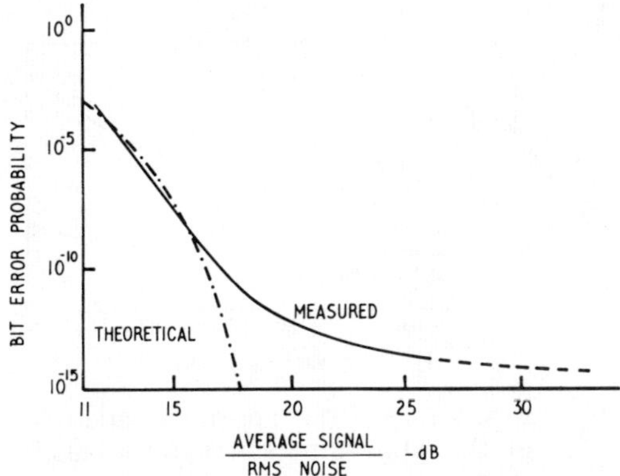

Figure 7.18 Measured and theoretical probabilities of a bit error for the ASK 30 GHz communication link. (Copyright © Controller HMSO, London, 1975)

The system has a number of features designed to enhance its simplicity. It uses incoherent ASK modulation, rather than biphase or quadriphase PSK. The penalty incurred in choosing ASK is a reduction in capacity compared with QPSK, but at the system bit rate of 120 Mbaud this is not a limitation. There is a 3 dB loss in mean power at the output of the ASK modulator, compared to the c.w. oscillator input, which would not be introduced by an ideal QPSK modulator; however, this is offset by the added complexity and loss of a practical QPSK modulator. Coherent ASK detection allows only a 0.5 dB reduction in signal-to-noise ratio for the same error rate compared with incoherent ASK, and so the added complexity for carrier recovery is not justified.

Each end of the link transmits at a frequency which differs from the transmitted frequency at the other end of the link by the IF. Thus, a single oscillator in each module can be used as both trasmitter and local oscillator. The system performance is degraded by AM noise from the silicon impatt oscillator at a frequency removed from the carrier equal to the IF, but this effect can be removed by placing a filter in the local oscillator channel and by using a balanced mixer.

The modulator uses a *p-i-n* diode which has an insertion loss of 0.5 dB and a switching time of less than 1 ns.

Tests with a 2 km link showed close agreement between the theoretical and measured bit error probabilities for error probabilities greater than 10^{-8}, as shown in Figure 7.18.

7.4.4 Waveguide communication

The demand for greater capacity communication systems has led to the investigation and development of millimetric waveguide systems in a number of countries.[131,165–170]

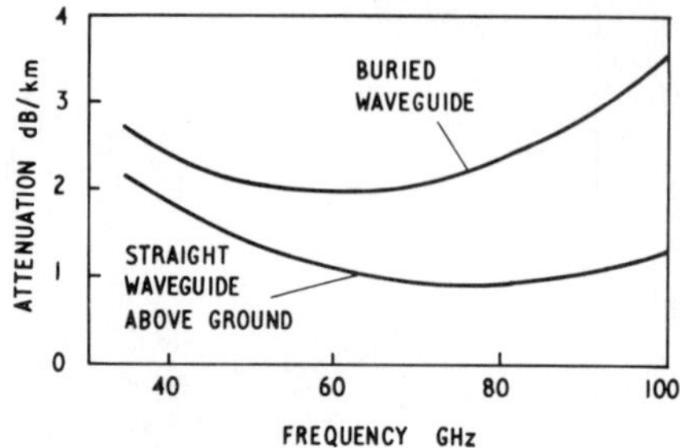

Figure 7.19 The measured variation of the TE_{01} attenuation for straight waveguide above ground and for buried waveguide. (From Sushi and Shimba,[172] reproduced by permission of the Institution of Electrical Engineers)

Line-of-sight links will suffer severe atmospheric attenuation at millimetre wavelengths, as well as severe fading due to rain. The solution chosen to circumvent these problems is to transmit the millimetre wave signal along a circular waveguide filled with dry nitrogen using the TE_{01} mode. The TE_{01} mode is a relatively low loss mode whose attentuation decreases with frequency.[171] In practice, the loss reaches a minimum and then rises with increasing frequency due to waveguide imperfections causing conversion from the TE_{01} mode to other propagating modes. The wall of the waveguide consists of a finely-wound metal-wire helix designed to attenuate the power in the unwanted modes. Figure 7.19 shows the measured variation of the TE_{01} attenuation over the range 30 GHz to 100 GHz reported by Sushi and Shimba[172] for waveguide above ground and for buried waveguide. Figure 7.20 shows the cross-section of a circular waveguide developed for the British Post Office.[173] With this waveguide the extra

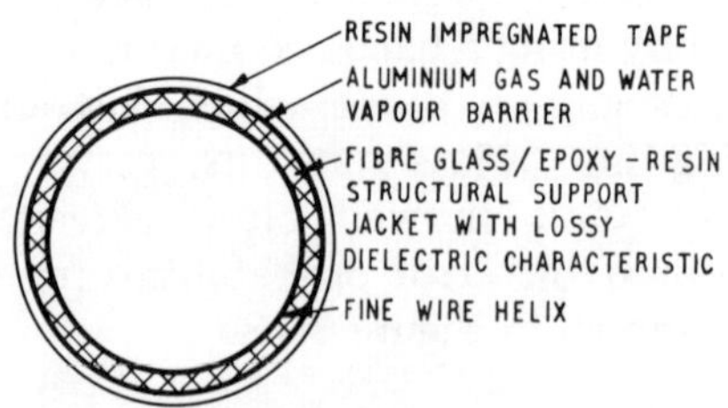

Figure 7.20 Cross-section of a circular waveguide. (From Ritchie,[173] reproduced by permission of the 1973 European Microwave Conference)

attenuation due to the curvature variations expected to be met in practice is calculated to be less than 0.1 dB/km at 30 GHz and about 0.5 dB/km at 100 GHz.

The maximum separation between repeaters is dependent on the bit rate and the net delay distortion due to the waveguide and delay compensation in a signal repeater. Lorek and Hanke[169] have shown that 640 Mbaud pulse patterns may be regenerated unambiguously after transmission down 5 km of waveguide. Evaluation of links of 14 km,[165] 15 km,[167] 20 km[131] and 30 km[166,169] are planned. The transmitter power required in each channel depends upon the length of the waveguide between repeaters, the attenuation per unit length and the minimum signal-to-noise ratio at the receiver in order to achieve an acceptable bit error rate. Transmitted powers of 10 mW or more are required for each channel. Impatt oscillators and amplifiers can provide these powers, with potentially good reliability, and heavy emphasis has been placed on their development at these frequencies.

7.4.5 Repeaters for digital systems

Repeaters and terminal equipment may either use direct baseband modulation of the transmission frequency or, alternatively, the baseband signal modulates an IF carrier which is upconverted to the transmission frequency band. On reception it is usual to downconvert to an intermediate frequency and then to demodulate to recover a baseband signal. Gallium arsenide diodes are the usual choice for making upconverters and downconverters at millimetre wave frequencies but recent work by Stover and Leedy[104] demonstrates the use of silicon devices in millimetre wave upconverters, downconverters, multipliers and varactors.

Figure 7.21 shows a block diagram of a transmitter—receiver developed for the British Post Office waveguide system for use in bandwidths of 0.5 GHz in the range 32 to 50 GHz.[103] Two input

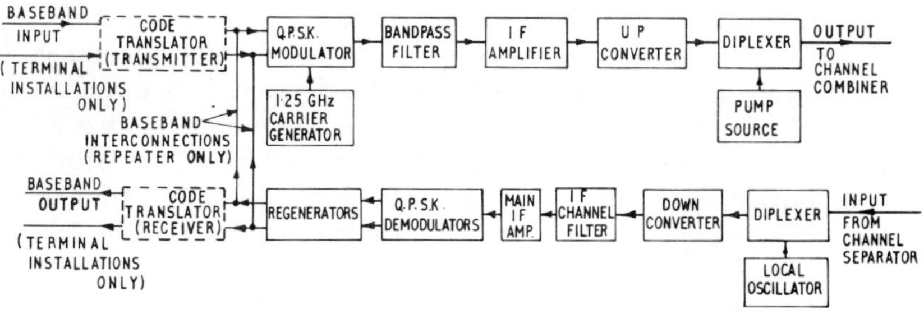

Figure 7.21 Block diagram of a transmitter—receiver for use with waveguide communication. (From Read,[103] reproduced by permission of the 1973 European Microwave Conference)

streams at 250 Mbaud drive the in-phase and quadrature channels of a QPSK modulator which modulates an IF of 1.25 GHz which has been generated by multiplication from a 125 MHz crystal oscillator. The quadrature phase accuracy is better than 2°. The gallium arsenide upconverter is pumped by a 100 mW Si impatt oscillator, with an upconversion loss of about 10 dB producing an output of about 10 mW. Branching and channelling introduces a further loss of about 5—6 dB before transmission down the waveguide. The mixer downconverter in the receiver is a hybrid 'm.i.c.' utilizing a beam lead Schottky barrier diode on a quartz substrate. The overall receiver noise figure is about 10 dB at 32 GHz and about 15 dB at 50 GHz.

The required short-term frequency stability Δf over one bit period of the downconverter local oscillator (and also pump source for the upconverter if a separate oscillator is used) for differential phase shift key systems is related to the tolerable phase shift $\Delta\phi$ in one bit period T by

$$\Delta f = \Delta\phi/(2\pi T) \tag{7.18}$$

Frequency stabilities of around 1 in 10^5 are required at millimetric frequencies for bit rates of hundreds of megabauds and phase shifts of one or two degrees.[102]

In order to increase the transmitted power in digital systems, millimetric amplifiers and injection-locked oscillators are being investigated as power stages.[175,176] The bandwidth of the stable amplifier or injection-locked oscillator must be sufficiently large in order not to increase the phase transition time of the phase modulated signal. Kuno[176] has shown, for impatt amplifiers and locked oscillators, that in order for there not to be an increase in the phase transition time, T_s,

$$B \geqslant 1/T_s \tag{7.19}$$

where B is the 3 dB bandwidth. If $BT_s < 1$ then for the same bandwidth, a faster output phase transition time is obtained with a stable amplifier than with a locked oscillator.

Paik and coworkers[175] have compared the error rate introduced by impatt stable amplifiers and injection-locked oscillators intended for use in an 11 GHz radio-relay system. The data rate was 40 Mbaud (two streams of 20 Mbaud) driving a differential QPSK modulator. The stable amplifier introduced insignificant additional error whereas the injection-locked oscillator introduced an error-rate floor at the 10^{-5} to 10^{-6} level.

Kita and coworkers[177] have described an 800 Mbaud QPSK repeater for a 30 GHz radio relay system. A 50 mW gallium arsenide impatt local oscillator is directly modulated using gallium arsenide Schottky barrier diodes. The modulator introduces a loss of 5 dB and it is followed by a

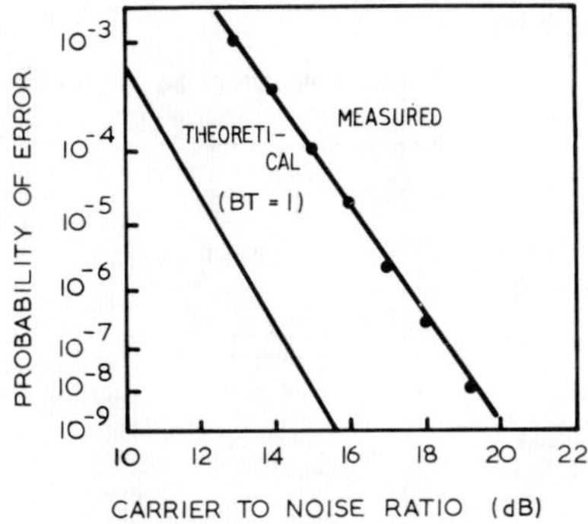

Figure 7.22 Measured and theoretical error probabilities for a 30 GHz 800 Mbaud QPSK repeater. (From Kita, Kaneko, Fujioka and Migitaka,[177] reproduced by permission of the 1973 European Microwave Conference)

gallium arsenide amplifier with a 100 mW output. The receiver downconverts to an IF of 1.7 GHz. Carrier recovery is achieved at IF using a phase locked loop which contains a voltage-controlled oscillator, a 4× multiplier and a phase comparator. Figure 7.22 shows the measured variation of the error rate with signal-to-noise ratio, compared with the theoretical variation for coherent QPSK modulation. The measured signal-to-noise ratios are about 4 dB worse than the theoretical values for the same error rate.

7.4.6 Optical communication

The demand for greater bandwidth communication systems and the advent of the laser have led to the investigation of systems using signals transmitted along optical fibres. A problem has been that of modulating semiconductor lasers at microwave frequencies. A possible solution is to pulse modulate the laser at repetition rates of a few gigahertz using a trapatt diode. Mackintosh[178,179] has shown how the output waveform from a trapatt oscillator depends upon the response of the cavity filter to the train of current pulses at the cavity filter plane. By terminating the cavity resistively, a pulse train output which is porportional to the filter plane current waveform is obtained. Kawamoto[180] has obtained 125 V pulses with a pulse width of 400 ps at a repetition rate of 1.2 GHz.

Carroll[181] has described a trapatt diode circuit which has been used successfully to pulse modulate a gallium arsenide laser. The trapatt diode laser modulator is shown in Figure 7.23. It consists of a conventional coaxial trapatt cavity to which has been added two

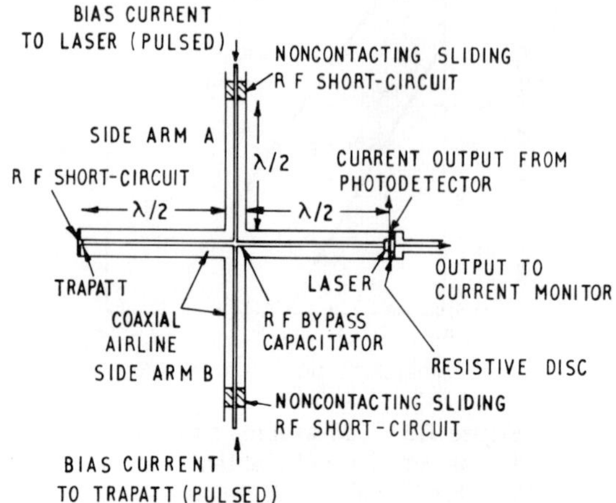

Figure 7.23 A pulse modulator for a gallium arsenide laser using a trapatt diode. (From Carroll,[181] reproduced by permission of the Institution of Electrical Engineers)

perpendicular side arms each approximately half a wave length long measured from the centre point of the cavity. The laser diode is mounted a further half wavelength from the centre point. The current pulses into the laser diode were measured and shown to have a rise time of about 100 ps and a peak current of 7 A. The optical output from the laser was measured with a silicon photodiode and found to have a rise time of 150 ps and a fall time of 300 ps. The pulse repetition rate was around 1 GHz.

An alternative approach is to drive a gallium arsenide laser with a Gunn diode logic element. Thim and coworkers[182] have described a modulator in which a Gunn diode operating as a bistable switch is in series with a double hetero-junction GaAs/GaAlAs laser diode. The current was switched between two levels above the laser threshold and optical pulses with a 200ps rise time and 400 ps fall time were produced at a repetition rate of 1.2 GHz. The modulation was controlled by trigger pulses of less than 1 V magnitude.

For comparison, Andrews[184] has shown that modulation of a GaAs laser with an avalanche transistor results in optical pulses as narrow as 110 ps. The repetition rate was 50 kHz.

The techniques discussed so far operate by modulating the current bias to the laser. Levinshtein[183] has suggested that trapatt diodes may be used to modulate infrared radiation directly. During plasma trapping, incident radiation is attenuated by free-carrier absorption, and during depletion this absorption is absent. No experimental results are available.

7.5 SOLID STATE MICROWAVE DEVICES IN RADAR SYSTEMS

In this section we consider the present and potential roles of solid state devices in radar systems. As is the case with communication systems, solid state microwave devices are presently used extensively in receivers. The use of solid state local oscillators has become widespread, either with crystal controlled oscillators followed by multipliers, or with stable fundamental oscillators. The limits on tolerable noise levels from local oscillators is discussed at the end of the section. Parametric amplifiers are being used at receiver front ends for increased range performance where otherwise expensive increases in transmitter powers would be required. Low noise transistor amplifiers will have an increasing role where losses are present between the antenna and the receiver. A small head amplifier placed behind the antenna can preserve the system noise figure in the presence of cable loss between antenna and receiver.

It is in the transmitter stages that solid state amplifiers and oscillators are starting to make inroads. Table 7.8 gives transmitter parameters for some examples of radar systems.[185–192] Except for the last one in the table, these radars are ones for which solid state transmitters either have been used, or could be used, or are under investigation. The satellite surveillance radar has a 5184 module transmitting array and 4660 module receiving array. However, high power microwave tubes are unlikely to be supplanted by solid state devices for the longer range radars unless low cost solid state arrays become practicable. We return to this question in Section 7.7.

Radar systems use a variety of transmitted waveforms. Table 7.9 gives a classification of waveforms with their type of modulation and an

Table 7.8
Examples of radar systems

Type	Range (km)	Frequency (GHz)	Mean power (W)	Peak power	Pulse (μs)
Intruder alarm	0.03	10.5	0.005	c.w.	
Short-range vehicle radar	0.06—0.8	10.5	0.1	c.w.	
Radar altimeter	1.6	4.87	0.15—1.0	c.w.	
Ground surveillance radar	15	9.3	4	2 kW	0.25
Airborne multifunction radar (mapping, weather, etc.)		9.3—9.5	80	1.6 kW	0.2—100
Airborne multifunction radar (mapping, weather, air-to-air, etc.)	100—200	10	65	52 kW	7.5
Ground-based 3-D radar		3.3	6×10^3— 6×10^4	60 kW— 600 kW	25
Satellite surveillance		0.442			1—250

Table 7.9
Examples of radar waveforms

Waveform	Type of modulation	Duty cycle (%)
Simple pulse	Rectangular amplitude modulation	0.01 to 1
Pulse compression	Linear FM	0.1 to 10
	In-pulse phase coding	
High duty Doppler	Rectangular AM	30 to 50
FM continuous wave	Linear FM	100
	Sinusoidal FM	
	Phase coding	
Continuous wave		100

indication of the duty cycle. The duty cycles are typical and are not to be interpreted as constraints. The main point to be made from Table 7.9 is that the radar designer makes use of a variety of wave forms and a wide range of duty cycles.

Radar range performance depends primarily on the mean power transmitted. The basis for this statement is as follows. The maximum range of a radar to a single target is given by[193]

$$R_{max} = \left[\frac{(P_T A_R) G_T \sigma}{(4\pi)^2 P_R L} \right]^{1/4} \tag{7.20}$$

where P_T = transmitter power
G_T = transmitting antenna gain
σ = radar cross-sectional area of target
A_R = receiving antenna effective area
P_R = power at input to receiver front end
L = loss factor.

The loss factor includes losses due to a variety of causes, e.g. feeder losses at the transmitting and receiving antennas, attentuation and fading in the atmospheric transmission path and a factor to take into account in-service loss. The minimum receiver power is given by

$$P_R = SFkTB \tag{7.21}$$

where F is the receiver noise factor and B is the receiver bandwidth. S is the signal-to-noise ratio which is the minimum set by the receiver threshold detector which in turn is set by the acceptable levels of false alarm rate and probability of missed targets. S includes parameters concerned with the noise statistics and the number of integrated pulses. Note that $kT = 4 \times 10^{-21}$ W for $T = 290$ K. Hence, for a 5 MHz bandwidth, a 10 dB noise figure and $S = 20$ dB, P_R is 2×10^{-11} W.

The transmitter power appearing in Equation 7.20 is the peak power or, in the case of a pulse compression radar, the effective peak power.

The mean power is given by

$$P_M = P_T \, \tau \, f_R \tag{7.22}$$

where f_R is the pulse repetition frequency and τ is the pulse length, or compressed pulse length with pulse compression. The product τf_R is the duty cycle. The maximum range may be recast as

$$R_{max} = \left[\frac{P_M A_R G_T \sigma}{(4\pi)^2 SFkT(B\tau) f_R} \right]^{1/4} \tag{7.23}$$

The receiver bandwidth will be such that $B\tau \simeq 1$. If the design fixes the antenna and S, then the range depends upon the energy P_M/f_R transmitted in one pulse repetition interval. For constant f_R, the range performance is therefore determined by the mean power transmitted.

The mean power dependence allows a certain measure of flexibility in the choice of pulse length and duty cycle for the same range performance. However, with a single antenna system leakage between transmitter and receiver will rule out the possibility of c.w. operation for medium and high power radars, and pulse lengths have to be restricted to comply with the minimum range specification for the radar. Thus, just as existing radars utilize a variety of duty cycles, future radars will do the same, and the radar designer demands that solid state devices be available to match these varied needs.

LSA devices can provide high peak power at low duty cycle; transferred-electron transit time devices give lower peak powers but with longer pulse lengths and higher duty cycles. Figures 7.24 and 7.25

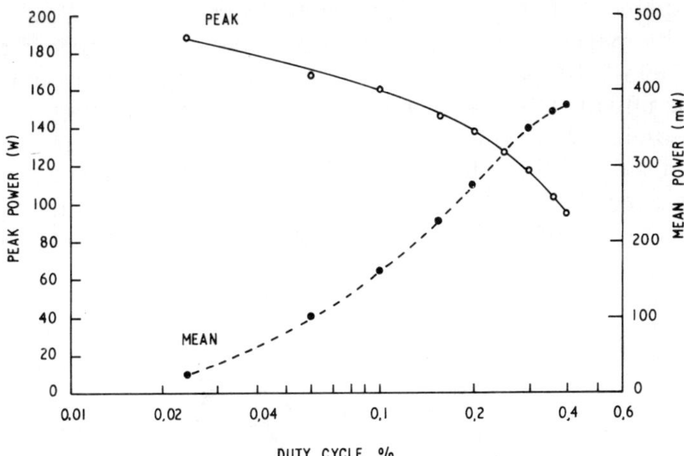

Figure 7.24 The measured variation of peak and mean powers with duty cycle for an 8.4 GHz LSA oscillator. (From Mun and Heeks,[194] reproduced by permission of Standard Telecommunication Laboratories Limited)

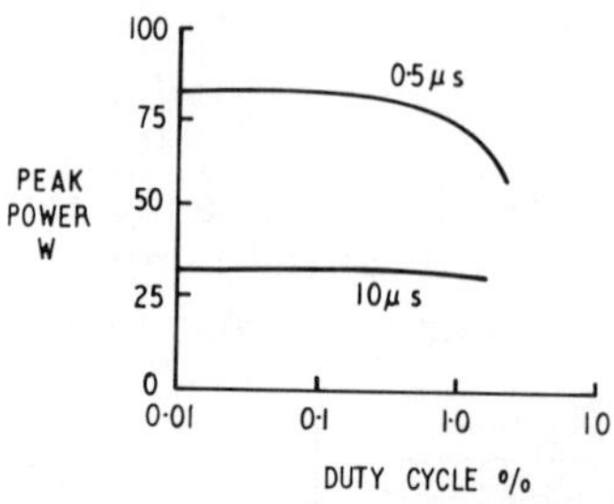

Figure 7.25 The measured variations of peak power with duty cycle for a 3 GHz Gunn diode oscillator. (From Cooke, Crisp, Conlon and Heeks,[195] reproduced by permission of the Institution of Electrical Engineers)

show how peak power degrades with increasing duty cycle for an 8.4 GHz LSA oscillator[194] and for a 3 GHz Gunn diode oscillator.[195] Trapatt and impatt devices are favoured for high duty, high mean power applications, and GaAs impatt diodes hold the present record for c.w. performance around 10 GHz. Transistors are favoured for c.w. and high duty cycle performance at lower frequencies. GaAs FETs are increasingly being advocated for power roles up to 10 GHz.

7.5.1 Elimination of temperature-induced chirp

The frequency of a pulsed solid state microwave oscillator changes during the pulse owing to the rise in device temperature during the pulse. Table 7.10 shows that the frequency change is around 0.2% to 0.3% for impatt, Gunn and trapatt oscillators, but is much greater for an LSA oscillator. In the case of the LSA oscillator entry in Table 7.10, the magnitude of the effect has been estimated from the rate of change of frequency with ambient temperature and an assumed in-pulse temperature rise of 50 C. In making this estimate, the contribution to

Table 7.10
Pulsed oscillator chirp

Oscillator	Frequency (GHz)	Peak power (W)	Pulse (μs)	Frequency (MHz)	Change (%)	Reference
Gunn	15.5—18	3	0.5	30	0.2	196
LSA	9	150		150	1.6	197
Impatt	16	0.5	1.7	30	0.2	198
Trapatt	9	5	0.25	25	0.3	199

Table 7.11
Variation of pulsed oscillator frequency with temperature

Oscillator	Frequency (GHz)	Temperature range (°C)	$\left\lvert \frac{df}{d\theta} \right\rvert$ (MHz °C^{-1})	$\left\lvert \frac{1}{f}\frac{df}{d\theta} \right\rvert$ (°C^{-1})	Reference
Gunn	15.5–18	–20 to +50	0.7	4×10^{-5}	196
LSA	9	–25 to +100	3.8	4×10^{-4}	197
Trapatt	1.3	–50 to +100	0.04	1.3×10^{-5}	197

the frequency change by the circuit is small and has been neglected and any effect on frequency due to differences in temperature profile in the device at the same mean temperature has been ignored.

A frequency variation during the pulse may also be present due to bias variation, and additionally the effect of a temperature change may be to induce a bias change resulting in a frequency change. For example, with an impatt oscillator a temperature rise of the device is accompanied by a rise in breakdown voltage, and a high impedance bias supply is required to minimize bias current droop.

It is necessary to reduce or eliminate chirp in order that the received radar signal falls within the receiver bandwidth. In a simple pulsed radar the receiver bandwidth is approximately equal to the reciprocal of the transmitted pulse duration, that is about 2 MHz for a 0.5 μs pulse, and an increase in receiver bandwidth beyond this figure will reduce the signal-to-noise ratio.

Similarly, it is desirable that a frequency change due to a change in ambient temperature will not take the received signal outside the receiver bandwidth. Table 7.11 compares the rates of change, and fractional rates of change, of frequency with ambient temperature, together with the temperature range to which the data applies, for Gunn, LSA and trapatt oscillators. The LSA oscillator shows the greatest fractional frequency change and the trapatt oscillator the least. Rather than to attempt to temperature-compensate a transmitter over the required operating temperature range, the IF signal can be kept within the IF bandwidth by an electronically-tunable local oscillator and an AFC loop.

Chirp compensation may be achieved by control of the bias level during the pulse, by injection locking,[196,198,200,204] by incorporating a varactor in the oscillator cavity, or by using an amplifier rather than an oscillator[201] Bias level chirp control has been used for LSA,[28,200,203] and impatt[3] oscillators, and it may be also used for ambient temperature compensation with LSA oscillators.[197]

Varactor compensation of a pulsed Gunn oscillator has been demonstrated by Bullimore and coworkers.[196] Their cavity is shown in Figure 7.26. The Gunn diode is mounted in the centre of a waveguide

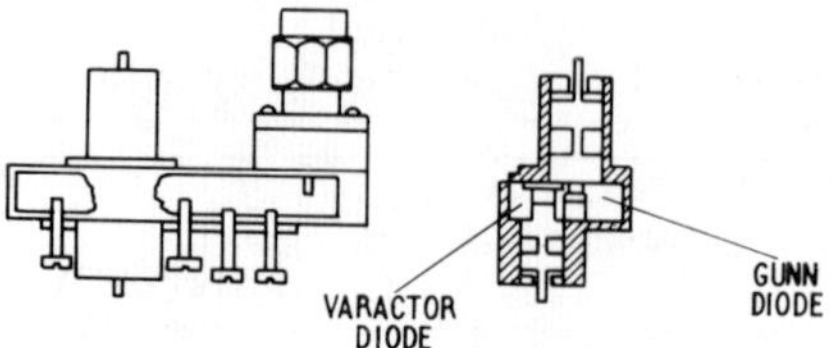

Figure 7.26 A pulsed Gunn diode oscillator with a varactor for temperature-induced chirp compensation. (From Bullimore, Downing and Myers,[196] reproduced by permission of the Institution of Electrical Engineers)

cavity and the varactor is mounted close to the sidewall. A ramp voltage derived from the Gunn diode bias voltage pulse is applied to the varactor. The chirp was reduced from 30 MHz to 2 MHz with a 17 GHz 3 W oscillator. Ambient temperature compensation was provided by a thermistor in the varactor control circuit.

The problems of temperature-induced chirp and starting transients may be avoided by using a c.w. oscillator followed by a switch. This technique has the advantage of eliminating the oscillator modulator, but clearly results in a poor overall efficiency. However, inefficiency may be tolerated in low power stages, and the technique has been used for a driver stage to a TWT in a pulsed Doppler radar.[202]

7.5.2 Pulse Compression Radar,

In a simple pulsed radar, the range resolution dictates the required pulse length, and the maximum range determines the pulse repetition frequency. Examination of the data of Figure 7.2 will reveal whether there is a solid state device which can provide the necessary peak power. If insufficient peak power is available, the necessary mean power may be attainable by increasing the energy in each pulse from the solid state device by lengthening the pulse. At the same time, range resolution may be preserved by exploiting the radar technique known as pulse compression. A pulse compression radar uses a modulated transmitted waveform for which the product of the modulation bandwidth, B, and the transmitted pulse length, τ, is much greater than unity. On reception, a matched filter compresses the pulse to a compressed pulse length, τ_e, such that $B\tau_e \simeq 1$. It is τ_e that determines the radar range resolution. The modulation may for example be linear FM or phase coded.

Pulse compression in a modern radar system is performed by a surface acoustic wave (SAW) filter.[205] Although little work has been reported to date, the combination of solid state microwave devices and

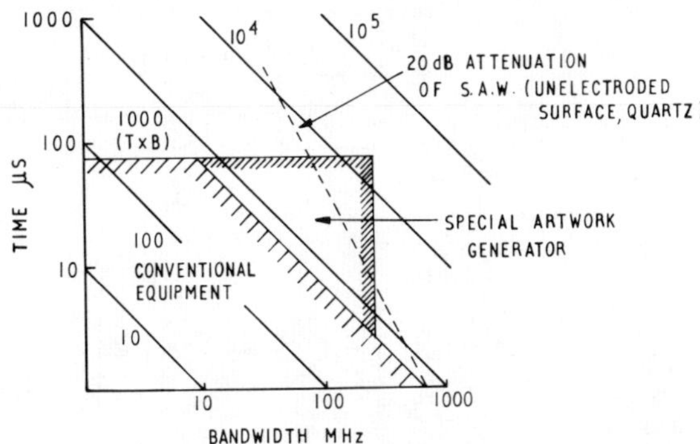

Figure 7.27 Fabrication limits on pulse length, bandwidth and time-bandwidth products for surface acoustic wave pulse compression filters. (Copyright © Controller HMSO, London, 1975)

SAW devices in pulse compression radars is a potentially fruitful area. Figure 7.27 shows the limits on time–bandwidth products of present surface acoustic wave devices. Time–bandwidth products of hundreds are possible, with pulse lengths ranging up to tens of microseconds and bandwidths of a few hundred megahertz.

Pulse lengths of 50 μs have been obtained with trapatt diodes by Chang and coworkers;[201] this was achieved by increasing the thermal capacity of the diode by increasing the thickness of the top gold contact, allowing the top contact to act effectively as a heat sink for long pulses. Amplifiers using these diodes have given 80 W of peak power at 3 GHz with 5 dB gain and trapatt amplifiers with 150 MHz bandwidth at this frequency are reported. Gunn diodes have achieved pulse lengths of 10 μs at 3 GHz with a peak power of 30 W.[195] Bandwidths attained in this frequency range were around 50 MHz.

An achievable specification for bipolar transistor amplifiers at 3.3 GHz is 20 W peak with a 25 μs pulse at 10% duty cycle with a bandwidth of 330 MHz and an overall efficiency of 20%.[191]

A gain ripple of less than 0.2 dB and a departure from phase linearity of less than 1° to 2° over the bandwidth is required in order to achieve a 40 dB range sidelobe level.[206]

Figure 7.28 shows a block diagram of a pulse compression radar system. FM modulated pulses of IF are applied to an upconverter which is pumped by a local oscillator. The upconverted signal is amplified and transmitted. On reception the signal is downconverted to IF, amplified and passed through a SAW pulse compression filter. The SAW filter has built-in weighting in order to reduce the time sidelobes on the compressed pulse. The signal is finally detected by phase sensitive detectors giving in-phase and quadrature bipolar video outputs.

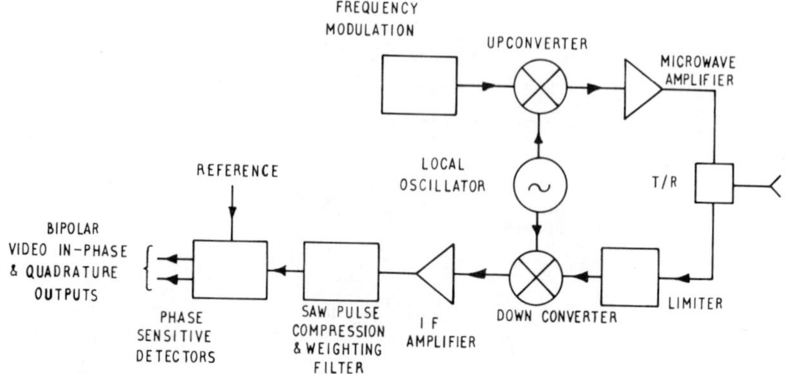

Figure 7.28 Block diagram of a pulse compression radar system

The local oscillator may be either a crystal controlled oscillator followed by a multiplier or a free-running Gunn or impatt oscillator. However, if the phase stability requirement is a stringent one the latter sources may be inadequate. A more detailed discussion on noise and stability is to be found later in the chapter.

At present the power amplifier stages are the domain of tubes. However, LSA locked oscillators, transistor, trapatt, impatt and transferred-electron amplifiers could all find application in pulse compression radar transmitters either as driver stages in high power systems or as output stages in modest mean power systems.

7.5.3 Solid state Doppler radars

Established radar systems exploit the Doppler frequency shift from moving targets in a variety of system configurations.[193,207,208] They include low duty cycle, short-pulse, ground surveillance radars, high duty cycle, pulsed Doppler radars for airborne operation, and c.w. and FM c.w. radars for airborne Doppler navigators.

Solid state devices are finding application as local oscillators and driver stages, and, increasingly, as final stage transmitting devices in all-solid-state radars. In this section we consider pulsed Doppler and c.w. Doppler solid state systems for use in ground surveillance. Later sections discuss FM c.w. radars and noise considerations.

Short-range (up to about 1 km) Doppler radars are used to provide detection of moving targets against a background of clutter returns from the ground and other stationary targets. A man walking at 1 m/s either towards or away from the radar will give a frequency shift of 67 Hz at 10 GHz, and a vehicle at 20 m/s will give a shift of 1.33 kHz.

The presence of a return from clutter is exploited in a clutter reference radar. Moving targets may be detected by listening to the audible Doppler beat note between the moving target return and

Figure 7.29 Photograph of the Shrimp radar. (Copyright © Controller HMSO, London, 1975)

returns from adjacent clutter. The use of a pulsed waveform allows ranging in addition to velocity measurement.

A simple hand-held clutter reference system (SPRAT) using a Gunn diode transmitter has been developed at RRE by Skinner.[209] The Gunn diode gave an output peak power of 3 W with a 250 ns pulse and a p.r.f. of 40 kHz corresponding to a mean power of 30 mW. The system used a Gunn diode local oscillator and consumed less than 5 W from its battery. A 550 m range gate was used to measure range between 25 m and 625 m. A more advanced version (SHRIMP) of this radar is shown in Figure 7.29 and a block diagram of the radar is shown in Figure 7.30.

An advantage of the clutter reference system is that it is tolerant of thermal chirp on the transmitter pulse, so long as the chirp does not take the downconverted received signal outside the IF bandwidth. A

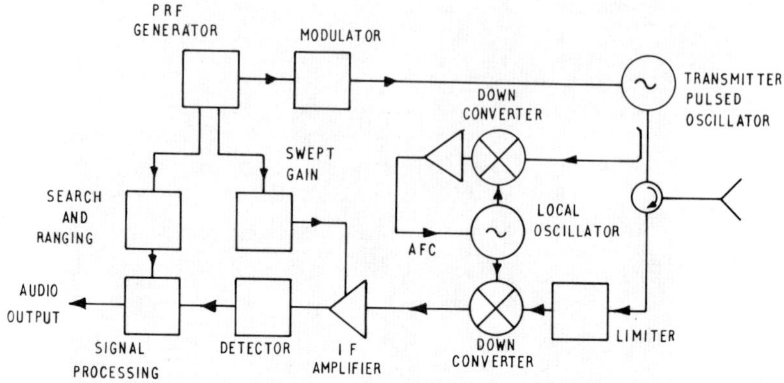

Figure 7.30 Block diagram of the Shrimp radar. (Copyright © Controller HMSO, London, 1975)

possible disadvantage is that performance may be degraded if there is little clutter adjacent to the target.

Brown and coworkers[210] have described a system in which an internal reference is provided by a surface acoustic wave delay line. Figure 7.31 shows a block diagram of the system. The single-port delay line contains an array of reflectors arranged to match the inverse of the IF amplifier swept gain variation with time. Breakthrough from the pulsed Gunn diode transmitter is downconverted to IF and enters the delay line and is successively reflected to give a reference signal of the required duration. Range and Doppler information is obtained from envelope detection of the sum of the target return and the delay line reference signal. With a Gunn diode system, the internal reference gave a 10 dB increase in return from a man at 250 m

Prior and Warren[211] have described a man-portable ground surveillance radar which operates in two modes — a surveillance mode and a ranging mode. In the surveillance mode, the Gunn diode local

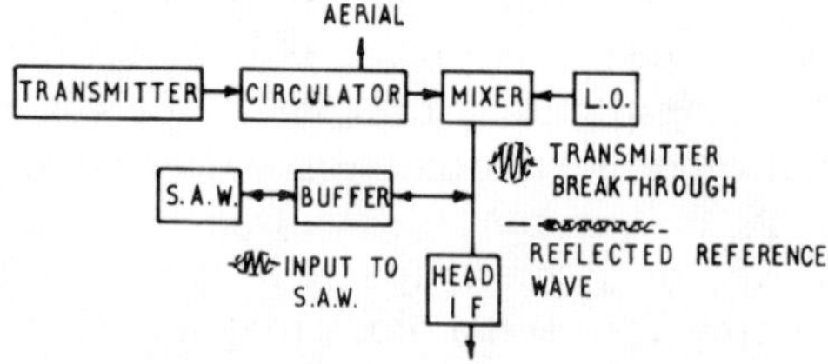

Figure 7.31 Block diagram of a clutter reference radar with an internal surface acoustic wave delay line. (From Brown, Hannis, Skinner and Turton.[210] Copyright © Controller HMSO, London, 1975)

oscillator acts as the transmitter for sinusoidal FM c.w. operation. The IF is at the modulation frequency. In the ranging mode, the local oscillator acts as a priming source for a pulsed Gunn diode transmitter, as well as providing the pump for the downconverter. Priming has the effect of bringing a p.r.f. sideband of the transmitter frequency into coincidence with the local oscillator frequency, so that the IF is a multiple of the p.r.f. This technique eliminates the need for upconversion. A priming effect has also been observed with impatt diodes.[212]

A major application of solid state microwave devices is with c.w. Doppler radar heads, for use as intruder alarms, for vehicle velocity measurement and as velocity sensors in industrial control applications. Gunn oscillators,[185,209,213,219,229] impatt oscillators[186,219] and baritt oscillators[221,222] are all suited to this role. Commercial systems operate at frequencies in the vicinity of 10.5 to 10.7 GHz.

Short-range Doppler radar heads may take a number of forms, as shown in Figure 7.32. They may use separate diodes in the transmitter and for the mixer in the receiver, or alternatively a single diode may combine transmitting and receiving functions and operate as a self-detecting oscillator.[186,214–219,221,222] When a separate mixer diode is used, the system may have individual antennas for transmission and reception, or the system can be monostatic with a circulator-coupled single antenna. A further alternative is to place the transmitting diode and the mixer diode across the same waveguide cavity and a simple horn is used as the common transmitting and receiving antenna.

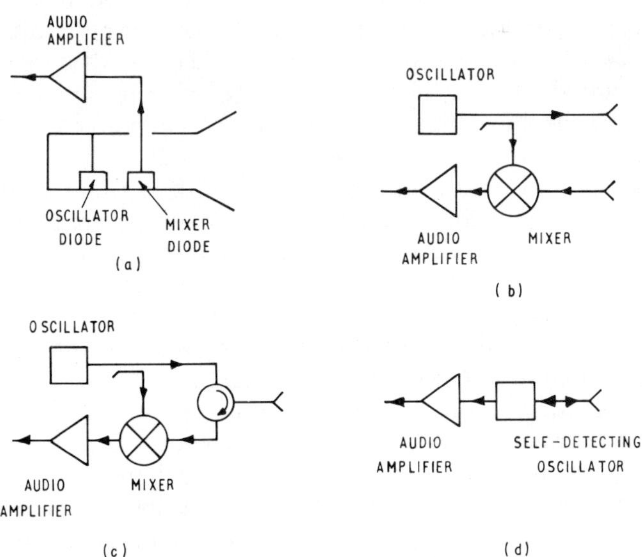

Figure 7.32 Short-range Doppler radars

Theoretical treatments of self-detecting oscillators have been given by a number of authors. They are based upon the examination of the interaction of the returned signal from the target on the oscillation in purely electrical terms. In contrast, the treatment by Schuck[214] ascribes the self-detection with an impatt oscillator to a bias voltage change due to a temperature change of the device induced by power returning from the target altering the power dissipated in the device.

Baritt and Gunn oscillators working in the self-detecting mode have noise figures which fall with increasing Doppler frequency reaching a plateau above about 10 kHz.[221] Baritts achieve noise figures as low as 25 dB,[221] but Gunn oscillators are reported as having noise figures in the range 40—70 dB.[215,220,221]

Range performance with human targets extends from a few metres with low power (~5 mW) self-detecting oscillators using a low gain antenna to hundreds of metres with higher power oscillators (~100 mW) using a bistatic high gain antenna configuration. A self detecting 100 mW impatt system with an 18 dB gain antenna gives a range of 30 m, and with a separate mixer should give about 150 m. A self-detecting 7 mW Gunn oscillator system gives a range of about 8 m with a 16 dB antenna,[213] whereas a bistatic system should give a range of about 300 m.[220]

7.5.4 FM c.w. radar

We have already seen, from Equation 7.23, that the maximum range of a radar depends upon the mean power transmitted. Continuous wave radars are of interest because the powers available from c.w. solid state sources are comparable to the maximum mean powers available from pulsed sources. Continuous wave sources have a clear advantage over low duty cycle pulsed sources, but it is to be noted that higher duty cycle operation can produce high mean powers. For example, Chang and coworkers[201] report a mean power of 7.5 W from a silicon trapatt oscillator at 3.2 GHz with an efficiency of 30%, a pulse width of 1μs and with 7.5% duty cycle. Pfund and coworkers[3] report a mean power of 4 W from a silicon double-drift impatt oscillator at 10 GHz with an efficiency of 12%, a pulse width of 800 ns and with 25% duty cycle. Figure 7.33 compares this experimental data with the c.w. data of Figure 7.1 together with curves given by Camp and coworkers[28] on the mean power capability of gallium arsenide LSA devices.

Continuous wave radar operation has the advantages that the problems of thermal transients encountered with pulsed operation are avoided and the need for a pulse modulator is eliminated. The major disadvantage of a c.w. radar is that separate transmitting and receiving antennas are required. With a monostatic radar, unless the transmitter power is very low, the leakage of power from the transmitter to the receiver front end would saturate the receiver (that is, take it out of its linear range) and introduce intolerable intermodulation products. The

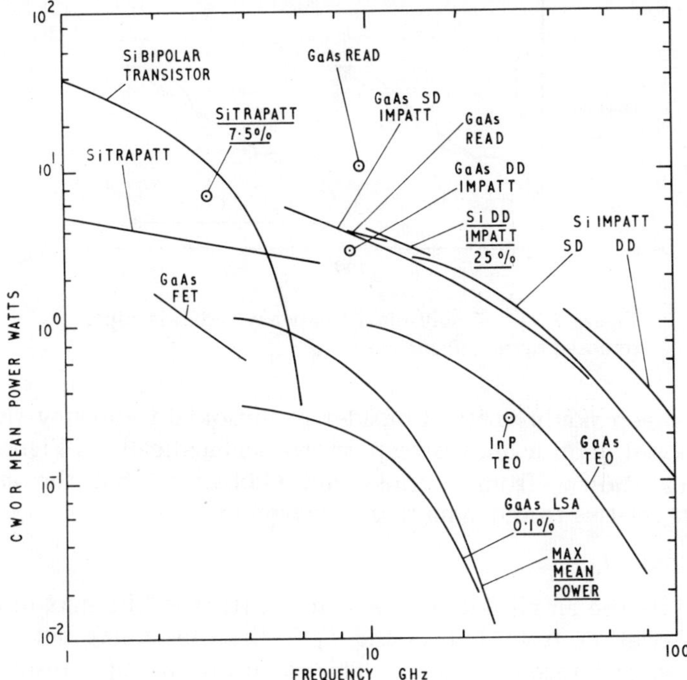

Figure 7.33 Best laboratory microwave solid-state c.w. power performance compared with mean power performance. The mean power results have underlined titles and the duty cycle is specified as a percentage

leakage will occur mainly because of antenna mismatch, but also to some extent by leakage through the circulator which separates the transmit and receive channels.

With mean powers up to 10 W it can be seen from Table 7.8 that ranges of tens of kilometres are possible and that single c.w. devices or a combination of a small number of devices can provide these powers up to at least 10 GHz.

Linear FM

In order to extract range information, the transmitted signal is required to be modulated. We consider the case of linear FM. The transmitter frequency is a sawtooth function of time in which linear frequency sweeps between frequencies f_1 and f_2 alternate with sweeps between f_2 and f_1 The difference between the transmitted and received frequencies is proportional to target range, and so spectral analysis of the beat signal provides a measure of returned signal versus range.

Nonlinearity in the frequency sweep will produce spurious range sidelobes on either side of a target return. This means that a specification of the tolerable range sidelobe level leads to an allowable

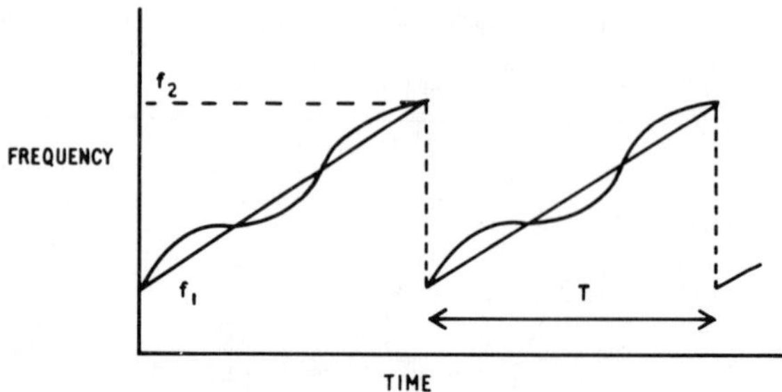

Figure 7.34 A sinusoidal frequency variation superimposed upon a linear sweep

level of sweep nonlinearity. Consider a sinusoidal frequency variation superimposed upon a linear sweep, shown schematically in Figure 7.34. It can be shown, from Denniss and Gibbs,[223] that the sidelobe amplitude relative to the main target is given by

$$S_m = \pi a_m t_d \tag{7.24}$$

where a_m is the amplitude of the sinusoid (that is, the maximum frequency deviation from linear) and t_d is the time delay between the transmitted and received signals. This is an approximate result which results in less than a few per cent error provided that

$$\frac{m\pi t_d}{T} \leq \frac{\pi}{10}, \quad \pi a_m t_d \leq 0.15 \tag{7.25}$$

where T is the period of the sawtooth, and T/m is the period of the sinusoid where m is an integer. For example, for the sidelobes to be 20 dB down for a range of 100 m the maximum deviation, from Equation 7.24, is approximately 40 kHz. The non-sinusoidal nature of the nonlinearity may be taken into account by reducing this frequency deviation by a factor of 2.[223] The required range resolution determines the sweep bandwidth f_2-f_1 For a 1 m resolution the bandwidth is 300 MHz, and the required linearity is 0.007% where the percentage linearity is defined by

% linearity = (frequency deviation/frequency sweep) x 100

and where the frequency deviation is the departure from the best straight line.

The period of the swept frequency waveform will be in the region of a few hundred microseconds. Thus the requirement is for high linearity, wide bandwidth and fast tuning rate voltage-controlled oscillators.

YIG tuned oscillators are highly linear but have slow tuning rates, whereas varactor-tuned oscillators require linearization by external means but have high tuning rates. Examples of performance of tunable

Table 7.12
Electronically-tunable sources

Source	Tuning method	Tuning range (GHz)	Linearity (%)	Reference
Transistor	YIG	1–2	±0.3	224
Transistor	Varactor	1–2	±10 (±1)*	225
TEO	YIG	8–12.4	±0.34	224
TEO	Varactor	13.5–17.5	±15	226
Impatt	YIG	8–10	±7	227
Impatt	Varactor	4.2–4.5	±7	228
Impatt	Ferrite substrate	10.3–10.7	±2	229
		9.8–10.7	±10	

*With linearizer.

sources is given in Table 7.12. It is evident that available linearity performance can be inadequate and that additional system complexity is needed to achieve the required performance. When the powers available from present electronically tunable sources are not sufficient to provide the required c.w. transmitter power, the solution is to follow the oscillator with a power amplifier.

Sinusoidal FM

Sinusoidal frequency modulation affords a method for eliminating the effects of antenna mismatch in a monostatic antenna system;[207] and extensive use has been made of this type of modulation in Doppler navigators and in radar altimeters. It has been used in ground surveillance as discussed previously. Solid state systems using transistor driven varactor multiplier chains have already been developed, and impatt and Gunn oscillators are presently being investigated for these applications. The effect of AM noise on performance is discussed in the next section.

Phase coding

Higgins and Baranowski have reported using a c.w. impatt oscillator as the transmitter in a phase coded radar.[202] Although little else has been reported on the use of solid state devices in these types of radar, the combination of phase coding and the extraction of information from the received signal using new signal processing techniques is likely to be a growing area.

7.5.5 Stability and Noise Specifications

In this section we discuss the various demands that radars make in regard of local oscillator and transmitter stability and noise performance.

Our starting point is to consider the origin of noise sidebands on a received signal. The sidebands arise from FM and AM sidebands on the transmitted signal and from FM and AM sidebands on the local oscillator which drives the receiver downconverter. We shall assume that AM noise from the local oscillator is cancelled by use of a balanced downconverter. We discuss FM noise first of all, and we shall restrict the discussion to the case of common FM in which there is identical modulation on the local oscillator and transmitted signals. This will occur, for example, where the transmitted signal is obtained by upconversion and a single local oscillator drives both upconverter and downconverter.

Owing to the time delay, t_d, for a signal to travel to the target and back to the receiver, the modulation spectrum in the receiver after downconversion is a modified version of the local oscillator modulation spectrum. If the instantaneous phase of the local oscillator is $\phi(t)$, then that of the received signal is $\phi(t-t_d)$ and the phase of the downconverted signal is given by

$$\Delta\phi(t) = \phi(t) - \phi(t-t_d) \tag{7.26}$$

We shall describe the local oscillator modulation spectrum by the frequency deviation 'power' spectral density function $S_{\dot{\phi}}(\omega_m)$ where $\omega_m = 2\pi f_m$ is the modulation angular frequency.[74] The downconverted frequency deviation 'power' spectral density function, $S_{\Delta\phi}(\omega_m)$, is given by[230]

$$S_{\Delta\dot{\phi}}(\omega_m) = 4S_{\dot{\phi}}(\omega_m)\sin^2(\omega_m t_d/2) \tag{7.27}$$

Equation 7.27 may be recast in terms of r.m.s. frequency deviations in the form[231]

$$\Delta f'(f_m) = 2\Delta f(f_m)\,|\sin(\pi f_m t_d)| \tag{7.28}$$

where $\Delta f(f_m)$, $\Delta f'(f_m)$ are the local oscillator and downconverted r.m.s. frequency deviations respectively in the same specified bandwidth. The ratio of $\Delta f'(f_m)$ to $\Delta f(f_m)$ is shown plotted in Figure 7.35 as a function of the product of the modulation frequency and time delay. The dashed lines indicate a simple practical approximation.

The frequency deviation spectrum of the local oscillator has to be multiplied by the function shown in Figure 7.35 in order to obtain the received downconverted spectrum. For small values of $f_m t_d$, that is for short ranges or low modulation frequencies, there is considerable suppression of the local oscillator FM noise in the downconverted signal owing to the correlation of the transmitted and returning FM noise. For large values of $f_m t_d$, the downconverted deviation can be twice that of the local oscillator.

The variance of the phase of the downconverted signal is given by[74]

$$\sigma^2[\Delta\phi(t)] = \frac{1}{\pi}\int_0^\infty S_{\Delta\phi}(\omega_m)d\omega_m \tag{7.29}$$

where $S_{\Delta\phi}(\omega_m)$ is the downconverted phase 'power' spectral density

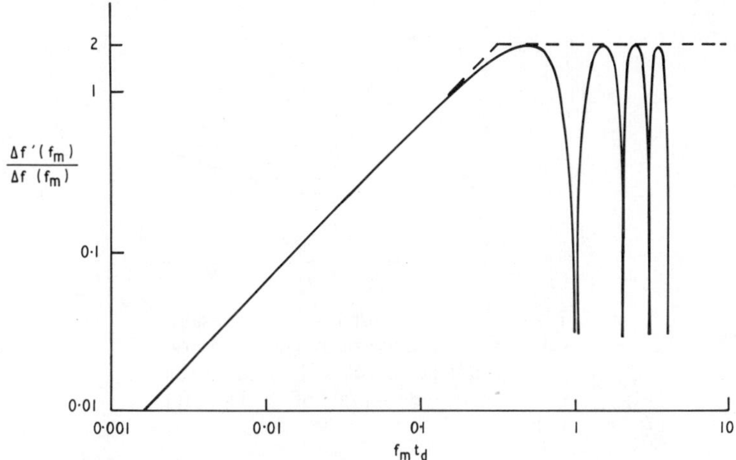

Figure 7.35 Common FM noise modification spectrum

function and is related to $S_{\Delta\dot{\phi}}(\omega_m)$ by

$$S_{\Delta\phi}(\omega_m) = \frac{1}{\omega_m^2} S_{\Delta\dot{\phi}}(\omega_m) \tag{7.30}$$

The square root of the variance of the downconverted phase is the r.m.s. phase error introduced into the received signal by the local oscillator FM noise. It may be related to the r.m.s. frequency deviation in a 1 Hz bandwidth of the local oscillator as follows.

We note that[74]

$$[\Delta f(f_m)]^2 = 2\pi[\Delta f(\omega_m)]^2 = \frac{1}{2\pi^2} S_{\dot{\phi}}(\omega_m) \tag{7.31}$$

and combining Equations 7.27, 7.29, 7.30 and 7.31 we obtain

$$\sigma^2[\Delta\phi(t)] = 4\pi^2 t_d^2 \int_0^\infty [\Delta f(\omega_m)]^2 \left(\frac{\sin(\omega_m t_d/2)}{\omega_m t_d/2}\right)^2 d\omega_m \tag{7.32}$$

When the local oscillation frequency deviation is independent of the modulation frequency, as is approximately the case for impatt oscillators, we put $\Delta f = \Delta f(f_m)$ and obtain from Equation 7.32 the following result for the r.m.s. phase error, in radians, in the downconverted signal,

$$\Delta\phi = \pi\Delta f\sqrt{2t_d} \tag{7.33}$$

For a range of 10 km, t_d is 67μs, and a deviation $\Delta f = 1$ Hz in a 1 Hz bandwidth will produce a phase error of approximately 4×10^{-2} radians or about 2°.

This phase deviation in the downconverted signal limits the performance of a moving target indicator (MTI) radar. Stationary target

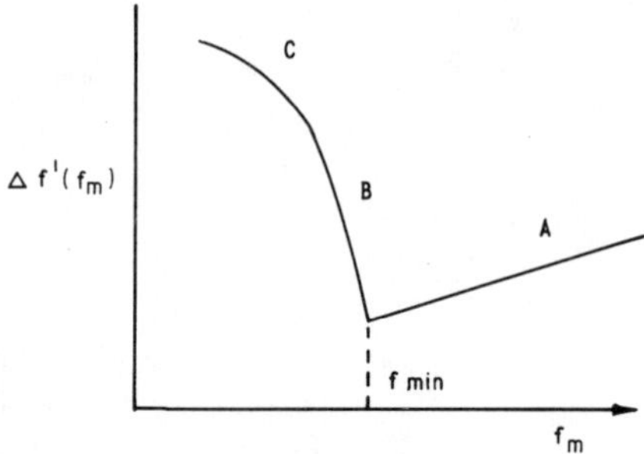

Figure 7.36 Schematic representation of the maximum specified frequency deviation for FM noise in the downconverted signal of a Doppler radar

returns are cancelled by subtracting the received signal from the received signal delayed by one interpulse period. Phase detection is used, and phase errors limit the sub-clutter visibility. For an r.m.s. phase error of $\Delta\phi$ radians, the maximum target to clutter improvement is $1/(\Delta\phi)^2$.[232] Thus, $\Delta\phi = 2 \times 10^{-2}$ radians corresponds to a sub-clutter visibility of 34 dB.

Doppler radars specify a maximum FM noise spectrum of the form shown schematically in Figure 7.36.[208,230,231] Segment A arises from the requirement that the noise sidebands on stationery clutter do not mask wanted returns from moving targets. The ratio of carrier to noise sideband power within the Doppler filter bandwidth must be greater than the required sub-clutter visibility. This means that beyond the cut-off frequency, f_{min}, of the clutter rejection filter, the maximum tolerable frequency deviation rises linearly with frequency. Below f_{min} the deviation spectrum must be less than segment B in order that higher-order FM sidebands do not contribute unacceptably to the power spectrum above f_{min} Segment C ensures that signal losses on passage through the Doppler filter and on integration are kept below a tolerable level, say 0.1 dB.

With a clutter reference Doppler radar the effects of FM noise on both the transmitter and local oscillator are suppressed considerably. Whereas we have assumed previously that the transmitted FM noise is identical to the local oscillator noise by use of an upconverter, we shall assume that the clutter reference transmitter is an independent source. The instantaneous FM noise deviation on the received downconverted signal is given by the sum of the instantaneous frequency deviation introduced by the transmitter and that introduced on downconversion by the local oscillator. The Doppler information is contained in the frequency difference between the return from the target and returns from clutter within half a pulse length of the target. The FM noise deviations on these two signals are correlated with a time delay of up to τ, where τ is the pulse duration, in contrast to the conventional radar case where the delay time is that between transmission and reception.

When a surface acoustic wave delay line is used to provide an internal reference, suppression of transmitter FM noise is achieved with a decorrelation time of τ as before. However, the suppression of local oscillator FM noise is not reduced as effectively as before. This is because the reference signal is downconverted on transmission, but the received signal is downconverted after the transmit—receive time delay, t_d, and therefore the decorrelation time is t_d.

AM noise on the transmitted signal is a limiting factor on the performance of c.w. radars. For example, with Doppler navigator radars, owing to FM noise suppression due to the correlation effect with common FM, AM noise, and not FM noise, limits performance. Antenna mismatch in a monostatic antenna system, or strong clutter, will give returns with AM noise sidebands which can mask wanted Doppler returns. The ratio of carrier to noise sideband power within the Doppler filter bandwidth must be greater than the required sub-clutter visibility.

With sinusoidal FM c.w. radars, an IF is chosen at a harmonic of the modulation frequency. The AM noise is required to be reduced to an acceptable level in the vicinity of the IF. By so moving up in frequency, advantage is being taken of reduced AM noise of solid state sources and of reduced receiver flicker noise. However, with impatt oscillators, once the flat region of the noise spectrum has been reached, there is no further advantage in increasing the IF.

7.6 SECONDARY RADAR AND MICROWAVE CONTROL

The unifying theme of this section is the application of solid state microwave devices in radar-type systems, often using secondary radar, for the control of aircraft, road vehicles, rail vehicles and shipping.

With aircraft, two well-established systems are altimeters and Doppler navigators, and solid state microwave devices provide solutions for both of them. Another possible area of application is with microwave landing systems.[233] Silicon bipolar transistors and silicon trapatt diodes[234] are contenders for airborne transponders for use with civil air traffic control and military IFF. This latter application requires the transmission of bursts of pulses of about 500 W peak power at a frequency of 1.09 GHz.

7.6.1 Microwave control of vehicles

A number of organizations throughout the world are developing control systems for the safe operation of vehicles in congested road conditions. The simplest system uses a sensor to detect an impending collision. Automatic application of the brakes will either avoid a collision, or automatic triggering of the inflation of an airbag can reduce injury to the vehicle's occupants. Headway control is a more sophisticated approach which requires range, relative velocity and possibly acceleration data in order to achieve safe following distances between vehicles, and so avoid collision situations occuring. An even more sophisticated system for motorway driving involves control of headway and lane changing, and also control of vehicles joining and leaving the motorway.

The sensor for collision avoidance should have an all-weather capability, be cheap and reliable, be able to detect obstacles such as crashed vehicles, and not require a target vehicle to carry specialized equipment.[235] The false alarm rate due to unwanted returns from spurious targets should be very low. Primary radar systems can satisfy these requirements, except that they require features to reduce the false alarm rate to acceptable levels. Secondary radar systems give low false alarm rates but require cooperative targets. Solid state components are natural candidates for these sensors.

Kiyoto and coworkers have described a solid state pulsed Doppler dual-antenna system.[236] A c.w. 10 GHz Gunn oscillator is pulse modulated by a gallium arsenide Schottky barrier diode switch at a repetition rate of 250 kHz and with a pulse duration of 10 ns. A portion of the transmitted pulse is used to drive a downconverter in a homodyne receiver. The receiver only operates for the duration of the transmitted pulse, thus restricting the range to about 1.5 m ahead of the vehicle, and so reducing the probability of false alarms. If the measured relative velocity is greater than a preset value the sensor will trigger an airbag.

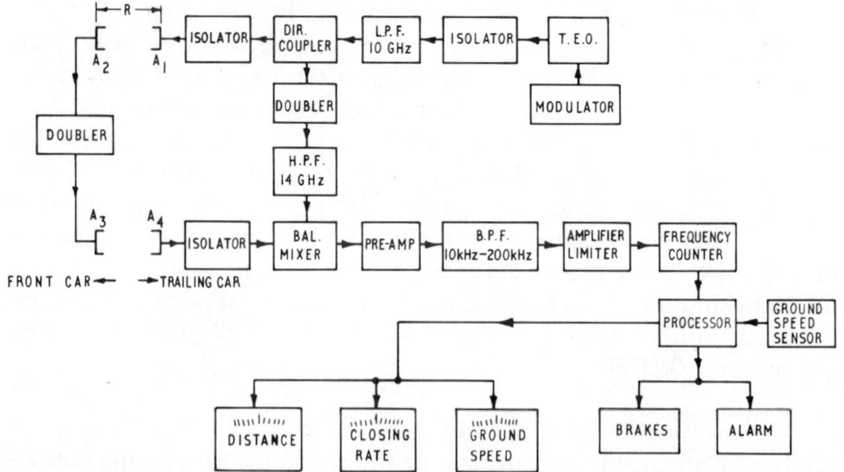

Figure 7.37 Block diagram of an experimental collision avoidance system which uses a harmonic reflector. (From Shefer, Klensch, Kaplan and Johnson,[239] reproduced by permission of Wireless World)

The Sperry BARDS system also uses separate transmitting and receiving antennas.[237] The transmitter utilizes a step recovery diode which generates unipolar 200 V peak triangular pulses of 200 ps duration. The pulse spectrum extends from 100 MHz to 20 GHz. The pulse length allows the use of three range gates out to a range of a few metres. Successive appearances of a target return in the three range gates with a closing velocity exceeding the threshold value triggers an airbag.

The RCA system achieves immunity from false returns by requiring the target vehicle to carry a rear-mounted harmonic reflector.[238,239] Figure 7.37 shows a block diagram of the system. The sensor transmits a linear FM signal from a modulated c.w. Gunn oscillator. The harmonic reflector on the target vehicle uses a silicon Schottky barrier diode to passively double the frequency of the incoming signal and it radiates back to the trailing vehicle on a separate antenna. The downconverter in the sensor's receiver is driven by a frequency-doubled sample of the transmitted power.

7.6.2 Railway applications

There is a requirement for the automation of the circulation of railway wagons, which involves the automatic reading of the number on each wagon. Becker[240] has described a microwave system in which each wagon is equipped with a transponder which acts as a passive frequency-dependent reflector. The output from a linearly-frequency-modulated (3.1 GHz to 4.2 GHz) transistor oscillator is used to feed a reader antenna placed between the tracks. The frequencies of the absorption dips in the received signal contain the required information in coded form.

Another application is the use of solid state Doppler radars for speedometers on railway trains. An accurate measure of distance may be obtained by integration.

7.6.3 Marine applications

In this section we describe a ship berthing system which has been developed by RRE.[241,242] It is a secondary radar system with Gunn diode transmitters.

Possible berthing systems could be based upon sonar, lasers or conventional radar. Sonar is susceptible to effects from turbulence, laser systems do not have an all-weather capability and conventional radar does not achieve the required accuracy of ±0.25 m. The secondary radar system achieved an accuracy of ±0.15 m.

Two interrogating radars with fixed omnidirectional antennas are mounted on the ship and two or more transponders are mounted on the shore. The ship radars measure the distances to the shore-based transponders. A computer then solves the complete geometry of the

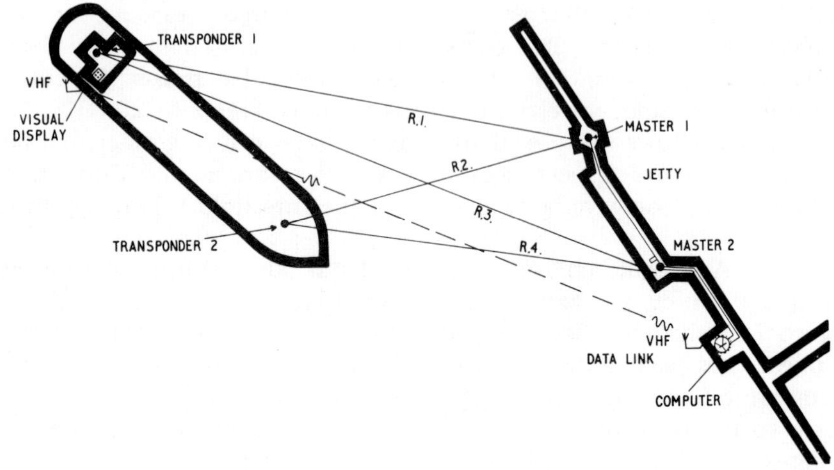

Figure 7.38 A radar transponder berthing system. (Copyright © Controller HMSO, London, 1975)

ship's position and aspect, and outputs the information to a display; from successive measurements of position the ship's speed is deduced. An alternative system places the radar and computer on shore with the transponders on the ship as shown in Figure 7.38. A radio link is required to drive the display on the ship.

The radars and the transponders used 20 ns pulsed Gunn diodes on frequencies in the band 13.5 GHz to 14.3 GHz with a peak power of 5 W. The transponders transmit on a different frequency to that of the primary transmission. The system has received extensive trials with supertankers at Milford Haven and gave coverage out to about 1 km.

Other possible marine applications of solid state microwave sources are in navigation beacons. A marine radar beacon transmits a coded signal that enhances a radar echo. A solid state radar beacon is available[243] that has an output power of 100 mW over a 20 μs pulse. The output frequency is 9.3 GHz to 9.5 GHz, and the effective range is up to eight nautical miles.

7.7 PHASED ARRAYS

With some types of microwave system it is an advantage to have some form of inertialess electronic scanning, rather than mechanical scanning with a rotating or mechanically-manoeuvrable antenna. At the present time, electronically-scanned antennas are expensive and their use is only warranted where electronic scanning offers a real advantage over mechanical scanning, rather than just an alternative.

There are a number of methods for electronic scanning which include swept frequency scanning, beam switching and beam steering using arrays of time delay elements or phase shifters. A major application of electronic scanning is in radar systems which can exploit to advantage multi-beam operation, interlaced scanning, simultaneous search and track, the steering of reception nulls, an adaptive polar diagram capability, or attitude correction of the platform on which the radar is mounted. An application in communication systems would be for the correction in pointing direction required to keep an antenna in a communications satellite positioned on a chosen ground station.

In this section we devote our attention to the phased array, paying especial attention to the use of solid state devices in active array antennas for radar systems. With a phased array antenna, beam steering is achieved by controlling the individual phases of radiating elements in the plane of the aperture of the antenna. Arrays may be linear in which the elements are arranged in a one-dimensional pattern, or planar with a two-dimensional pattern.

The essential features of a phased array are shown in Figure 7.39. On transmission, the power splitter distributes the power fed into the antenna into a number of channels. A phase shifter in each channel controls the phase of the radiating element at the channel output. On reception, the incoming signal to each element is phase shifted, and the individual signals are combined by the power splitter working in reverse as a power combiner.

In a passive array, active elements are absent; the phase shifters have to handle the full transmitted channel power, and the loss in the phase shifters attenuates the received signal prior to it reaching the receiver front end. In an active array, the transmitted signal is amplified by power sources that are placed between the phase shifters and the radiating elements. On reception, possible configurations include the use of a low noise amplifier prior to the signal passing through the phase shifter. Alternatively, the incoming signal is downconverted in

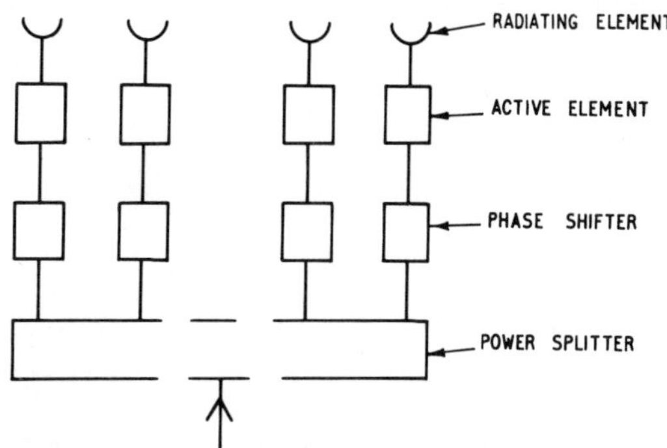

Figure 7.39 Schematic layout of an active phased array antenna

each element; the local oscillator signal can either carry the appropriately shifted phase, or the phase shifting can be performed in subsequent lower frequency stages.

An active array is attractive for solid state devices because it naturally combines a number of devices, and combinations of devices are required for other than low mean power radars. The output from devices combined in an active array is radiated, thus eliminating the need to channel the output powers from the individual devices into a single-port output. However, although attractive for solid state devices, this does not constitute a case by itself for an active array.

The case for an active array is based upon increased reliability and compactness compared with a passive array. Failure of the microwave power source which drives a passive array results in total system failure whereas an active array would still have usable performance in the event of failure of a number of its active elements. There is transmission loss in the power splitter and phase shifters and this means that a passive array requires more power at the input to the array than the active array has to provide at the output. An active array can make use of phase shifters with a lower power rating and can tolerate phase shifters with a higher transmission loss. Finally, there is the ultimate goal of a single compact module which integrates the solid state elements in the phase shifters, the transmitter power amplifier and the receiver.

7.7.1 Antenna gain and beamwidth

The array element spacing is determined by the requirement that grating lobes do not appear in visible space. For a maximum scan angle θ_{max} (see Figure 7.40) off boresight, for a linear array, we have the condition[244]

$$\frac{D}{\lambda} \leqslant (1 + |\sin \theta_{max}|)^{-1} \tag{7.34}$$

where D is the element spacing.

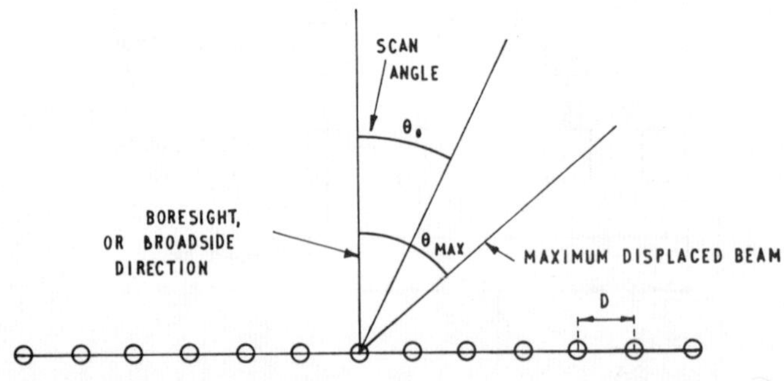

Figure 7.40 Parameters concerned with an electronically-scanned array antenna

This condition is always satisfied for $D = \lambda/2$. Spacings greater than half a wavelength are acceptable depending upon θ_{max}. A similar expression holds for planar arrays where D is now the spacing along a principal axis in the plane of the array, and θ_{max} is the maximum scan angle in the plane defined by the boresight and the chosen principal axis.[245] The gain of an antenna is

$$G_0 = \frac{4\pi A}{\lambda^2} \tag{7.35}$$

where A is the area and λ is the wavelength. For a planar array with $\lambda/2$ spacing and N elements, the broadside gain is given by

$$G_0 = \pi N \tag{7.36}$$

This means that a specification of the antenna gain and radiated power determines the number of elements and the power from each element. For example, a 1000-element 1 kW mean power airborne radar antenna would have a gain of 35 dB and would require a mean power of 1 W per element.

For a linear array with a vertical aperture dimension of M half wavelengths, and N elements spaced by $\lambda/2$, the broadside gain is

$$G_0 = \pi NM \tag{7.37}$$

When scanning to an angle θ_0 the gain is reduced such that $G(\theta_0) = G_0 \cos\theta_0$. The 3 dB beamwidth of a linear array is given by

$$\theta_B = \frac{K\lambda}{ND} \tag{7.38}$$

where K is the beamwidth coefficient.[244] θ_B is the azimuthal beamwidth for a horizontally mounted array. K is around 50° at broadside for a uniformly illuminated array. Watkins and Crossley[188] have described a 25-element linear array with a spacing of $5\lambda/8$ giving a beamwidth of 3° at a frequency of 9.3 GHz. This array is intended for use in a ground surveillance radar with a peak power of 2 kW and a pulse length of 250 ns. A peak power of 80 W per element would be required for an active array (neglecting amplitude tapering).

Amplitude tapering of the illumination in the aperture is required to reduce the sidelobe level. This is more important for reception than transmission in a jamming environment. An alternative on transmission with active elements of the same nominal output power is to thin the element density towards the array edges.

The far out sidelobe level may be set by random errors. For a planar array with half-wavelength element spacing, the variance of the amplitude variation σ_a^2 and the variance of the phase variation σ_p^2 must satisfy the condition derived by Allen[244,246]

$$\sigma_p{}^2 + \sigma_a{}^2 \leq \frac{1}{10\pi} G_0 r \tag{7.39}$$

where r is the required ratio of average sidelobe level to mainbeam gain. For the case $G_0 r \sim 1$, $\sigma_a = 0.18$, corresponding to a power variation of ± 1.4 dB, assuming no phase errors.

7.7.2 Phase control

The change in the phase increment between adjacent elements in order to move the beam by one beamwidth is approximately $2\pi/N$ radians where N is the number of elements for a linear array. This phase change is the same for all phase increments and is independent of the beam pointing direction.

Phase shifters are usually controlled in a digital fashion; an n-bit phase shifter will allow phase increments of $2\pi/(2^n)$. The maximum quantization phase error is therefore $\pm\pi/(2^n)$ and hence the r.m.s. phase error is given by

$$\alpha = \frac{\pi}{\sqrt{3}\, 2^n} \tag{7.40}$$

Quantization phase errors reduce the gain of the antenna and increase the sidelobe level. The ratio r of r.m.s. sidelobe level to the main beam gain is given by[247,248]

$$r \triangleq \frac{\alpha^2}{0.63\, N} \triangleq \frac{5}{2^{2n} N} \tag{7.41}$$

where N is the total number of elements. An r.m.s. sidelobe level of 31 dB down on the mainbeam with a 25-element linear array will be achieved with a 4-bit phase shifter.

Phase shifters are either based on ferrite or p–i–n diodes. Each type has a wide variety of different configurations. The reader is referred to the reviews by Ince and Temme,[244] Stark, Burns and Clark,[249] and Ince[250] for details. We note that power handling is up to 100 kW for ferrite phasers and 50 kW for p–i–n diode phasers. Insertion losses may be as low as 0.5 dB for ferrite phasers and 0.75 dB for p–i–n diode phasers.

A number of alternative schemes for producing the required phase variation across an array have been suggested. They include the harmonic locking of active elements,[251] interpolation locking of arrays of oscillators,[252] phase interpolation using frequency multiplication[253] and the switching of the phase of the output from impatt diode reflection amplifiers by control of the bias current.[254]

7.7.3 Phase stability of active elements

We have seen that in order not to degrade the antenna sidelobe level there is a maximum tolerance on the phase error in each channel, and a specification on the number of phase bits for the phase shifter. Typical specifications of r.m.s. phase errors are in the range ±5° to ±20°. Active elements must ideally be phase stable with temperature, or at least track in phase with temperature, and they must track in phase with frequency over the system bandwidth. It may be unnecessary to require that each element have identical insertion phase if the same requirement has not been placed on the phase shifters and if some form of phase trimming is available in each channel.

Yarrington and Hawkins have measured the phase change with temperature of gallium arsenide impatt amplifiers.[255] They found that at a given frequency the phase varied linearly with temperature over the range 11°C to 70°C. The rate of change of phase with temperature varied from −5.06 deg/°C at 9.0 GHz to −4.27 deg/°C at 9.6 GHz. A 1—2 GHz transistor amplifier is reported to have a phase variation with temperature of less than 0.35 deg/°C and a phase sensitivity to bias changes of less than 4.8 deg/V.[256] The bias voltage was 28 V to 30 V and the output peak power was 130 W with up to 10% duty cycle. Chang and coworkers[201] have shown that two trapatt amplifiers with a 12% bandwidth at just below 3 GHz will track in gain over the band to within 1 dB, and in phase to within 12°.

For pulsed oscillators and amplifiers, there are the additional problems of phase changes produced by in-pulse heating and bias variations during the pulse. Consider the case of a locked oscillator whose free-running frequency f_0 is a function of its bias voltage, V_0. The phase difference, between a locking signal at frequency f_L and the output of the locked oscillator, is a function of the frequency difference $(f_0 - f_L)$. For a constant locking frequency, the phase difference is a function of f_0. Hence the phase change introduced by a change in bias voltage ΔV_0 is given by

$$\Delta\phi = \left(\frac{d\phi}{df_0}\right)\left(\frac{df_0}{dV_0}\right)\Delta V_0 = -\left(\frac{d\phi}{d(f_L - f_0)}\right)\left(\frac{df_0}{dV_0}\right)\Delta V_0 \qquad (7.42)$$

A pulsed LSA oscillator at 9.25 GHz when biased at 624 V gave an unlocked power of 148 W, and 100 W when locked with a 10 W injected signal, and had $(df_0/dV_0) = 6$ MHz/V and $d\phi/d(f_L - f_0) = 1°$/MHz.[257] In order to achieve a phase stability of ±6°, the bias needs to be stable to ∓1 V or ∓0.16%.

7.7.4 Active modules

Active solid state modules which incorporate phase shifters, transmitter power stages, duplexer and receiver stages are required to meet a number of stringent requirements. They are to supply the required transmitter power with high efficiency in a small volume with low weight, achieving the required phase and amplitude stability and tracking requirements, with high reliability at low cost. These problems are more severe with planar arrays than with linear arrays.

Active modules have been considered for use at a number of frequencies, but principally at around 3 GHz and around 10 GHz. The alternatives that have been investigated for 3 GHz ground-based phased array radars are bipolar transistors and trapatt diodes. Belohoubek and coworkers[258] have investigated a power module which combines a number of 1.5 GHz bipolar transistors followed by a 2x multiplier to give 15 W c.w. output at 3 GHz. Tsuda and Stitt[191] have described an active module at 3.3 GHz with a 20 W peak output and which contains a 5-bit phase shifter, a 3-stage bipolar transistor amplifier, a *p–i–n* diode duplexer and a low noise amplifier. Chang and coworkers[201] advocate the use of trapatt amplifiers. Trapatt amplifiers may be operated under 'class-C' conditions in a manner similar to transistor amplifiers.[259] In this mode of operation, the trapatt diode is biased below the breakdown voltage and draws negligible current from the supply in the absence of an input signal. In the presence of the signal, the diode voltage drops, a large current is taken from the supply, and an amplified microwave output signal is produced. On cessation of the input signal the amplifier returns to its quiescent state.

The Texas Instruments MERA system[260] entailed a planar array of 604 modules for airborne use at 9 GHz, providing terrain following, terrain avoidance, ground mapping and air-to-ground ranging. Each module contained 4-bit phase shifters, a pulsed transistor power amplifier at 2.25 GHz followed by a 4x multiplier, a mixer and a 4x multiplier to provide a local oscillator signal. The output was 0.58 W peak with a 19.5 μs pulse and 6 MHz chirp.

The subsequent RASSR system[261] is a solid state 1648 element active array intended for an airborne multifunction radar which includes mapping, weather avoidance and beacon roles. Each module contains two active array elements; the block diagram of a single element is shown in Figure 7.41. Phase shifting is performed at 2.3 GHz to 2.375 GHz followed by a transistor power amplifier and a 4x multiplier. The peak output power is 1 W with a maximum duty cycle of 5% and a maximum pulse compression ratio of 100. The amplitude distribution is uniform on transmit and tapered on receive. The overall r.m.s. phase deviation is 10° and the r.m.s. power deviation is 0.8 dB.

7.8 MICROWAVE MEASUREMENTS AND WIDEBAND COMPONENTS

The theme of this section is the use of solid state microwave devices in equipment for spectral measurements. Our considerations include

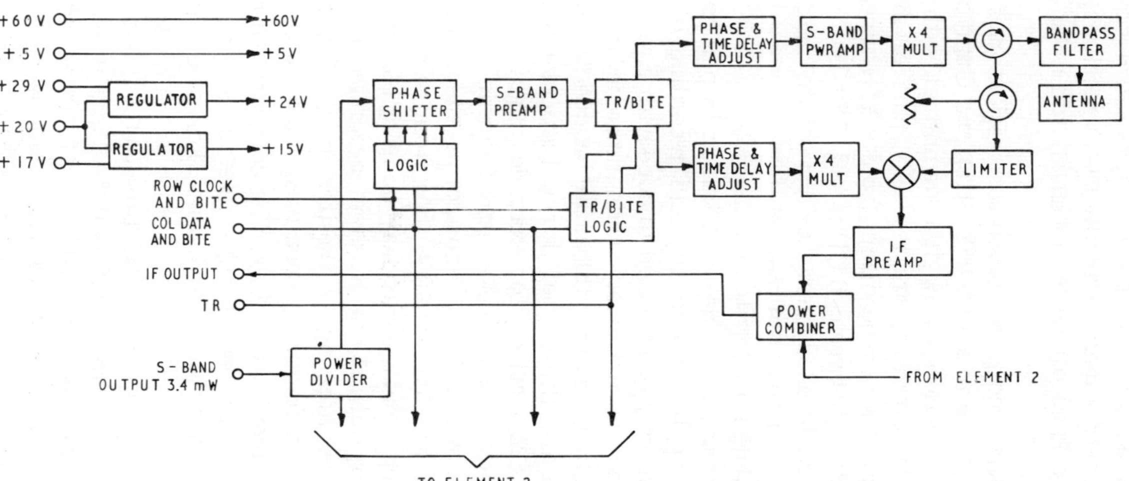

Figure 7.41 The block diagram of a single element of the RASSR solid state array. (From Collins and Harwell,[261] reproduced by permission of Microwaves (August 1972)

laboratory sweepers, receiving systems, microwave spectroscopy and electron spin resonance spectroscopy.

A basic component of a number of these types of measurement is an electronically-tunable oscillator. The use of this type of oscillator has been discussed in connection with FM c.w. radars. They are also used with laboratory sweepers, superheterodyne swept receivers, frequency--agile radars, tracking local oscillators in magnetron radars and as FM oscillators.

Laboratory sweepers incorporate transistor oscillators at the lower microwave frequencies (up to about 4—8 GHz), and Gunn oscillators at higher frequencies.[262] Octave bandwidths are normal, with YIG-sphere tuning elements. Output powers are typically tens of milliwatts. A power of about 10 mW is now available from a YIG tuned transistor oscillator which tunes between 6 GHz and 12 GHz.[263] Sweeper outputs are normally levelled using *p—i—n* diode modulators. At millimetre wavelengths, impatt sources are now being used in sweeper equipment.[264] Levelling is by ferrite modulator.

Sophisticated receiver systems are called upon to perform a number of tasks which involve the monitoring and recording of microwave emissions. The system may be expected to cope with c.w., AM, FM and pulsed emissions over an extremely wide range in frequency. The system may be digitally tuned over the entire bandwidth, or swept over sections of the bandwidth, or set for a particular frequency. A requirement of an ECM receiver, for example, might be the measurement of the frequency of a pulsed emission during a single received pulse. This situation will occur in monitoring the emission from a frequency-agile radar which frequency hops from pulse to pulse. The local oscillator of such a radar either may be using electronic selection from a number of crystal references, or incorporates a phase locked loop with a frequency hopping synthesizer input. Techniques using SAW devices can give a frequency measurement on a single received pulse with a high intercept probability. The conventional receiver, for this and other tasks, uses swept superheterodyne reception. Such systems require wideband local oscillators. For example, the band 0.5 GHz to 18 GHz has been covered with six YIG tuned oscillators.[265] These systems also exploit YIG preselector filters to eliminate spurious responses, wideband *p—i—n* switches, and wideband mixers with good intermodulation performance. The YIG filter and YIG tuned oscillator may well be packaged into single unit with good tracking between filter and oscillator over a wide temperature range.[266]

Voltage-tuned sources are also required for FM noise exciters in ECM noise jamming transmitters. Baseband inputs, with varactor tuned oscillators, are typically in the range d.c. to 50 MHz, and with deviations over 1 GHz up to 18 GHz.[267] Wide-bandwidth transferred-electron amplifiers are expected to establish themselves in ECM systems.

We finally note that impatt oscillators are finding a place as sources in microwave electron spin resonance spectrometers[268,269] and that their use is proposed in the mapping of terrestrial surface atmospheric pressure from satellites by measuring the two-path attenuation between the satellite and the surface due to oxygen absorption at 60 GHz.[270]

ACKNOWLEDGEMENTS

It is a pleasure to acknowledge many valuable discussions with colleagues at the Royal Radar Establishment.

REFERENCES

1. M. Gilden and M. Moroney, 'High power pulsed avalanche diode oscillators for microwave frequencies', *Proc IEEE*, 55, 1227—1228 (1967).
2. S. G. Liu, 'High-power punch-through avalanche-diode microwave oscillators', *IEEE J. Solid State Ccts.*, SC-3, 213—217.
3. G. Pfund, A. Podell and U. Tarakci, 'Pulsed silicon double-drift IMPATTs for microwave and millimetre wave applications', *Electronics Letters*, 518—520 (1973).
4. C. B. Swan, 'The importance of providing a good heat sink for avalanching transit-time oscillator diodes', *Proc IEEE*, 55, 451—452 (1967).
5. T. E. Seidel, R. E. Davis and D. E. Iglesias, 'Double-drift-region ion-implanted millimetre wave IMPATT diodes', *Proc IEEE*, 59, 1222—1227 (1971).
6. R. Edwards, D. Ciccolella, T. Misawa and D. E. Iglesias, 'Millimetre wave silicon IMPATT diodes', *Int. Electron Devices Mtg, Washington, USA, October 1969.*
7. W. C. Niehaus, 'Double-drift IMPATT diodes near 100 GHz', *IEEE Trans. Electron Devices*, 20, 765—771 (1973).
8. W. G. Matthei, 'State of the art of GaAs IMPATT diodes', *The Microwave J.*, 16, 29—36 (1973).
9. T. Watanabe, H. Kodera and M. Migitaka, 'GaAs 50 GHz Schottky barrier IMPATT diodes', *Electronics Letters*, 10, 7—8 (1974).
10. R. E. Goldwasser and F. E. Rosztoczy, *Appl. Phys. Lett.*, 25, 92—94 (1974).
11. F. Hasegawa and Y. Aono, 'Efficiency versus doping in GaAs Read type Impatt diodes', *Proc IEEE*, 62, 641—643 (1974).
12. M. Omori, F. Rosztoczy and R. Hayashi, 'High power GaAs double-drift IMPATT devices', *1973 Conference on Electron Device Techniques, New York, USA*, pp. 17—20, 1—2 May 1973.

13. M. I. Grace, H. Kroger and J. Telio, 'Improved performance of X-Band TRAPATTS', *Proc IEEE*, **61**, 1443—1444 (1972).
14. N. W. Cox, K. E. Gsteiger, G. N. Hill and C. T. Rucker, 'X-band CW TRAPATT oscillators using ring diodes on diamond heat spreaders', *Electronics Letters*, **9**, 269—270 (1973).
15. J. Risko, J. Thomas, H. J. Prager and K. K. N. Chang, 'I band (8—10 GHz) TRAPATT diode sources', *Electronics Letters*, **9**, 572—573 (1973).
16. H. J. Prager, K. K. N. Chang and S. Weisbrod, 'High-power, high-efficiency silicon avalanche diodes at ultra high frequencies', *Proc. IEEE*, **55**, 586—587 (1967).
17. M. Yeou Ta, 'High power subharmonic oscillations in silicon avalanche diodes at Ku band', *Proc. IEEE*, **57**, 2054—2055 (1969).
18. W. J. Evans and R. L. Johnston, 'Improved performance of CW silicon TRAPATT oscillators', *Proc. IEEE*, 58, 845—846 (1970).
19. Ho-chung Huang and L. A. MacKenzie, 'A Gunn diode operated in the hybrid mode', *Proc. IEEE*, **56**, 1232—1233 (1968).
20. J. F. Reynolds, B. E. Berson and R. E. Enstrom, 'High-efficiency transferred electron oscillators', *Proc. IEEE*, **57**, 1692—1693 (1969).
21. H. C. Huang, R. E. Enstrom and S. Y. Narayan, 'High efficiency operation of transferred electron oscillators in the hybrid mode', *Electron. Letts.*, 8, 271—273 (1972).
22. S. Y. Narayan and J. P. Paczkowski, 'Integral heat sink transferred electron oscillators', *RCA Review*, **33**, 752—765 (1972).
23. T. G. Rutlan and R. E. Brown, 'High frequency Gunn oscillators', *Int. Electron Devices Mtg., Washington, USA, December 1972.*
24. J. McGeeham and F. A. Myers, 'High power Q-band Gunn oscillators', *IEE specialist seminar on microwave semiconductors, Aviemore, Scotland, October 1972.*
25. F. E. Rosztoczy, R. E. Goldwasser and T. G. Ruttan, 'GaAs Gunn diodes', *The Microwave J.*, **16**, 51—54 (1973).
26. S. Christensson, D. W. Woodward and L. F. Eastman, 'High peak-power LSA operation from epitaxial GaAs', *IEEE Trans. Electron Devices*, **ED-17**, 732—738 (1970).
27. B. Jeppson and P. Jeppesen, 'A high power LSA relaxation oscillator', *Proc. IEEE*, **57**, 1218—1219 (1969).
28. W. O. Camp, L. F. Eastman, J. S. Bravman and D. W. Woodward, 'You can rely on LSA diodes', *MicroWaves*, **11**, 44—45 (1972).
29. D. J. Colliver, 'Progress with InP transferred electron devices', *Proceedings Fourth Biennial Cornell Electrical Engineering Conference*, pp. 11—19, August 1973.
30. D. J. Colliver, L. D. Irving, J. E. Pattison and H. D. Rees, 'High

efficiency InP transferred electron oscillators', *Electronics Letters*, **10**, 221—222 (1974).

31. D. M. Brookbanks and P. M. White, 'Design and performance of indium phosphide millimetre wave transferred electron oscillators', *Conference Proceedings, 4th European Microwave Conference, Montreux* pp. 227—231, September 1974.
32. L. S. Napoli, R. E. DeBrecht, J. J. Hughes, W. F. Reichert, A. Dreeben and A. Triano, 'High power GaAs FET amplifier — a multigate structure'. *IEEE Int. Solid State Ccts. Conference, Digest of Technical Papers, Philadelphia, USA*, pp. 82—83, February 1973.
33. M. Fukuta, T. Mimura, I. Tujimura and A. Furumoto, 'Mesh source type microwave power FET', *IEEE Int. Solid State Ccts. Conference, Digest of Technical Papers, Philadelphia, USA*, pp. 84—85, February 1973.
34. W. E. Poole and D. Renkowitz, 'M/W power transistors and MIC amplifiers: state-of-the-art', *The Microwave J.*, **15**, 23—30 (1972).
35. H. Cooke, 'Microwave transistors, Part 1: Power devices', *Micro-Waves*, **8**, 46 — 52 (1969).
36. Microwave Semiconductor Corporation Advertisement, *Micro-Waves*, **12**, 15, (1973).
37. R. S. Harp and H. L. Stover, 'Power combining of X-band IMPATT circuit modules', *IEEE Int. Solid State Ccts. Conference, Digest of Technical Papers, Philadelphia, USA*, pp. 118—119, February 1973.
38. S. G. Liu, '2000-W-GHz Complementary TRAPATT diodes', *IEEE Int. Solid State Ccts. Conference, Digest of Technical Papers, Philadelphia, USA*, pp. 124—125, February 1973.
39. R. Stevens, D. Tarrant and F. A. Myers, '40 watt 16 GHz pulsed Gunn diode oscillator', *Conference Proceedings, 4th European Microwave Conference, Montreux*, pp. 257—261, September 1974.
40. R. M. Henry, 'Stacked varactors with four snap-off diodes in series for very high power multipliers (5 watts at 16 GHz)', *1973 European Microwave Conference, Proceedings*, vol. 1, p. A.4.5, Brussels, Belgium, September 1973.
41. P. Staeker, 'K_A-band IMPATT diode reliability', *1973 International Electron Devices Meeting Technical Digest, Washington, DC, USA*, 3—5 December 1973, pp. 493—496.
42. R. A. Murphy, W. T. Lindley, D. F. Peterson and P. Staecker, 'Performance and reliability of K_a-band GaAs IMPATT diodes', *Digest of Technical Papers, International Microwave Symposium, Atlanta, Georgia, USA*, pp. 315—317, June 1974.
43. J. R. Black, 'Electromigration — a brief survey and some recent results', *IEEE Trans. Electron Devices*, **ED—16**, 338—347 (1969).

44. M. Flashie, 'Reliability and MTF — the long and short of it', *MicroWaves*, **11**, 36 — 44, (1972).
45. I. Bazovsky, *Reliability theory and practice*, Prentice—Hall, Englewood Cliffs, New Jersey, 1961.
46. *Reliability Handbook*, (Ed. W. G. Ireson), McGraw—Hill, New York, 1966.
47. A. E. Green and A. J. Bourne, *Reliability technology*, Wiley—Interscience, London, 1972.
48. Ministry of Defence, *Defence Specification DEF133*, 'Climatic, shock and vibration testing of service equipment', London, Her Majesty's Stationery Office 1969 (reprinted August 1971).
49. US Department of Defense Military specifications, 'Electronic Equipment, Ground, General requirements for', MIL-E-4158E; 'Electronic Equipment, Airborne, General specification for', MIL-E-5400P; 'Electronic Equipment, Missiles, boosters and allied vehicles', MIL-E-8189G; 'Electronic Equipment, airborne extended space environment', MIL-E-8983B; 'Electronic Equipment, Naval, ship and shore, general specification (Navy)', MIL-E-16400F.
50. British Standard 9300:1969, 'Specification for semiconductor devices of assessed quality: generic data and methods of test', British Standards Institution.
51. H. C. Okean and P. P. Lombardo, 'Noise performance of m/w and mm-wave receivers', *Microwave J.*, **16**, 41—46, 48, 50 (1973).
52. R. T. Davis, 'FET's edging out bipolar transistors above C-band', *MicroWaves*, **12**, 48—51 (1973).
53. S. Kakihana, 'Current status and trends in high frequency transistors', *Conference Proceedings, Microwave 73, Brighton*, pp. 258—267, June 1973.
54. *Microwave Journal Engineers Handbook*, p. 64, January 1974.
55. P. A. Levine and V. W. Chan, 'A comparative study of IMPATT noise properties', *Proc. IEEE*, **60**, 745—746 (1972).
56. H. J. Kuno, J. R. Collard and A. Gobat, 'Microwave amplification with GaAs avalanche diodes', *Electron Lett.*, 4, 540—542 (1968).
57. A. Rabier and R. Spitalnik, 'Diode characterization and circuit optimization for transferred electron amplifiers', *Proceedings European Microwave Conference, Brussels*, Paper A.6.1, September 1973.
58. P. W. Braddock and K. W. Gray, 'Low-noise wideband indium-phosphide transferred-electron amplifiers', *Electron Lett.*, 9, 36—37 (1973).
59. S. Baskaran and P. N. Robson, 'Noise performance of InP reflection amplifiers in Q-band', *Electron Lett.*, 8, 137—138 (1972).

60. E. F. Scherer, 'A multistage high-power avalanche amplifier at X-band', *IEEE J. Solid-State Circuits*, SC-4, 396—399 (1969).
61. C. N. Dunn, B. L. Morris, C. L. Paulnack, T. E. Seidel and L. J. Smith, 'Improved fabrication and noise performance of silicon double-drift millimeter-wave IMPATT diodes', *IEEE 1973 International Electron Devices Meeting Technical Digest*, Washington, pp. 486—488, December 1973.
62. D. Neuf, D. Brown and B. Jaracz, 'Multi-octave double-balanced mixer', *Microwave J.*, 16, 13—14 (1973).
63. A. H. Solomon, 'Microwave components for civil satellite communication systems', *Conference Proceedings, Microwave '73, Brighton*, pp. 12—21, June 1973.
64. Y. Anand and W. J. Moroney, 'Microwave mixer and detector diodes', *Proc. IEEE*, 59, 1182—1190 (1971).
65. H. E. Elder and V. J. Glinski, 'Detector and mixer diodes and circuits', Chapter 12 in *Microwave semiconductor devices and their circuit applications* (edited by H. A. Watson), McGraw—Hill, 1969.
66. T. H. Oxley and G. H. Swallow, 'Planar germanium backward diodes as broadband detectors', *Proceedings European Microwave Conference, IEEE Conference Publication Number 58*, pp. 468—471, September 1969.
67. S. March, 'Microwave micromin bibliography', *MicroWaves*, 8, 59—64 (1969).
68. S. March, 'Microwave micromin bibliography', *MicroWaves*, 9, 53—56, 122, 124 (1970).
69. T. H. Oxley, 'The application of integrated circuits to microwave receivers', *Proceedings 1973 European Microwave Conference*, vol. 1, Paper A.14.1, September 1973.
70. P. Chavas, 'An integrated 6 GHz receiver front end for high--capacity radio-relay systems', *IEEE Int. Solid State Ccts. Conference, Digest of Technical Papers*, pp. 90—91, February 1971.
71. D. A. Gray, 'High performance thin-film CARS band receiver', *Microwave '73 Conference Proceedings, Brighton*, pp. 278—282, June 1973.
72. B. Glance and W. W. Snell, 'Low-noise integrated millimeter-wave receiver', *Proceedings 1973 European Microwave Conference*, vol. 1, Paper A.15.1, September 1973.
73. R. P. G. Allen and G. R. Antell, 'Monolithic mixers for 60—80 GHz', *Proceedings 1973 European Microwave Conference*, vol. 1, Paper A.15.3, September 1973.
74. L. S. Cutler and C. S. Searle, 'Some aspects of the theory and measurement of frequency fluctuations in frequency standards', *Proc. IEEE*, 54, 136—154 (1966).

75. P. A. Levine, H. C. Huang and H. Johnson, 'IMPATTs shoot for Gunn noise levels', *MicroWaves*, **11**, 54—56 (1972).
76. *Hewlett—Packard Application Note 935*.
77. J. R. Ashley and F. M. Palka, 'Measured FM noise reduction by injection phase locking', *Proc. IEEE*, **58**, 155—157 (1970).
78. E. F. Scherer, 'Investigation of the noise spectra of avalanche oscillators', *IEEE Trans. Microwave Theory Tech.*, **MTT-16**, 781—788 (1968).
79. H-J. Thaler, G. Ulrich and G. Weidmann, 'Noise in IMPATT diode amplifiers and oscillators', *IEEE Trans Microwave Theory Tech.*, **MTT-19**, 692—705 (1971).
80. M. Ohtomo, 'Experimental evaluation of noise parameters in Gunn and avalanche oscillators', *IEEE Trans. Microwave Theory Tech.*, **MTT-20**, 425—437 (1972).
81. P. A. Levine and V. W. Chan, 'A comparative study of IMPATT noise properties', *Proc. IEEE*, **60**, 745—746 (1972).
82. W. J. Evans, 'Noise spectra of a CW silicon TRAPATT oscillator', *Proc. IEEE*, **60**, 125—126 (1972).
83. J. R. Ashley and F. M. Palka, 'Transmission cavity and injection stabilization of an X-band transferred electron oscillator', *Digest of Technical Papers, 1973 IEEE—G—MTT Microwave Symposium*, pp. 181—182, June 1973.
84. M. Omori, 'Gunn diodes and sources', *The Microwave J.*, **17**, 57—62 (1974).
85. J. Helmcke, M. Herbst, M. Claasen and W. Harth, 'FM noise and bias fluctuations of a silicon Pd—n—p^+ microwave oscillator', *Electronics Letters*, **8**, 158—159 (1972).
86. M. E. Hines, J-C. R. Collinet and J. G. Ondria, 'FM noise suppression of an injection phase locked oscillator', *IEEE Trans. Microwave Theory Tech.*, **MTT-16**, 738—742 (1968).
87. H. Mager, S. L. Johnson and D. A. Calder, 'Noise spectrum properties of low noise microwave tube and solid-state signal sources', *IEEE—NASA symposium on short-term frequency stability*, November 1964.
88. F. Diamond, 'Low noise silicon IMPATT structure', *Electronics Letters*, **9**, 405—406 (1973).
89. G. A. Swartz, Y. S. Chiang, C. P. Wen and A. Young, 'FM noise measurements on p-type and n-type silicon IMPATT oscillators', *Electronics Letters*, **9**, 578—580 (1973).
90. C. T. Rucker, 'A multiple-diode high-average-power avalanche-diode oscillator', *IEEE Trans. Microwave Theory. Tech.*, **MTT-17**, 1156—1158 (1969).
91. C. H. Chien and G. C. Dalman, 'Subharmonically injected phase-locked IMPATT-oscillator experiments', *Electronics Letters*, **6**, 240—241 (1970).

92. S. Rupp, 'Phase-locked signal-source design', *The Microwave J.*, 17, 45, 46, 48 (1974).

93. M. F. Lewis, 'The design, performance and limitations of SAW oscillators', *IEE International Specialist Seminar on Component performance and system application of surface acoustic wave devices, Aviemore*, September 1973.

94. M. F. Lewis, 'Surface acoustic wave devices and applications, 6-Oscillators — the next successful surface acoustic wave device', *Ultrasonics*, 1115—1123 (1974).

95. K. Akada, K. Nishio, S. Yamamoto, Y. Sawayama and K. Hirai, 'A new stabilized Gunn oscillator of double resonance coupled type', *Proceedings Fourth Biennial Cornell Electrical Engineering Conference*, pp. 185—194, August 1973.

96. K. Wilson, A. J. Tebby and D. W. Langdon, 'A novel high stability high power IMPATT oscillator', *1971 European Microwave Conference*, vol. 1, p. A6/2:1, The Royal Swedish Academy of Engineering Sciences (IVA) Stockholm, Sweden.

97. G. Hodowanec, 'Microwave transistor oscillators', *Microwave J.*, 17, 39—42 (1974).

98. J. Miller, 'An upper sideband upconverter using MIS-varactors', *IEEE Int. Solid-State Ccts. Conference, Digest of Technical Papers, Philadelphia, USA*, pp. 28—29, February 1971.

99. M. Uenohara and J. W. Gewartowski, 'Varactor applications', Chapter 8 of *Microwave Semiconductor Devices and their circuit applications* (edited by H. A. Watson), McGraw—Hill, 1969.

100. D. R. Hill and R. R. Thomas, 'High power X-band parametric upconverter, STL, Harlow, Unpublished work for MOD(PE).

101. M. J. Ahmed, 'Avalanche-diode amplifier upconverter', *Electronics Letters*, 9, 490—491 (1973).

102. S. Kita, Y. Fukatsu and N. Kanmuri, 'Millimeter-wave solid-state circuits of trial 4-phase PSK millimeter-wave repeaters', *Rev. Electrical Comm. Labs.*, 21, 77—86 (1973).

103. M. B. Read, 'The realisation and performance of terminal equipment for circular waveguide TE_{01} mode trunk communication systems', *Proceedings 1973 European Microwave Conference, Brussels*, Paper B.13.3, September 1973.

104. H. L. Stover and H. M. Leedy, 'Solid-state devices and components for a 60 GHz communication system', *Conference Record, National Telecommunications Conference, Atlanta, Georgia*, pp. 23A-1—23A-7, November 1973.

105. J. D. Pearson and K. S. Lunt, 'A broad-band balanced idler circuit for parametric amplifiers', *Radio Electron Engr*, 27, 333—351 (1964).

106. J. F. Gittins, J. C. Vokes and C. S. Whitehead, 'A broad-band parametric amplifier', *Int. J. Electron*, 24, 333—351 (1968).

107. P. Bura, R. Camison, W. Y. Pan, S. Yuan and A. Block, 'Design considerations for an integrated 1.8 GHz parametric amplifier', *IEEE Trans. Microwave Theory Tech*, MTT-16, 424—428 (1968).
108. T. Kudo, M. Fukui and K. Kurachi, '4 GHz uncooled parametric amplifier', *FUJITSU Scientific Technical J.*, 9, 1—23 (1973).
109. D. M. Clunie, C. A. Tearle and L. D. Clough, 'Q band IMPATT oscillators for parametric amplifier pumps', *1972 European Solid State Device Research Conference, Invited papers and abstracts of contributed papers, Lancaster, England*, pp. 195—196, September 1972.
110. A. D'Ambrosio, 'Realization of a KU band uncooled parametric amplifier for spacecraft application', *Microwave '73 Conference Proceedings, Brighton*, pp. 29, 495—499, June 1973.
111. J. Whelehan, E. Kraemer and H. Paczkowski, 'Millimeter-wave paramp with a solid-state pump source', *Microwave J.*, 16, 35—38 (1973).
112. P. Penfield and R. P. Rafuse, *Varactor Applications*, MIT Press, Cambridge, Mass., 1962, p. 273.
113. P. Penfield and R. P. Rafuse, *Varactor Applications*, MIT Press, Cambridge, Mass., 1962, pp. 198—200.
114. J. Clarke, 'Effects of pump noise in parametric amplifiers', to be published.
115. M. Dean and J. Clarke, 'Pump noise coupling in parametric amplifiers', *Conference Proceedings 4th European Microwave Conference, Montreux*, September 1974, pp. 338—342.
116. R. J. Wagner, W. W. Gray and A. Luber, 'The use of IMPATT generated power for parametric amplifier pumps', *IEEE International Solid State Ccts. Conference, Digest of Technical Papers*, pp. 22—23, 194, February 1971.
117. G. H. Swallow, T. H. Oxley and A. M. Hansom, 'Recent advances in the burnout properties of low noise gallium arsenide Schottky barrier mixer diodes', *Proceedings 1973 European Microwave Conference, Brussels*, Paper A.7.6, September 1973.
118. Y. Anand and C. Howell, 'A burnout criterion for Schottky-barrier mixer diodes', *Proc. IEEE*, 56, 2098 (1968).
119. Y. Anand and C. Howell, 'The real culprit in diode failure', *MicroWaves*, 9, 36—38 (1970).
120. G. E. Morris and G. A. Hall, 'RF burnout of K_a-band mixer diodes', *IEEE Trans Microwave Theory Tech.*, MTT-22, 745—746 (1974).
121. *Diode and Transistor Designers Catalog*, Hewlett Packard Components, May 1974.
122. M. R. Barber, K. F. Sodomsky and A. Zacharias, 'Microwave switches, limiters and phase shifters', Chapter 10 in *Microwave*

Semiconductor Devices and their Circuit Applications (edited by H. A. Watson), McGraw—Hill, 1969.

123. I. H. Macdonald, S. A. J. Matykiewicz and D. G. Pulley, 'Pin limiter for microstrip circuits', *Microwave J.*, 16, 52, 54, 56 (1973).
124. 'Broadband limiter modules, ML5200 series', *Microwave Associates, Bulletin L/0113.*
125. D. L. LaCombe, 'Broadband MIC limiter-detectors . . . design through production', *Conference Proceedings, Microwave '73, Brighton*, June 1973, pp. 138—142.
126. D. C. Hogg, 'Millimeter-wave communication through the atmosphere', *Science*, 159, 39—45 (1968).
127. J. W. Mink, 'Rain-attenuation measurements of millimetre waves over short paths', *Electronics Letters*, 9, 198—199 (1973).
128. P. F. Panter, *Communication systems design — line-of-sight and tropo-scatter systems*, McGraw — Hill, New York, 1972.
129. H. Carl. *Radio relay systems*, MacDonald, London, 1966.
130. A. H. Solomon, 'Components for modern microwave communications systems — an overview', *IEEE 1973 Communications Conference* 11—13 June 1973, *Conference Record*, pp. 27-1—27-7.
131. B. Oguchi, 'Microwave and millimeter-wave communications in Japan', *IEEE Trans. Comms.*, **COM-20**, 717—725 (1972).
132. *CCIR Documents of the XIth Plenary Assembly*, Vol. IV, 'Radio-relay systems, space systems and radio astronomy', International Telecommunication Union Oslo, 1966.
133. B. Standal and B. Målsnes, 'Transistorized high capacity microwave radio link systems', *Microwave 73 Conference Proceedings*, June 1973, *Brighton*, Published by Microwave Exhibitions and Publishers Ltd., Surbiton, Surrey, England.
134. A. A. Sweet, 'Factors limiting the signal-to-noise ratio of negative conductance amplifiers and oscillators in FM/FDM communications systems', *IEEE Trans. Microwave Theory Tech.*, MTT-22, 146—149 (1974).
135. H-J. Thaler, 'Excess noise in stable and injection-locked IMPATT diode amplifiers at high-power levels', *1973 International Solid-State Circuits Conference, Digest of Technical Papers, Philadelphia*, February 1973, pp. 116—117, 206.
136. H. B. Wood, 'Injection-locked IMPATT oscillators applied to FDM microwave transmission', *Electronics Letters*, 9, 141—142 (1973).
137. R. A. Brockbank and C. A. A. Wass, 'Non-linear distortion in transmission systems', *J. IEE*, 92, 45—56 (1945).
138. S. Fedida and D. S. Palmer, 'Some design considerations for links

carrying multichannel telephony, Part II', *The Marconi Review*, 19, 1—46 (1956).

139. L. P. Yeh, 'Consideration of non-linear noise and its testing in frequency division multiplex voice UHF radio communication systems', *IRE Trans. Comm. Syst.*, CS-9, 115—129 (1961).
140. H. C. Bowers and W. H. Lockyear, 'IMPATT amplifiers for communication systems', *1973 Wescon Technical Papers, Western Electronic Show and Convention, San Francisco*, vol. 17, pp. 15/1—5, September 1973.
141. J. G. deKoning, R. E. Goldwasser, R. J. Hamilton and F. E. Rosztoczy, 'Gunn amplifiers come of age', *MicroWaves*, 13, 52, 54, 56—57 (1974).
142. R. J. Trew, N. A. Masnari and G. I. Haddad, 'Intermodulation characteristics of X-band IMPATT amplifiers', *Digest of Technical Papers, IEEE 1972 International Microwave Symposium*, pp. 182—184.
143. H. Komizo, Y. Daido, H. Ashida, Y. Ito and M. Honma, 'Improvement of nonlinear distortion in an IMPATT stable amplifier', *IEEE Trans Microwave Theory Tech*, MTT-21, 721—728 (1973).
144. P. D. Lubell, W. B. Denniston and R. F. Hertz, 'Linearizing amplifiers for multi-signal use', *MicroWaves*, 13, 46, 48, 50 (1974).
145. H. Komizo, Y. Ito, H. Ashida and M. Shinoda, 'A 0.5 W CW IMPATT diode amplifier for high capacity 11 GHz FM radio-relay equipment', *IEEE J. Sol. St. Circuits*, SC-8, 14—20 (1973).
146. D. C. Hanson and W. W. Heinz, 'Integrated electrically tuned X-band power amplifier utilizing Gunn and IMPATT diodes', *IEEE J. Solid State Circuits*, SC-8, 3—14 (1973).
147. G. Endersz, 'High-linearity FM Gunn oscillator', *Proceedings 1973 European Microwave Conference, Brussels*, Paper A.3.3, September 1973.
148. T. Inatomi, Y. Ito and K. Ikai, '11 GHz all solid state radio relay equipment using Gunn-diode oscillators and IMPATT-diode stable amplifier', *FUJITSU Scientific Tech. J.*, 25—53 (1973).
149. K. Tada, M. Oguchi and T. Kikuchi, '11 GHz fully solid-stated short-haul microwave system', *Japan Telecommunications Review*, 138—142 (April 1973).
150. E. M. Hicken, '12 GHz solid state links for television outside broadcast use', *IEE International Broadcasting Convention London*, pp. 141—146, September 1972.
151. G. A. R. Teesdale and E. M. Hicken, 'Microwave links for television outside broadcasts', *The Royal Television Society J.*, 14, 107—112 (1972).
152. C. L. Cuccia, 'Progress towards million channel satellite com-

munication systems', *Conference Proceedings, Microwave '73, Brighton*, pp. 3—11, June 1973.

153. G. Crippa, F. Riva and G. Savino, '13 GHz radio relay link for 35 M bit/s digital signals', *Conference Proceedings, Microwave '73, Brighton*, pp. 107—111, June 1973.
154. T. R. Rowbotham, 'Short-hop radio-relay system work at 20 GHz', *Conference Proceedings, Microwave '73, Brighton*, pp. 112—116, June 1973.
155. A. B. Carlson, *Communication Systems*', McGraw—Hill, New York, 1968.
156. R. W. Lucky, J. Salz and E. J. Weldon, *Principles of Data Communication*, McGraw—Hill, New York, 1968.
157. J. A. Betts, *Signal Processing, Modulation and Noise*, The English Universities Press Ltd, London 1970.
158. P. F. Panter, Modulation, Noise and Spectral Analysis, McGraw—Hill, New York, 1965.
159. D. J. Torrieri and J. N. O'Connor, *Naval Research Lab, Washington, Report 7609*, September 1973.
160. S. E. Miller, 'Millimeter waves in communications', *Proceedings Symposium on Millimeter Waves, Polytechnic Institute of Brooklyn*, March 31, April 1, 2, 1959, pp. 25—43.
161. M. I. Skolnik, 'Millimeter and submillimeter wave applications', *Proceedings Symposium on Submillimeter Waves, Polytechnic Institute of Brooklyn*, March 31, April 1, 2, 1970, pp. 9—25.
162. C. E. White, 'Microwave technology for the US Navy', *The Microwave J.*, 17, 14, 16, 18 (1974).
163. J. J. Taub, 'The future of millimeter waves', *The Microwave Journal*, 16, 6, 8 (1973).
164. P. A. Volckman, S. E. Gibbs and P. Denniss, '30 GHz high-speed portable digital data link', *International Communications Conference*, June 1974.
165. D. W. Morris, I. A. Ravenscroft and R. W. White, 'Millimetric waveguide system research in the British Post Office', *Proceedings 1973 European Microwave Conference, Brussels*, p. B.12.4.
166. W. D. Waters, 'A high-capacity millimeter waveguide transmission system — concepts and implementation plans', *Conference Record, National Telecommunications Conference, Atlanta, Georgia, USA*, November 1973, p. 35A-1.
167. Ph. Dupuis and M. Goloubkoff, 'Comparison of different communication systems', *Proceedings 1973 European Microwave Conference, Brussels*, p. B.12.1.
168. V. F. Suchkov, N. P. Kerzhentseva and V. A. Aron, 'Helix filters for waveguide communication systems', *Proceedings 1973 European Microwave Conference, Brussels*, p. B.11.4.
169. W. Lorek and G. Hanke, 'Waveguide transmission experiments at

bit rates of up to 640 Mbits/s', *Proceedings 1973 European Microwave Conference, Brussels*, p. B.13.5.

170. G. C. Corazza, 'Microwave measurements in a telecommunication system by circular waveguides', *Conference Proceedings, Microwave 73, Brighton*, pp. 213—217, June 1973.
171. S. E. Miller, 'Millimeter waves in communication', *Proceedings Symposium on Millimeter Waves, Polytechnic Institute of Brooklyn*, March 1959, pp. 25—43.
172. N. Sushi and M. Shimba, 'Transmission characteristics of buried millimetre waveguide lines', *Conference on Trunk Communication by Guided Waves, London, IEE Conference Publication Number 71*, pp. 27—32, 29 September—2nd October 1970.
173. W. K. Ritchie, 'Design and performance of a TE_{01}-mode helix waveguide', *Proceedings 1973 European Microwave Conference, Brussels*, p. B.13.1, September 1973.
174. H. J. Kuno and D. L. English, 'Nonlinear and large-signal characteristics of millimeter-wave IMPATT amplifiers', *IEEE Trans. Microwave Theory Tech.*, MTT-21, 703—706 (1973).
175. S. F. Paik, P. J. Tanzi and D. J. Kelley, 'IMPATT-diode power amplifiers for digital communication systems', *IEEE Trans. Microwave Theory Tech.*, MTT-21, 716—720 (1973).
176. H. J. Kuno, 'Analysis of nonlinear characteristics and transient response of IMPATT amplifiers', *IEEE Trans. Microwave Theory Tech.*, MTT-21, 694—702 (1973).
177. Y. Kita, Y. Kaneko, A. Fujioka and M. Migitaka, '800 Mbits/s 4-phase PSK all solid-state repeater for millimeter-wave radio relay system', *Proceedings 1973 European Microwave Conference, Brussels*, p. B.11.2.
178. I. W. Mackintosh, 'Circuit model for TRAPATT-oscillator cavities', *Proc. IEE*, 119, 1529—1537 (1972).
179. I. W. Mackintosh, 'A circuit mode chart for high-efficiency avalanche diode oscillators', *IEEE J. Solid State Circuits*, SC-8, 44—53 (1973).
180. H. Kawamoto, 'Gigahertz-rate 100 V pulse generator', *IEEE J. Solid-State Circuits*, SC-8, 63-66 (1973).
181. J. E. Carroll, 'Short pulse modulation of gallium-arsenide lasers with TRAPATT diodes', *Electronics Letters*, 9, 166—167 (1973).
182. H. W. Thim, L. R. Dawson, J. V. DiLorenzo, J. C. Dyment, C. J. Hwang and D. L. Rode, 'Subnanosecond PCM of GaAs lasers by Gunn-effect switches', *IEEE Solid-State Circuits Conference, Digest of Technical Papers*, pp. 92—93, February 1973.
183. M. E. Levinshtein, 'The TRAPATT mode in high-frequency modulation of infrared radiation', *Sov. Phys. Semicond.*, 7, 832 (1973).

184. J. R. Andrews, 'Inexpensive laser diode pulse generator for optical waveguide studies', *Rev. Sci. Instrum.*, 45, 22—24 (1974).
185. J. S. T. Charters, 'Electronics and security systems', *Electronics and Power*, 18, 266—268 (1972).
186. M. Cowley and S. Hamilton, 'Cut the costs of Doppler radars', *Electronic Design*, 48—53 (1971).
187. W. K. Saunders, 'CW and FM radar', Chapter 16, *Radar Handbook*, McGraw—Hill, New York, 1970.
188. C. D. Watkins and G. E. Crossley, 'A 9 GHz phase-scanned linear array antenna', *International Conference on Radar — present and future, London*, October 1973, *IEE Conference Publication Number 105*, pp. 81—87.
189. J. R. Collins, 'RASSR array comes of age', *MicroWaves*, 11, 36—42 (1972).
190. J. D. Adams, 'Capability of a projected 1975 airborne solid-state phased-array radar', *The Microwave J.*, 14, 23—26, 28, 30, 32, 34, 48 (1971).
191. G. I. Tsuda and J. J. Stitt, 'Are we really ready for S-band solid-state arrays', *MicroWaves*, 13, 46, 48, 50, 52—53 (1974).
192. J. E. Reed, 'The AN/FPS-85 Radar system', *Proc. IEEE*, 57, 324—335 (1969).
193. M. I. Skolnik, *Introduction to Radar Systems*, McGraw—Hill, 1962, p. 4.
194. J. Mun and J. S. Heeks, 'Study of LSA devices for X-band', STL, Harlow, unpublished work for MOD(PE).
195. R. E. Cooke, J. J. Crisp, R. F. B. Conlon and J. S. Heeks, 'Transferred electron oscillators for phased array radar', *International conference on radar — present and future, London, IEE Conference publication Number 105*, October 1973, pp. 118—123.
196. E. D. Bullimore, B. J. Downing and F. A. Myers, 'Frequency-stable pulsed Gunn oscillators', *Electron. Letters*, 10, 220—221 (1974).
197. W. E. Wilson, 'Pulsed LSA and TRAPATT sources for microwave systems', *The Microwave J.*, 14, 33—41 (1971).
198. Y. Amblard, 'Emetteur état solide pour radar Doppler, bande ku', *Electron Fisc., Apli*, 16, 241—247 (1973).
199. R. J. Royds, Private communication.
200. W. L. Wilson, 'Precise frequency and phase control of LSA oscillators', *IEEE Trans Microwave Theory Tech*, MTT-21, 146—149 (1973).
201. K. K. N. Chang, H. Kawamoto, H. J. Prager, J. F. Reynolds and A. Rosen, 'TRAPATT amplifiers for phased-array radar systems', *The Microwave J.*, 16, 27—32 (1973).

202. V. J. Higgins and J. J. Baranowski, 'The utility and performance of avalanche transit time diode and transferred electron oscillators in microwave systems', *The Microwave J.*, 13, 37—42 (1970).
203. R. E. Keller, 'LSA transmitters', *The Microwave J.*, 14, 50, 53—54 (1971).
204. M. I. Grace, 'Injection locking of pulsed avalanche diode oscillators', *Proc. IEEE*, 55, 713—714 (1967).
205. J. D. Maines and E. G. S. Paige, 'Surface-acoustic-wave components, devices and applications', *IEE Reviews*, 120, 1078—1110 (1973).
206. E.C. Farnett, T. B. Howard and G. H. Stevens, 'Pulse compression radar', Chapter 20 in *Radar Handbook*, McGraw—Hill, New York, 1970.
207. W. K. Saunders, 'CW and FM Radar', Chapter 16 in *Radar Handbook*, McGraw—Hill, New York, 1970.
208. D. H. Mooney and W. A. Skillman, 'Pulse-doppler radar', Chapter 19 in *Radar Handbook*, McGraw—Hill, New York, 1970.
209. P. J. Bulman, G. S. Hobson and B. C. Taylor, *Transferred electron devices*, Academic Press, London, 1972.
210. M. A. C. S. Brown, W. J. Hannis, J. M. Skinner and D. K. Turton, 'Use of a surface-acoustic-wave delay line to provide pseudo-coherence in a clutter-reference pulse Doppler radar', *Electronics Letters*, 9, 17—18 (1973).
211. A. C. Prior and K. A. J. Warren, 'A man portable radar for use by patrols', *International Conference on radar — present and future, London, IEE Conference publication Number 105*, October 1973, pp. 384—394.
212. F. R. Domer and R. H. Kyle, 'Phase locking an IMPATT device in the pulsed mode', *Proc. IEEE*, 55, 1753—1754 (1967).
213. M. L. Nyss, 'Intruder detector using a Gunn effect oscillator', *Philips Electron Appl. Bull.*, 31, 28—36 (1972).
214. W. D. Schuck, 'Thernal effects of IMPATT diodes used as self-detecting oscillators in Doppler radar applications', *Nachrichtentechn Z.*, 26, 517—519.
215. S. Nagano, H. Ueno, H. Kondo and H. Murakami, 'Self-exited microwave mixer with a Gunn diode and its applications to Doppler radar', *Electron Commun. Japan*, 52, 112—114 (1969).
216. S. Nagano and Y. Akaiwa, 'Behaviour of Gunn diode oscillator with a moving reflector as a self-exited mixer and a load variation detector', *IEEE Trans. Microwave Theory Tech.*, MTT-19, 906—910 (1971).
217. Y. Takayama, 'Doppler signal detection with negative-resistance diode oscillators', *IEEE Trans. Microwave Theory Tech.*, MTT-21, 89—94 (1973).

218. M. S. Gupta, R. J. Lomax and G. I. Haddad, 'Noise considerations in self-mixing IMPATT-diode oscillators for short-range Doppler radar applications', *IEEE Trans. Microwave Theory Tech.*, MTT-22, 37—43 (1974).
219. C. T. Rucker and C. T. Nations, 'Some practical Doppler sensors', *Conference Proceedings Microwave '73, Brighton*, June 1973, pp. 298—303.
220. 'CL8960 X-band Doppler radar module', *Mullard New Product Information, TP1427*, Mullard Limited, London.
221. P. Weissglas, 'BARITT and Gunn diodes as local oscillators', *1973 European Microwave Conference Proceedings, Brussels*, September 1973, Paper A.2.1.
222. D. Delagebeaudeuf, 'Punch through injection structures for low voltage oscillation and low noise amplification', *Conference Proceedings, 4th European Microwave Conference, Montreux*, pp. 178—181, September 1974.
223. P. Denniss and S. E. Gibbs, 'Solid-state linear FM/CW radar systems — their promise and their problems', *IEEE G-MTT International Microwave Symposium*, June 1974.
224. Sivers Lab Advertisement, *The Microwave J.*, 17, 36C, (1974).
225. W. Heichel and K. Zublin, 'YIG-tuned transistor oscillators', *Watkins—Johnson Application Note, 100239*, August 1968.
226. B. Hoglund and A. Isaacs, 'Varactor-tuned oscillators', *Watkins —Johnson Application Note, 100465*, July 1973.
227. M. I. Grace, 'Magnetically tunable transit-time oscillator', *Proc. IEEE*, 56, 771—773 (1968).
228. M. I. Grace, 'Varactor-tuned avalanche transit-time oscillator with linear tuning characteristics', *IEEE Trans. Microwave Theory Tech.*, MTT-18, 44—45 (1970).
229. B. Glance, 'A magnetically tunable microstrip IMPATT oscillator', *IEEE Trans. Microwave Theory Tech.*, 21, 425—426 (1973).
230. R. S. Raven, 'Requirements on master oscillators for coherent radar', *Proc. IEEE*, 54, 237—243 (1966).
231. L. P. Goetz and W. A. Skillman, 'Master oscillator requirements for coherent radar sets', *IEEE—NASA symposium on short-term frequency stability, Goddard Space Flight Center, Greenbelt, Maryland, USA*, November 1964, pp. 19—27.
232. W. W. Shrader, 'MTI radar', *Radar Handbook* (edited by M. I. Skolnik), McGraw—Hill, New York, 1970.
233. 'MLS: is it really on the beam for 1976?', *Microwave Systems News*, 4, 30, 32, 35, 37, 38 (1974).
234. J. F. Reynolds, J. Assour and A. Rosen, 'A solid-state transponder source using high-efficiency silicon avalanche oscillators', *RCA Review*, 33, 344—356 (1972).
235. A. P. Ives, 'A comparison of microwave systems for the headway

control of automobiles', *Conference Proceedings, Microwave 73, Brighton*, June 1973, pp. 413—417.

236. M. Kiyoto, T. Kondoh, K. Ban and K. Shirahata, 'Radar sensor for automobiles', *Digest of Technical Papers, 1974 IEEE International Solid-state Circuits Conf., Philadelphia*, February 1974, pp. 74—75.
237. G. F. Ross, 'Automobile pulse radar reduces false alarms', *MicroWaves*, 13, 12 (1974).
238. H. Storas, J. Shefer, R. J. Klensch and H. C. Johnson, 'A radar sensor for automatic braking that is immune to blinding and clutter', *Conference Proceedings, Microwave 73, Brighton*, June 1973, pp. 408—412.
239. J. Shefer, R. J. Klensch, G. Kaplan and H. C. Johnson, 'Clutter-free radar for cars', *Wireless World*, 80, 117—122 (1974).
240. F. G. Becker, 'Identification of moving vehicles by microwave techniques', *Microwave 73, Brighton, Conference Proceedings*, pp. 403—407.
241. F. R. Albrow, R. Allard, P. J. Bulman, J. L. Clarke, J. E. Copeland, *UK Patent No. 1290916.*
242. C. J. Huxter, M. J. O'Brian, J. T. Pinson and D. F. Sampson, 'MANAV esturial and berthing sub-systems', *IEE International Conference on Advances in Marine Navigational Aids, London*, July 1972, *Conference Publication No. 87*, pp. 171—176.
243. 'Radar beacon for buoys', *Microwave J.*, 15, 59E (1972).
244. W. J. Ince and D. H. Temme, 'Phasers and time delay elements', *Advances in microwaves* (edited by L. Young), Vol. 4, Academic Press, New York and London, 1969.
245. N. Amitay, V. Galindo and C. P. Wu, *Theory and analysis of phased array antennas*, Wiley—Interscience, 1972, p. 11.
246. J. L. Allen, 'The theory of array antennas', *Massachusetts Institute of Technology, Lincoln Laboratory, Technical Report No. 323.*
247. T. C. Cheston and J. Frank, 'Array antennas', Chapter 11, *Radar Handbook* (edited by M. I. Skolnik), McGraw—Hill, New York, 1970.
248. C. J. Miller, 'Minimizing the effects of phase quantization errors in an electronically scanned array', *Proceedings of Symposium on electronically scanned array techniques and applications, RADC-TDR-64-225, vol. 1, pp. 17—38.*
249. L. Stark, R. W. Burns and W. P. Clark, 'Phase shifters for arrays', Chapter 12, *Radar Handbook* (edited by M. I. Skolnik), McGraw—Hill, New York, 1970.
250. W. J. Ince, 'Recent advances in diode and ferrite phaser technology for phased-array radars, Pt. I', *The Microwave J.*, 15,

36—38, 40, 42, 46 (1972), 'Pt. II', *The Microwave J.*, 15, 31—34, 36 (1972).

251. A. H. Al-Ani, A. L. Cullen and J. R. Forrest, 'Novel techniques in electronically steerable arrays', *International Conference on Radar — present and future, London*, October 1973, *IEEE Conference publication Number 105*, pp. 88—93.
252. A. L. Cullen, 'Phase interpolation locking: new technique for beam-steered active antennas', *Electronics Letters*, 10, 81—2 (1974).
253. R. J. Mailloux and P. R. Caron, 'A class of phase interpolation circuits for scanning phased arrays', *NEREM Record 1969, Boston Section IEEE*, pp. 200—201.
254. P. Brook, L. D. Clough and K. G. Hambleton, 'Microwave phase shifting with gain using IMPATT diodes', *Electronics Letters*, 8, 489—491 (1972).
255. L. I. Yarrington and P. W. Hawkins, 'Analysis of phase characteristics as a function of ambient temperature of IMPATT amplifiers', *IEEE Trans. Microwave Theory Tech.*, MTT-21, 728—730 (1973).
256. 'Phased array power amplifier', *The Microwave J.*, 17, 28 (1974).
257. M. Dixon and J. L. Glasper, Private communication.
258. E. Belohoubek, A. Presser, D. M. Stevenson, A. Rosen and D. Zieger, 'S-band CW power module for phased arrays', *The Microwave J.*, 13, 29—30, 32, 34 (1970).
259. A. Rosen, J. F. Reynolds, S. G. Liu and G. E. Theriault, 'Wideband class-C TRAPATT amplifiers', *RCA Review*, 33, 729—736 (1972).
260. G. C. Bandy, L. J. Hardeman and W. F. Hayes, 'MERA modules — how good in an array?', *MicroWaves*, 8, 39—49 (1969).
261. J. R. Collins and T. E. Harwell, 'RASSR array comes of age', *MicroWaves*, 11, 36—38, 40—42 (1972).
262. G. Sorger, J. Herrero and H. Verhoeven, 'Design of a broadband RF and M/W sweep generator', *Microwave J.*, 17, 51—57 (1974).
263. 'Fundamental transistor source tunes over entire X-band', *MicroWaves*, 12, 76 (1973).
264. H. J. Kuno and P. H. Pusateri, 'Use of solid-state components for mm-wave measurements', *Microwave J.*, 17, 35—36, 39—40 (1974).
265. J. C. Goggins, 'An EMC frequency synthesizer', *Microwave J.*, 17, 51—53 (1974).
266. R. H. Rector, 'High performance design-to-price surveillance receivers', *Conference Proceedings, Microwave 73, Brighton*, pp. 244—250, June 1973.
267. 'ECM solid-state oscillators to 18 GHz', Omni-spectra Inc, Advertisement, *MicroWaves*, 11, (1972).

268. R. D. Hogg, 'A low cost X-band IMPATT diode marginal oscillator for EPR', *Amer. J. Phys.*, 41, 224–229 (1973).
269. R. D. Hogg, 'Applications of IMPATT diodes as RF sources for microwave EPR spectroscopy', *Rev. Sci. Inst.*, 44, 582–587 (1973).
270. S. D. Smith, M. J. Colles and G. Peckham, 'The measurement of surface pressure from a satellite', *Quarterly J. Royal Meteorological Soc.*, 98, 431–433 (1972).
271. J. L. Fikart, 'AM and FM noise of BARITT oscillators', *IEEE Trans. Microwave Theory Tech.*, MTT-22, 517–523 (1974).
272. V. F. Kroupa, *Frequency Synthesis — Theory, Design and Applications*, Griffin, London 1973.
273. F. Hasegawa, Y. Aono and Y. Kaneko, 'Performance and characterisation of X-band GaAs Read-type IMPATT diodes', Gallium Arsenide and Related Compounds, Deauville, France, Inst. Phys. 61–70, Sept. 1974.

Subject Index